SPACECRAFT ELECTROMAGNETIC DOCKING AND SEPARATION

SPACECRAFT ELECTROMAGNETIC DOCKING AND SEPARATION

KEKE SHI
Xi'an University of Science and Technology, Xi'an, Shaanxi, China

CHUANG LIU
Northwestern Polytechnical University, Xi'an, Shaanxi, China

XIAOKUI YUE
Northwestern Polytechnical University, Xi'an, Shaanxi, China

MAHE SHU
Northwestern Polytechnical University, Xi'an, Shaanxi, China

ELSEVIER

Elsevier
Radarweg 29, PO Box 211, 1000 AE Amsterdam, Netherlands
50 Hampshire Street, 5th Floor, Cambridge, MA 02139, United States

ISBN: 978-0-443-49334-8

For Information on all Elsevier publications visit our website at https://www.elsevier.com/books-and-journals

Publisher: Matthew Deans
Acquisitions Editor: Glyn Jones
Editorial Project Manager: Tripti Rawat
Production Project Manager: Maria Shalini
Cover Designer: Gopalakrishnan Venkatraman

Typeset by Aptara, New Delhi, India

Contents

Preface

Spacecraft electromagnetic docking and separation technology has garnered increasing attention since the beginning of this century. This is not only because electromagnetic force serves as a renewable onboard energy source, but more critically, it enables contactless and gentle docking/separation operations between spacecraft, significantly reducing mechanical wear, plume contamination, and collision risks compared to traditional approaches. However, the strong nonlinearity of electromagnetic forces and the complex disturbances during docking make this technology extremely difficult to perfect. The challenges encountered in spacecraft electromagnetic docking and separation have driven us to investigate diverse advanced control strategies to achieve high-precision operations, balancing robustness and adaptability under complex disturbances.

The basic topic of this book is solving problems from spacecraft electromagnetic docking and separation control systems using different control approaches instead of the widely used proportional-integral-derivative (PID) control in the existing worldwide electromagnetic docking and separation missions. Parameter uncertainties, external disturbances, fault signals, time delay, and fragility will occur in on-orbit operation process of spacecraft, especially for the electromagnetic docking and separation, which will cause the control system performance degradation or even instability. To deal with the above-mentioned issues, in this book, we consider the problem of stability analysis and controller synthesis in spacecraft electromagnetic docking and separation control systems. We show that sliding mode control, active disturbance rejection control, disturbance observer-based control, model predictive control, and deep reinforcement learning-based control can effectively deal with the multisource disturbances and inherent uncertainties. Besides, we show that in some cases, nonfragile control can further enhance the control performance and robustness during electromagnetic docking and separation. The theorems developed in this text can be solved numerically in polynomial time, and so are tractable, at least in a theoretical sense.

This book is aimed at readers who are interested in the field of spacecraft electromagnetic docking and separation control, while it is also accessible for novice researchers in this field, as we have introduced some basic knowledge of magnetic field, electromagnetic force, and spacecraft relative motion dynamics to lower the barrier to comprehension. This book can be used as both a reference guide for practicing professionals and a textbook for undergraduate and graduate students in the field of on-orbit servicing and assembly or control engineering. Some contents of this book have been used to teach graduate students at Xi'an University of Science and Technology and Northwestern Polytechnical University, for the control technology of robot and on-orbit operation of spacecraft courses. Besides, a variety of control strategies, accompanied by stability proofs that integrate Lyapunov stability theory with sliding mode control theory, LMIs-based theorems, model predictive control theory, reinforcement learning theory and other frameworks are also introduced in this book, which can serve as first cut analysis for further research and are especially important in the initial design phase of a spacecraft electromagnetic operation system.

This book is organized as follows. Chapter 1 introduces some basic knowledge, all of which are used in the following chapters. Chapters 2–7 discuss different electromagnetic docking control strategies, and Chapters 8 and 9 discuss different electromagnetic separation control strategies, while in Chapter 10, an integrated control strategy for spacecraft electromagnetic docking and separation is discussed. In particular, Chapter 1 lays the theoretical foundation by introducing magnetic field concepts, presenting near-field and far-field models for electromagnetic force/torque calculation, establishing relative motion dynamics, and providing key mathematical lemmas. Chapter 2 investigates coupled orbit-attitude control for spacecraft electromagnetic docking, presenting a sliding mode controller with a planned trajectory. Chapter 3 develops an intermediate observer to estimate relative motion information and lumped disturbances, and proposes an antidisturbance controller with input constraints that demonstrates superior performance. Chapter 4 proposes an active disturbance rejection control scheme for spacecraft electromagnetic docking with time-varying delay, where an intermediate observer estimates relative motion information and lumped disturbances to ensure uniform ultimate boundedness. Chapter 5 develops a disturbance observer-based control (DOBC) strategy for spacecraft electromagnetic docking with input magnitude and rate constraints, estimating lumped disturbances via LMIs without requiring mass information. Chapter 6 presents an enhanced tube-based model predictive control (TMPC) strategy for spacecraft electromagnetic docking, establishing a liquid-solid-magnetic coupled dynamics model with constraint-tightening for multiobjective optimization. Chapter 7 introduces a deep reinforcement learning-based control approach using the deep deterministic policy gradient (DDPG) algorithm, designing a multiobjective reward function for high precision, disturbance rejection, and energy optimization. Chapter 8 investigates electromagnetic separation for mass-unknown spacecraft in elliptical orbits, designing a compliant trigonometric-function reference trajectory with an improved sliding mode controller ensuring fast convergence. Chapter 9 proposes an impedance control strategy optimized by integrated hippopotamus and whale algorithms, combining a time-optimal trajectory with a nonlinear disturbance observer to achieve high precision. Chapter 10 develops a dual hybrid nonfragile control strategy for integrated electromagnetic docking and separation, integrating a hybrid nonfragile disturbance observer with a nonfragile controller to address gain perturbations and ensure robust stability.

In this book, numerical simulation examples on spacecraft electromagnetic docking or separation systems considering multisource complex disturbances are given. Readers should note that the control strategies discussed in this work remain confined to theoretical analysis and numerical simulation, and potential challenges arising in real-world implementation—including but not limited to hardware latency, environmental disturbance, and cross-system interference—fall beyond the scope of this book.

This book represents the assemblage of long-term research outcomes by the authors working on spacecraft electromagnetic docking and separation control technologies. The completion of this book has greatly benefited from the assistance of our colleagues and students for their contributions to various aspects of this work. We owe a particular debt of gratitude to Prof. Zhaowei Sun and Prof. James D. Biggs for their steadfast support and invaluable scholarly guidance. Several students have provided excellent insights and recommendations to enhance the pedagogical value, as well as promote the completion of this book.

This work was supported in part by the National Natural Science Foundation of China under Grants 12572056, 62303374, and 62103336, and in part by the Young Elite Scientists Sponsorship Program by CAST under Grant 2022QNRC001.

CHAPTER 1

Introduction of basic knowledge

1.1 Basic knowledge of magnetic field

In this section, some definitions, properties, and theorems of magnetic field will be introduced. For relevant information on Sections 1.1 to 1.5, see Notaros (2010).

1.1.1 Flux and divergence of vector field

For a vector field $\boldsymbol{a}$, its surface integral over an oriented surface σ is called the flux of the vector field $\boldsymbol{a}$ through the surface σ on the side indicated by the surface normal.

$$\Phi = \iint_{\sigma} \boldsymbol{a} \cdot \boldsymbol{n} d\sigma = \iint_{\sigma} a_n d\sigma \tag{1.1}$$

where vector $\boldsymbol{n}$ is the unit normal vector of the surface, a_n is the projection of the vector $\boldsymbol{a}$ onto the surface normal. The physical significance of flux is the effective amount of the vector field passing through the surface. In Cartesian coordinates, let

$$\boldsymbol{a} = \begin{bmatrix} P(x, y, z) & Q(x, y, z) & R(x, y, z) \end{bmatrix}^{\mathrm{T}}$$

then Eq. (1.1) can be expressed as

$$\Phi = \iint_{\sigma} Pdydz + Qdxdz + Rdxdy \tag{1.2}$$

If σ is a closed surface, the flux represents the algebraic sum of the positive flow exiting the surface and the negative flow entering the surface, with the convention that quantities flowing out are positive and those flowing in are negative.

Spacecraft Electromagnetic Docking and Separation.
DOI: https://doi.org/10.1016/B978-0-443-49334-8.00001-X

The divergence of a vector field represents the flux exiting a unit volume at a point in the field, which is defined as

$$\mathrm{div}\boldsymbol{a} = \nabla \cdot \boldsymbol{a} = \lim_{V \to 0} \frac{\oint\!\!\oint_{\sigma} \boldsymbol{a} \cdot \boldsymbol{n} d\sigma}{V} \tag{1.3}$$

where V is the volume enclosed by the closed surface; $\nabla = \frac{\partial}{\partial x}\boldsymbol{i} + \frac{\partial}{\partial y}\boldsymbol{j} + \frac{\partial}{\partial z}\boldsymbol{k}$ is the Hamilton operator, with the coordinate axis unit direction vectors $\boldsymbol{i}$, $\boldsymbol{j}$ and $\boldsymbol{k}$. Divergence is used to measure the strength of sources or sinks at a certain point in a vector field, representing the distribution characteristics of the flux sources in the vector field. Divergence is a scalar and is a function of spatial coordinates.

1.1.2 Circulation and curl of vector field

For a vector field $\boldsymbol{a}$, consider a closed oriented curve l in the vector field. The integral of the vector field $\boldsymbol{a}$ along the closed oriented curve l is referred to as the circulation of the vector field $\boldsymbol{a}$ along the curve l.

$$\Gamma = \oint_l \boldsymbol{a} \cdot d\boldsymbol{r} = \oint_l \boldsymbol{a} \cdot \boldsymbol{\tau} ds \tag{1.4}$$

where $\boldsymbol{\tau}$ is the unit direction vector of curve l, and s is the curve arc length. The physical meaning of circulation is the cumulative amount of the vector along a closed curve. If the circulation of the vector field is zero at all points in a certain region, then the vector field is referred to as an irrotational field (or conservative field, or potential field). In Cartesian coordinates, Eq. (1.4) can be expressed as

$$\Gamma = \oint_l Pdx + Qdy + Rdz \tag{1.5}$$

The circulation density of a vector field $\boldsymbol{a}$ is defined as the limit of the circulation of the field along a closed curve as the area enclosed by the curve approaches zero. Mathematically, it can be expressed as:

$$\lim_{\Delta S \to 0} \frac{\Delta \Gamma}{\Delta S} = \lim_{\Delta S \to 0} \frac{\oint_l \boldsymbol{a} \cdot d\boldsymbol{r}}{\Delta S} \tag{1.6}$$

The curl of the vector field $\boldsymbol{a}$ is represented as rot$\boldsymbol{a}$. Stokes' theorem states that the circulation of the vector field $\boldsymbol{a}$ along a closed curve l is equal to the flux of the curl of the vector field through the surface enclosed by the curve. Using Stokes' theorem, the circulation density can be rewritten as:

$$\lim_{\Delta S \to 0} \frac{\Delta \Gamma}{\Delta S} = \lim_{\Delta S \to 0} \frac{\oiint_{\sigma} \mathrm{rot}\boldsymbol{a} \cdot d\boldsymbol{\sigma}}{\Delta S} = \mathrm{rot}\boldsymbol{a} \cdot \boldsymbol{n} = \nabla \times \boldsymbol{a} \cdot \boldsymbol{n} \tag{1.7}$$

Through Eq. (1.7), it is obvious that the curl of the vector field $\boldsymbol{a}$ projected in the direction of the surface element is the circulation density. The curl describes the strength of the vorticity source of the vector field $\boldsymbol{a}$ at a given point. In Cartesian coordinates, the curl can be computed using the following method:

$$\begin{aligned} \mathrm{rot}\boldsymbol{a} = \nabla \times \boldsymbol{a} &= \begin{vmatrix} \boldsymbol{i} & \boldsymbol{j} & \boldsymbol{k} \\ \frac{\partial}{\partial x} & \frac{\partial}{\partial y} & \frac{\partial}{\partial z} \\ A_x & A_y & A_z \end{vmatrix} \\ &= \left(\frac{\partial A_z}{\partial y} - \frac{\partial A_y}{\partial z}\right)\boldsymbol{i} + \left(\frac{\partial A_x}{\partial z} - \frac{\partial A_z}{\partial x}\right)\boldsymbol{j} + \left(\frac{\partial A_y}{\partial x} - \frac{\partial A_x}{\partial y}\right)\boldsymbol{k} \end{aligned} \tag{1.8}$$

1.1.3 Gauss's theorem for steady magnetic fields and Ampere's circuital law

The integral form of Gauss's theorem for steady magnetic fields is:

$$\oiint_{\sigma} \boldsymbol{B} \cdot \boldsymbol{n} d\sigma = 0 \tag{1.9}$$

where $\boldsymbol{B}$ is the magnetic induction intensity. Eq. (1.9) indicates that the magnetic flux of the magnetic induction intensity through any closed surface in the magnetic field is constant and equal to zero. For any point in the magnetic field, the differential form of Gauss's theorem is:

$$\mathrm{div}\boldsymbol{B} = \nabla \cdot \boldsymbol{B} = 0 \tag{1.10}$$

Gauss's theorem states that the divergence of the magnetic induction intensity in a steady magnetic field is always zero, and the magnetic field lines are closed curves. The magnetic field is a source-free field (a vortex field), and magnetic poles always occur in pairs.

The integral form of Ampere's circuital law for a steady magnetic field in a vacuum is:

$$\oint_{l} \boldsymbol{B} \cdot d\boldsymbol{r} = \mu_0 \sum I \tag{1.11}$$

where μ_0 is the vacuum permeability, and I is the current enclosed by the closed path l. When the direction of the loop follows the right-hand rule with respect to the current direction, the current is considered positive, otherwise, it is considered negative. This indicates that in the magnetic field generated by a steady closed current, the magnetic induction intensity along any closed loop is directly proportional to the algebraic sum of the currents enclosed by that loop. For any point in the magnetic field, the differential form of Ampere's circuital law is:

$$\mathrm{rot}\boldsymbol{B} = \nabla \times \boldsymbol{B} = \mu_0 \boldsymbol{J} \tag{1.12}$$

where $\boldsymbol{J}$ is the current density vector at that point, whose direction is the direction of the current at that point, and whose magnitude is the amount of charge passing through a unit perpendicular cross-section at that point per unit time. Ampere's circuital law indicates that a steady magnetic field is a nonconservative force field and is a field with circulation.

1.1.4 Magnetic vector potential

Since a steady magnetic field is a field with circulation, it cannot be described by a potential function like an electrostatic field. However, a steady magnetic field is a source-free field, and the following identities exist for its curl and divergence.

$$\nabla \cdot (\nabla \times \boldsymbol{A}) \equiv 0 \tag{1.13}$$

Thus, a vector function $\boldsymbol{A}$ can be introduced satisfying

$$\boldsymbol{B} = \nabla \times \boldsymbol{A} \tag{1.14}$$

The vector function $\boldsymbol{A}$ is called the magnetic vector potential (or vector magnetic potential) of a steady magnetic field. In the International System of Units, its unit is weber per meter (Wb/m).

In a vacuum, replacing magnetic induction intensity in Eq. (1.12) by magnetic vector potential, one has

$$\nabla \times \nabla \times \boldsymbol{A} = \mu_0 \boldsymbol{J} \tag{1.15}$$

Through vector identities

$$\nabla \times \nabla \times \boldsymbol{A} = \nabla \cdot (\nabla \cdot \boldsymbol{A}) - \nabla^2 \boldsymbol{A} \tag{1.16}$$

where $\nabla^2 = \frac{\partial^2}{\partial x^2} + \frac{\partial^2}{\partial y^2} + \frac{\partial^2}{\partial z^2}$.

Eq. (1.12) turns into

$$\nabla \cdot (\nabla \cdot \boldsymbol{A}) - \nabla^2 \boldsymbol{A} = \mu_0 \boldsymbol{J} \tag{1.17}$$

In a vector field, once a vector is determined, its divergence and curl are uniquely determined as well; conversely, to determine a vector, both its divergence and curl must be known. Therefore, to uniquely determine from Eq. (1.17), the divergence must be specified. For simplicity, let

$$\nabla \cdot \boldsymbol{A} = 0 \tag{1.18}$$

Eq. (1.18) is called the Coulomb criterion. Now Eq. (1.17) turns into

$$\nabla^2 \boldsymbol{A} = -\mu_0 \boldsymbol{J} \tag{1.19}$$

This indicates that the magnetic vector potential $\boldsymbol{A}$ satisfies the vector form of Poisson's equation. When the current is distributed in a finite space, and the value of the magnetic vector potential at infinity is set to zero, integrating Eq. (1.19) yields

$$\boldsymbol{A} = \frac{\mu_0}{4\pi} \iiint\limits_V \frac{\boldsymbol{J}}{r} dV \tag{1.20}$$

where r is the magnitude of the position vector.

For a current distribution in the form of a current-carrying conductor, the induced magnetic vector potential is

$$\boldsymbol{A} = \frac{\mu_0}{4\pi} \oint\limits_l \frac{Ni}{r} d\boldsymbol{l} \tag{1.21}$$

where i is the current in the wire; N is the coil turns.

1.1.5 Biot-Savart law

The Biot-Savart law gives the magnetic field intensity produced by a current source at any point in space. The Biot-Savart law can be derived from the magnetic vector potential, and the derivation process is presented later.

After obtaining the magnetic vector potential, the magnetic field intensity can be calculated using Eq. (1.14). For a planar current-carrying coil, the formula for calculating its magnetic field intensity is

$$\boldsymbol{B} = \nabla \times \frac{\mu_0 Ni}{4\pi} \oint\limits_l \frac{1}{r} d\boldsymbol{l} \tag{1.22}$$

In Cartesian coordinates, since the integrand $1/rd\boldsymbol{l}$ and its first-order partial derivatives with respect to x, y, z are continuous functions, by applying the theorem of interchange of differentiation and integration, one can obtain

$$\boldsymbol{B} = \frac{\mu_0 Ni}{4\pi} \oint\limits_l \nabla \times \frac{1}{r} d\boldsymbol{l} \tag{1.23}$$

Considering that the vector cross product and the calculation of curl satisfy the following rules:

$$\boldsymbol{m} \times \boldsymbol{n} = -\boldsymbol{n} \times \boldsymbol{m} \tag{1.24}$$

$$\nabla \times (u\boldsymbol{m}) = u\nabla \times \boldsymbol{m} + \nabla u \times \boldsymbol{m} \tag{1.25}$$

where u is a scalar function; $\boldsymbol{m}$, $\boldsymbol{n}$ are vector functions; ∇u is the gradient of u; r is the magnitude of the vector pointing from the origin to that position. Therefore, Eq. (1.23) can be transformed into:

$$\boldsymbol{B} = \frac{\mu_0 Ni}{4\pi} \oint_l \frac{\nabla \times d\boldsymbol{l}}{r} + \nabla\left(\frac{1}{r}\right) \times d\boldsymbol{l} \tag{1.26}$$

Expressing the integral curve l in parametric form with respect to the parameter t, that is, l: $x = x(t)$, $y = y(t)$, $z = z(t)$, the integral variable $d\boldsymbol{l}$ can be represented as

$$d\boldsymbol{l} = \begin{bmatrix} x'(t) & y'(t) & z'(t) \end{bmatrix} dt \tag{1.27}$$

Now, it is easy to find out that the second term of the integral in Eq. (1.26) is zero. Ultimately, Eq. (1.26) becomes:

$$\boldsymbol{B} = \frac{\mu_0 Ni}{4\pi} \oint_l \frac{d\boldsymbol{l} \times \boldsymbol{r}}{r^3} \tag{1.28}$$

Eq. (1.28) is the famous Biot-Savart law. The schematic of a closed-loop current-carrying coil is illustrated in Fig. 1.1.

1.1.6 Magnetic force, magnetic torque, and magnetic moment of a planar current-carrying coil

The force acting on a planar current-carrying coil under the influence of an external magnetic field is referred to as the magnetic field force. The torque relative to the center of the coil is called the magnetic torque. Their calculation methods are as follows:

$$\boldsymbol{F}_{\mathrm{e}} = \oint_L Ni d\boldsymbol{l} \times \boldsymbol{B} \tag{1.29}$$

$$\boldsymbol{\tau}_{\mathrm{e}} = \oint_l \boldsymbol{R} \times (Ni d\boldsymbol{l} \times \boldsymbol{B}) = NiS(\hat{\boldsymbol{n}} \times \boldsymbol{B}) \tag{1.30}$$

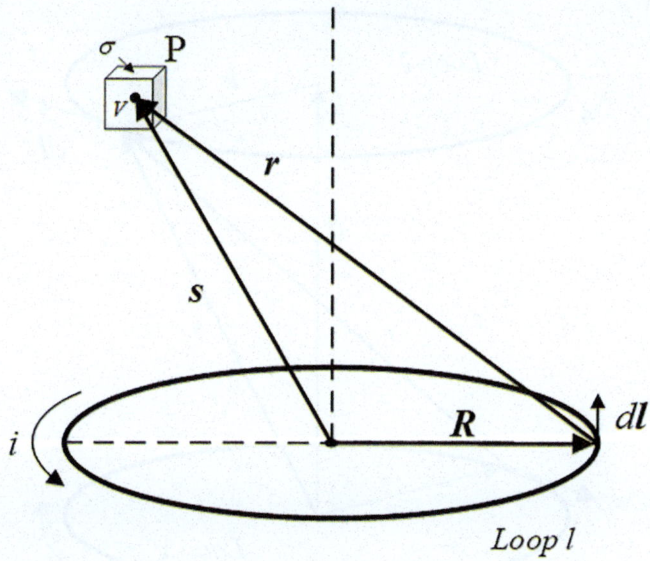

Figure 1.1 Schematic diagram of a closed-loop current-carrying coil. A current-carrying coil produces a magnetic field in space.

where $\boldsymbol{r}$ represents the position vector from any point on the coil to point P, $r = ||\boldsymbol{r}||$. Clearly, at most positions in space, the line integral is not easy to solve, making it difficult to obtain an analytical solution for the magnetic field intensity at those points.

Define magnetic moment as

$$\boldsymbol{\mu} = \frac{1}{2}Ni\oint_{l} \boldsymbol{R} \times d\boldsymbol{l} = NiS\hat{\boldsymbol{n}} \tag{1.31}$$

where N is the number of turns of the coil, i is the current flowing through the coil, S is the area of the coil, and $\hat{\boldsymbol{n}}$ is the normal vector of the coil.

When the magnetic induction distribution in the space is obtained using Eq. (1.28), the Ampere force on the planar current-carrying coil can be calculated using Eqs. (1.29) and (1.30). Considering the interaction between two electromagnetic coils in the space, the Ampere force and the Ampere moment exerted on coil 2 in the magnetic field generated by coil 1 are given by:

$$\boldsymbol{F}_{e2} = \frac{\mu_0 N_1 N_2 i_1 i_2}{4\pi}\oint_{l_2} d\boldsymbol{l}_2 \times \left(\oint_{l_1} \frac{\boldsymbol{r} \times d\boldsymbol{l}_1}{r^2}\right) \tag{1.32}$$

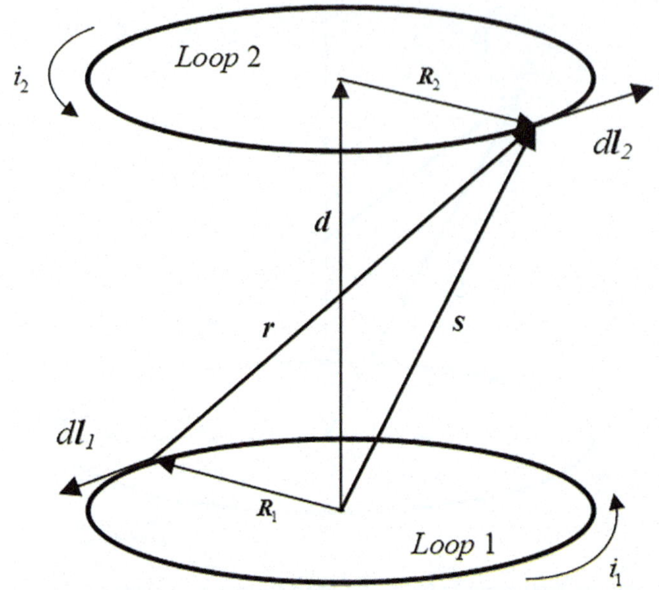

Figure 1.2 Electromagnetic force between closed-loop current-carrying coils. A current-carrying coil experiences an electromagnetic force when placed in the magnetic field generated by another current-carrying coil.

$$\boldsymbol{\tau}_{e2} = \frac{\mu_0 N_1 N_2 i_1 i_2}{4\pi} \oint_{l_2} \boldsymbol{h_2} \times \left(d\boldsymbol{l}_2 \times \left(\oint_{l_1} \frac{\boldsymbol{r} \times d\boldsymbol{l}_1}{r^2} \right) \right) \tag{1.33}$$

where $\boldsymbol{F}_{e2}$ is the Ampere force experienced by coil 2 under the magnetic field of coil 1; N_1 and N_2 represent the numbers of turns in coil 1 and coil 2, respectively; i_1 and i_2 represent the currents flowing through coil 1 and coil 2, respectively, with their directions shown in Fig. 1.2. $\boldsymbol{r}$ represents the position vector from any point on coil 1 to any point on coil 2; $\boldsymbol{s}$ represents the position vector from the center of coil 1 to any point on coil 2; L_1 and L_2 are the integral loops of coil 1 and coil 2; $d\boldsymbol{l}_1$ and $d\boldsymbol{l}_2$ are the length vector differentials on coil 1 and coil 2; $\boldsymbol{h}_2$ is a vector pointing from the centroid to any point on coil 2.

The double curve integral in Eqs. (1.32) and (1.33) pose challenges for numerical solution. By introducing parametric equations and elliptic integrals, the double curve integral can be transformed into a double definite integral for computation (Babic & Akyel, 2012).

Define a global coordinate $G{:}OXYZ$. In the global coordinate, coil 1 is placed on the surface λ_1: $a_1x + b_1y + c_1z + D_1 = 0$

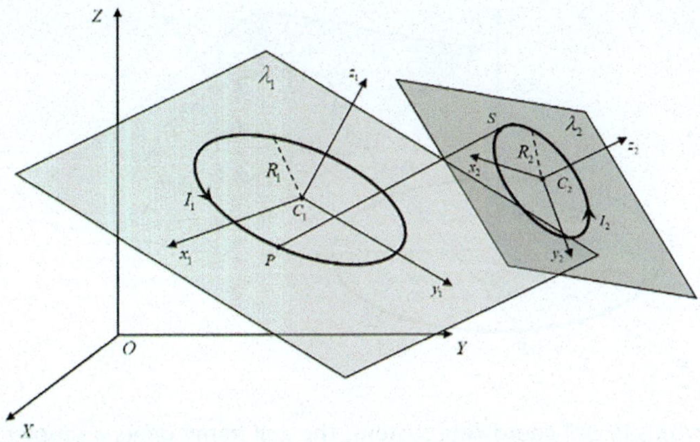

Figure 1.3 Coils in global coordinate system. Electromagnetic forces and torques often need to be transformed into the reference frame associated with the object under study.

with its center ${}^{G}C_1(x_{C1}, y_{C1}, z_{C1})$; coil 2 is placed on the surface λ_2: $a_2x + b_2y + c_2z + D_2 = 0$ with its center ${}^{G}C_2(x_{C2}, y_{C2}, z_{C2})$. Their schematic diagram is shown in Fig. 1.3.

Besides the global coordinate, coil 1 coordinate $C_1x_1y_1z_1$ is defined. The origin is located at the center of coil 1. The z-axis is perpendicular to the plane of the coil 1. The x and y axes lie within the plane of coil 1 and are mutually perpendicular. Similarly, the coil 2 coordinate $C_2x_2y_2z_2$ is also established. The coordinate rotation matrix from the global coordinate system to coil 1 coordinate system is:

$$\boldsymbol{R}_{G-1} = \begin{bmatrix} \boldsymbol{u}_1 \\ \boldsymbol{v}_1 \\ \boldsymbol{w}_1 \end{bmatrix} \tag{1.34}$$

where $\boldsymbol{w}_1$ is the unit normal vector of the surface $\lambda_1, \boldsymbol{u}_1, \boldsymbol{v}_1$ are the other two-unit orthogonal basis vectors of coil 1 coordinate. When $a_1 \neq 0$ and $b_1 \neq 0$, a unit orthonormal basis is formed by

$$\begin{cases} \boldsymbol{u}_1 = \dfrac{\begin{bmatrix} -a_1b_1 & \sqrt{a_1^2 + c_1^2} & -b_1c_1 \end{bmatrix}}{\sqrt{(a_1^2+b_1^2+c_1^2)(a_1^2+c_1^2)}} \\ \boldsymbol{v}_1 = \dfrac{\begin{bmatrix} -c_1 & 0 & a_1 \end{bmatrix}}{\sqrt{a_1^2+c_1^2}} \\ \boldsymbol{w}_1 = \dfrac{\begin{bmatrix} a_1 & b_1 & c_1 \end{bmatrix}}{\sqrt{a_1^2+b_1^2+c_1^2}} \end{cases} \tag{1.35}$$

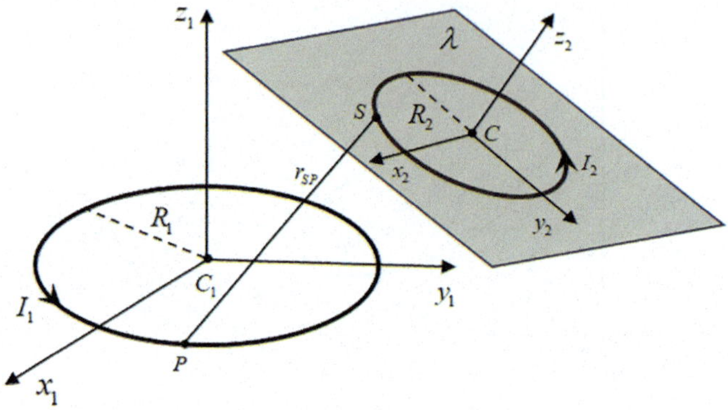

Figure 1.4 Coils in coil coordinate system. The coil frame offers a simpler description of the electromagnetic interactions between coils, featuring rotational symmetry about its z-axis. *Figure 1.4 in Babic & Akyel (2012), reproduced courtesy of The Electromagnetics Academy.*

Therefore, in two coordinates, the point and vector have the following relationship:

$$\begin{bmatrix} x & y & z \end{bmatrix}^{\mathrm{T}} = \boldsymbol{R}_{G-1}\left(\begin{bmatrix} X & Y & Z \end{bmatrix}^{\mathrm{T}} - \begin{bmatrix} x_{C1} & y_{C1} & z_{C1} \end{bmatrix}^{\mathrm{T}}\right) \tag{1.36}$$

$$\boldsymbol{vec} = \boldsymbol{R}_{G-1}\boldsymbol{VEC} \tag{1.37}$$

where $\begin{bmatrix} x & y & z \end{bmatrix}^{\mathrm{T}}$ and $\boldsymbol{vec}$ represent point and vector in coil 1 coordinate, $\begin{bmatrix} X & Y & Z \end{bmatrix}^{\mathrm{T}}$ and $\boldsymbol{VEC}$ represent point and vector in global coordinates. Now, coil 2 can be expressed in coil 1 coordinate, which is placed on the surface λ: $ax + by + cz + D = 0$ with the center $C(x_C, y_C, z_C)$. At this point, the problem has been transformed from the global coordinate system to the coil 1 coordinate system for analysis, which is shown in Fig. 1.4.

The parametric equation for an arbitrary point P on coil 1 is:

$$\begin{cases} x_P = R_1 \cos t \\ y_P = R_1 \sin t \ (t \in [0, 2\pi]) \\ z_P = 0 \end{cases} \tag{1.38}$$

where t is the parameter. The differential line element on coil 1 is given by:

$$d\boldsymbol{l}_1 = R_1(-\sin t\boldsymbol{i} + \cos t\boldsymbol{j})dt \quad (t \in [0, 2\pi]) \tag{1.39}$$

where $\boldsymbol{i}, \boldsymbol{j}$ are the unit vectors of the x and y axes.

Similar to Eq. (1.35), the unit orthonormal basis of coil 2 coordinate is

$$\begin{cases} \boldsymbol{n} = \begin{bmatrix} n_x & n_y & n_z \end{bmatrix}^{\mathrm{T}} = \begin{bmatrix} \frac{a}{L} & \frac{b}{L} & \frac{c}{L} \end{bmatrix}^{\mathrm{T}} \\ \boldsymbol{u} = \begin{bmatrix} u_x & u_y & u_z \end{bmatrix}^{\mathrm{T}} = \begin{bmatrix} -\frac{ab}{lL} & \frac{l}{L} & -\frac{bc}{lL} \end{bmatrix}^{\mathrm{T}} \\ \boldsymbol{v} = \begin{bmatrix} v_x & v_y & v_z \end{bmatrix}^{\mathrm{T}} = \begin{bmatrix} -\frac{c}{l} & 0 & \frac{a}{l} \end{bmatrix}^{\mathrm{T}} \end{cases}$$

The parametric equation for an arbitrary point S on coil 2 is:

$$\boldsymbol{x}_S = \boldsymbol{x}_C + R_2(\cos\phi\boldsymbol{u} + \sin\phi\boldsymbol{v}) \quad \phi \in [0, 2\pi] \tag{1.40}$$

or the component form

$$\begin{cases} x_S = x_C + R_2 u_x \cos\phi + R_2 v_x \sin\phi \\ y_S = y_C + R_2 u_y \cos\phi + R_2 v_y \sin\phi \; \phi \in [0, 2\pi] \\ z_S = z_C + R_2 u_z \cos\phi + R_2 v_z \sin\phi \end{cases}$$

The differential line element on coil 2 is given by:

$$d\boldsymbol{l}_2 = R_2(-\sin\phi\boldsymbol{u} + \cos\phi\boldsymbol{v})d\phi \quad \phi \in [0, 2\pi] \tag{1.41}$$

Based on the Biot-Savart law Eq. (1.28) and combined with Eq. (1.38), the magnetic flux density at an arbitrary point S in space, generated by coil 1, is given by:

$$\begin{cases} B_{1x} = \frac{\mu_0 I_1 R_1 z_{SP}}{4\pi} \int_0^{2\pi} \frac{\cos t}{r_{SP}^3} dt \\ B_{1y} = \frac{\mu_0 I_1 R_1 z_{SP}}{4\pi} \int_0^{2\pi} \frac{\sin t}{r_{SP}^3} dt \\ B_{1z} = \frac{\mu_0 I_1 R_1}{4\pi} \int_0^{2\pi} \frac{R_1 - \sqrt{x_S^2 + y_S^2}\cos(\gamma - t)}{r_{SP}^3} dt \end{cases} \tag{1.42}$$

where $\boldsymbol{r}_{SP} = \boldsymbol{r}_S - \boldsymbol{r}_P$, $\boldsymbol{r}_S = [x_S \quad y_S \quad z_S]^{\mathrm{T}}$, $\boldsymbol{r}_P = [x_P \quad y_P \quad z_P]^{\mathrm{T}}$, $\cos\gamma = \frac{x_S}{\sqrt{x_S^2 + y_S^2}}$, $\sin\gamma = \frac{y_S}{\sqrt{x_S^2 + y_S^2}}$.

Applying variable substitution $t - \gamma = \pi - 2\beta$, one gets

$$\cos t = \cos(\pi - 2\beta + \gamma) = -\cos\gamma\cos(2\beta) - \sin\gamma\sin(2\beta)$$

$$\sin t = \sin(\pi - 2\beta + \gamma) = \cos\gamma\sin(2\beta) - \sin\gamma\cos(2\beta)$$

$$R_1 - \sqrt{x_S^2 + y_S^2}\cos(\gamma - t) = R_1 + \sqrt{x_S^2 + y_S^2}\cos(2\beta)$$

$$\int_{0}^{2\pi}(\cdots)dt = 2\int_{\frac{\gamma-\pi}{2}}^{\frac{\gamma+\pi}{2}}(\cdots)d\beta$$

Since the integrands in Eq. (1.42) are periodic function of the parameter t with a period of 2π, one has

$$\int_{0}^{2\pi}(\cdots)dt = 2\int_{\frac{\gamma-\pi}{2}}^{\frac{\gamma+\pi}{2}}(\cdots)d\beta = \int_{0}^{2\pi}(\cdots)d\beta$$

Therefore, the integrands of the x-axis, y-axis, and z-axis component in Eq. (1.42) turn into

$$\int_{0}^{2\pi}\frac{\cos t}{r_{SP}^3}dt = -\cos\gamma\int_{0}^{2\pi}\frac{\cos(2\beta)}{r_{SP}^3}d\beta$$

$$\int_{0}^{2\pi}\frac{\sin t}{r_{SP}^3}dt = -\sin\gamma\int_{0}^{2\pi}\frac{\cos(2\beta)}{r_{SP}^3}dt$$

$$\int_{0}^{2\pi}\frac{R_1-\sqrt{x_S^2+y_S^2}\cos(\gamma-t)}{r_{SP}^3}dt = \int_{0}^{2\pi}\frac{R_1-\sqrt{x_S^2+y_S^2}+2\sqrt{x_S^2+y_S^2}\cos^2(\beta)}{r_{SP}^3}dt$$

To represent r_{SP}, through the Law of Cosines $r_{SP}^2 = \left(R_1^2+x_S^2+y_S^2-2R_1\sqrt{x_S^2+y_S^2}\cos(t-\gamma)\right)+z_S^2$, one has

$$r_{SP}^2 = \left(R_1+\sqrt{x_S^2+y_S^2}\right)^2 + z_S^2 - 4R_1\sqrt{x_S^2+y_S^2}\sin^2(\beta) \tag{1.43}$$

Define the following parameter

$$k^2 = \frac{4R_1\sqrt{x_S^2+y_S^2}}{\left(R_1+\sqrt{x_S^2+y_S^2}\right)^2+z_S^2} \tag{1.44}$$

Combining Eqs. (1.43) and (1.44), the expression of r_{SP}^3 is

$$r_{SP}^3 = \left(\left(R_1 + \sqrt{x_S^2 + y_S^2}\right)^2 + z_S^2\right)^{\frac{3}{2}} \left(1 - k^2\sin^2\beta\right)^{\frac{3}{2}} \tag{1.45}$$

Thus, Eq. (1.42) turns into

$$\begin{cases} B_{1x} = -\dfrac{\mu_0 I_1 R_1 z_{SP}\cos\gamma}{4\pi\left(\left(R_P+\sqrt{x_S^2+y_S^2}\right)^2+z_S^2\right)^{\frac{3}{2}}} \displaystyle\int_0^{2\pi} \frac{(2\cos^2\beta-1)}{\Delta^3} d\beta \\ B_{1y} = -\dfrac{\mu_0 I_1 R_1 z_{SP}\sin\gamma}{4\pi\left(\left(R_P+\sqrt{x_S^2+y_S^2}\right)^2+z_S^2\right)^{\frac{3}{2}}} \displaystyle\int_0^{2\pi} \frac{(2\cos^2\beta-1)}{\Delta^3} d\beta \\ B_{1z} = \dfrac{\mu_0 I_1 R_1}{4\pi\left(\left(R_P+\sqrt{x_S^2+y_S^2}\right)^2+z_S^2\right)^{\frac{3}{2}}} \displaystyle\int_0^{2\pi} \frac{R_1-\sqrt{x_S^2+y_S^2}+2\sqrt{x_S^2+y_S^2}\cos^2(\beta)}{\Delta^3} d\beta \end{cases} \tag{1.46}$$

where $\Delta = \sqrt{1 - k^2\sin^2\beta}$.

It can be observed that Eq. (1.46) exhibits a form analogous to the complete elliptic integral, where k represents the eccentricity.

The complete elliptic integral of the first kind $K(k)$ is defined and characterized by the following properties:

$$K(k) = \int_0^{\frac{\pi}{2}} \frac{d\beta}{\sqrt{1 - k^2\sin^2\beta}} = \int_0^{\frac{\pi}{2}} \frac{d\beta}{\Delta} \tag{1.47}$$

$$\frac{dK(k)}{dk} = \int_0^{\frac{\pi}{2}} \frac{k\sin^2\beta}{\Delta^3} d\beta \tag{1.48}$$

$$\frac{dK(k)}{dk} = \frac{E(k)}{k(1 - k^2)} - \frac{K(k)}{k} \tag{1.49}$$

The complete elliptic integral of the second kind $E(k)$ is defined and characterized by the following properties:

$$E(k) = \int_0^{\frac{\pi}{2}} \sqrt{1 - k^2\sin^2\beta}\, d\beta = \int_0^{\frac{\pi}{2}} \Delta\, d\beta \tag{1.50}$$

$$\frac{dE(k)}{dk} = \int_0^{\frac{\pi}{2}} \frac{-k\sin^2\beta}{\triangle} d\beta \tag{1.51}$$

$$\frac{dE(k)}{dk} = \frac{E(k) - K(k)}{k} \tag{1.52}$$

Based on Eqs.(1.47)– (1.52), the integrands of the x-axis, y-axis, and z-axis components in Eq. (1.46) turn into

$$\begin{aligned}\int_0^{2\pi} \frac{1-2\cos^2\beta}{\Delta^3} d\beta &= -4\left(\int_0^{\frac{\pi}{2}} \frac{1}{\Delta^3} d\beta - 2\int_0^{\frac{\pi}{2}} \frac{\sin^2\beta}{\Delta^3} d\beta\right) \\ &= -4\left(\frac{k^2-2}{k^2(1-k^2)} E(k) + \frac{2}{k^2} K(k)\right)\end{aligned}$$

$$\begin{aligned}&\int_0^{2\pi} \frac{\sqrt{x_S^2+y_S^2} - R_1 - 2\sqrt{x_S^2+y_S^2}\cos^2\beta}{\Delta^3} d\beta \\ &= \sqrt{x_S^2+y_S^2}\int_0^{2\pi} \frac{1-2\cos^2\beta}{\Delta^3} d\beta - R_1\int_0^{2\pi} \frac{1}{\Delta^3} d\beta \\ &= -\frac{8\sqrt{x_S^2+y_S^2}}{k^2} K(k) - 4\frac{-2\sqrt{x_S^2+y_S^2} + \left(R_1 + \sqrt{x_S^2+y_S^2}\right)k^2}{k^2(1-k^2)} E(k)\end{aligned}$$

Above all, the final expression of Eq. (1.46) is:

$$\begin{cases} B_x(x_S, y_S, z_S) = -\dfrac{\mu_0 I_1 z_S x_S k}{8\pi R_1^{\frac{1}{2}} \left(x_S^2+y_S^2\right)^{5/4}} L_0 \\ B_y(x_S, y_S, z_S) = -\dfrac{\mu_0 I_1 z_S y_S k}{8\pi R_1^{\frac{1}{2}} \left(x_S^2+y_S^2\right)^{5/4}} L_0 \\ B_z(x_S, y_S, z_S) = \dfrac{\mu_0 I_1 k}{8\pi R_1^{\frac{1}{2}} \left(x_S^2+y_S^2\right)^{3/4}} S_0 \end{cases} \tag{1.53}$$

with

$$L_0 = 2K(k) - \frac{2-k^2}{1-k^2} E(k)$$

$$S_0 = 2\sqrt{x_S^2 + y_S^2}K(k) - \frac{2\sqrt{x_S^2 + y_S^2} - \left(R_1 + \sqrt{x_S^2 + y_S^2}\right)k^2}{1 - k^2}E(k)$$

With the expression of the magnetic flux density, the components of the electromagnetic force, $\boldsymbol{F} = I_2 \oint_{l_2} d\boldsymbol{l}_2 \times \boldsymbol{B}_1$, can be expressed as

$$F_x = I_2 \int_0^{2\pi} \left[B_{1z}(x_S, y_S, z_S)dl_{2y} - B_{1y}(x_S, y_S, z_S)dl_{2z}\right]d\phi$$

$$= \frac{\mu_0 I_1 I_2}{8\pi R_1^{1/2}} \int_0^{2\pi} \left[\frac{k}{\left(x_S^2 + y_S^2\right)^{3/4}}\left(S_0 dl_{2y} + \frac{z_S y_S}{\left(x_S^2 + y_S^2\right)^{1/2}} L_0 dl_{2z}\right)\right]d\phi$$

$$F_y = I_2 \int_0^{2\pi} \left[B(x_S, y_S, z_S)dl_{2z} - B_{1z}(x_S, y_S, z_S)dl_{2x}\right]d\phi$$

$$= \frac{\mu_0 I_1 I_2}{8\pi R_1^{1/2}} \int_0^{2\pi} \left[-\frac{k}{\left(x_S^2 + y_S^2\right)^{3/4}}\left(\frac{z_S x_S}{\left(x_S^2 + y_S^2\right)^{1/2}} L_0 dl_{2z} + S_0 dl_{2x}\right)\right]d\phi$$

$$F_z = I_2 \int_0^{2\pi} \left[B_{1y}(x_S, y_S, z_S)dl_{2x} - B_{1x}(x_S, y_S, z_S)dl_{2y}\right]d\phi$$

$$= \frac{\mu_0 I_1 I_2}{8\pi R_1^{1/2}} \int_0^{2\pi} \left[\frac{z_S k}{\left(x_S^2 + y_S^2\right)^{5/4}}\left(x_S L_0 dl_{2y} - y_S L_0 dl_{2x}\right)\right]d\phi$$

The final expression of the electromagnetic force is

$$\begin{cases} F_x = \frac{\mu_0 I_1 I_2}{8\pi R_1^{\frac{1}{2}}} \int_0^{2\pi} I_x d\phi \\ F_y = \frac{\mu_0 I_1 I_2}{8\pi R_1^{\frac{1}{2}}} \int_0^{2\pi} I_y d\phi \\ F_z = \frac{\mu_0 I_1 I_2}{8\pi R_1^{\frac{1}{2}}} \int_0^{2\pi} I_z d\phi \end{cases} \tag{1.54}$$

with

$$\begin{cases} I_x = \frac{k}{\left(x_S^2+y_S^2\right)^{3/4}}\left(S_0 dl_{2y} + \frac{z_S y_S}{\left(x_S^2+y_S^2\right)^{1/2}} L_0 dl_{2z}\right) \\ I_y = -\frac{k}{\left(x_S^2+y_S^2\right)^{3/4}}\left(\frac{z_S x_S}{\left(x_S^2+y_S^2\right)^{1/2}} L_0 dl_{2z} + S_0 dl_{2x}\right) \\ I_z = \frac{z_S k}{\left(x_S^2+y_S^2\right)^{5/4}}\left(x_S dl_{2y} - y_S dl_{2x}\right) L_0 \end{cases}$$

Through electromagnetic forces, Eq. (1.54), the corresponding electromagnetic torque can be readily derived as

$$\boldsymbol{\tau} = I_2 \oint_{l_2} \boldsymbol{R}_2 \times (d\boldsymbol{l}_2 \times \boldsymbol{B}_1) = I_2 \oint_{l_2} \boldsymbol{R}_2 \times d\boldsymbol{F} \tag{1.55}$$

where $dF_x = \frac{\mu_0 I_1}{8\pi R_1^{\frac{1}{2}}} I_x d\phi$, $dF_y = \frac{\mu_0 I_1}{8\pi R_1^{\frac{1}{2}}} I_y d\phi$, $dF_z = \frac{\mu_0 I_1}{8\pi R_1^{\frac{1}{2}}} I_z d\phi$, $\boldsymbol{R}_2$ is the position vector from a fixed reference point x_r in space (e.g., the origin of the global coordinate system) to an arbitrary point x_S on coil 2, that is, $R_2 = x_S - x_r$.

Therefore, the final expression of the electromagnetic torque is

$$\begin{cases} \tau_x = \frac{\mu_0 I_1 I_2}{8\pi R_1^{\frac{1}{2}}} \int_0^{2\pi} \left[(y_S - y_r) I_z - (z_S - z_r) I_y\right] d\phi \\ \tau_y = \frac{\mu_0 I_1 I_2}{8\pi R_1^{\frac{1}{2}}} \int_0^{2\pi} \left[(z_S - z_r) I_x - (x_S - x_r) I_z\right] d\phi \\ \tau_z = \frac{\mu_0 I_1 I_2}{8\pi R_1^{\frac{1}{2}}} \int_0^{2\pi} \left[(x_S - x_r) I_y - (y_S - y_r) I_x\right] d\phi \end{cases} \tag{1.56}$$

At this point, the expressions for the electromagnetic force and torque have been obtained, Eqs. (1.54) and (1.56), that are well-suited for numerical computation and can be readily solved using computers. It should be noted that these equations calculate the electromagnetic force and torque in coil 1 coordinate. To obtain results in the global coordinate, a coordinate transformation, the inverse process of Eqs. (1.36) and (1.37) must be applied. To achieve fast and high-precision computation of the complete elliptic integrals, a method combining piecewise Taylor polynomial approximations with transformations based on Jacobi's nome can be utilized (Fukushima, 2009).

It is worth emphasizing that the near-field model presented in this section also introduces a key approximation: we approximate the closely wound solenoid with a single-layer coil. Consequently, the near-field model may also become inaccurate at very close distances, in which case it is necessary to resort to methods such as finite element analysis (Buck, 2013).

1.1.7 Magnetic dipole and its potential energy in an external field

A small loop carrying a constant current forms a structure similar to an electric dipole, referred to as a magnetic dipole. Here, "small" means that the size of the coil is much smaller than the distance from any point in the space to the center of the coil. The magnetic dipole is characterized by the magnetic moment.

The (mechanical) potential energy U of the current-carrying coil in an external field is equal to the work done by the external forces against the Ampere force while moving the coil, defined as follows:

$$U = -Ni\Phi_e \tag{1.57}$$

where Φ_e is the magnetic flux of the coil in an external magnetic field. According to the definition of potential energy, the relationship between the Ampere force and potential energy can be expressed as

$$\boldsymbol{F}_e = -\nabla U \tag{1.58}$$

For a magnetic dipole, since it can be considered a very small loop, its potential energy and the resulting Ampere force are given by:

$$U = -\boldsymbol{\mu}\boldsymbol{B}_e \tag{1.59}$$

$$\boldsymbol{F}_e = \nabla(\boldsymbol{\mu}\boldsymbol{B}_e) \tag{1.60}$$

where $\boldsymbol{B}_e$ is the magnetic flux density of the external magnetic field. Eqs. (1.29) and (1.60) can both be used to calculate the Ampere force, and they are equivalent for a magnetic dipole.

1.1.8 The far-field model of electromagnetic force and moment

Due to the difficulty in accurately solving Eqs. (1.32) and (1.33) despite the method Eq. (1.54), it is necessary to use a more straightforward approach to calculate the Ampere force and the Ampere moment, aiming to obtain a simplified electromagnetic force model applicable in engineering. Considering that the size of the coils on the spacecraft is generally much smaller than the relative distance between the two spacecraft, the electromagnetic docking device on the spacecraft is first simplified as a magnetic dipole. Then, the far-field model of electromagnetic force based on the magnetic dipole will be derived (Schweighart, 2006).

When $R \ll s$, since $\frac{R}{s} \approx 0$, one can perform a Taylor expansion of $f(\boldsymbol{r}) = \frac{1}{r}$ around $r = s$. Noticing that $f(\boldsymbol{r})$ is a ternary function, its first partial derivative is

$$\frac{\partial f}{\partial r_i} = -\frac{r_i}{r^3}(i = 1, 2, 3) \tag{1.61}$$

where r_i $(i = 1, 2, 3)$ are the components of $\boldsymbol{r}$ on the three coordinate axes. Thus, the Taylor expansion of $f(\boldsymbol{r})$ is

$$\begin{aligned} f(\boldsymbol{r}) &= f(\boldsymbol{s}) + \sum_{i=1}^{3} \frac{\partial f(\boldsymbol{s})}{\partial r_i}(r_i - s_i) + o\left(R^2\right) = \frac{1}{s} + \sum_{i=1}^{3} \frac{s_i R_i}{s^3} + o\left(R^2\right) \\ &= \frac{1}{s} + \frac{\boldsymbol{s} \cdot \boldsymbol{R}}{s^3} + o\left(R^2\right) \end{aligned} \tag{1.62}$$

where $s_i(i = 1, 2, 3)$ are the components of $\boldsymbol{s}$ on the three coordinate axes; $R_i(i = 1, 2, 3)$ are the components of $\boldsymbol{R}$ on the three coordinate axes.

Substituting Eq. (1.62) into (1.21) and retaining only the first two terms, noting that the integral of the first term in the Taylor expansion is zero, one has:

$$\boldsymbol{A} \approx \frac{\mu_0 N i}{4\pi s^3} \oint_{l} (\boldsymbol{s} \cdot \boldsymbol{R}) d\boldsymbol{l} \tag{1.63}$$

Using the vector triple product formula

$$(\boldsymbol{a} \times \boldsymbol{b}) \times \boldsymbol{c} = (\boldsymbol{a} \cdot \boldsymbol{c})\boldsymbol{b} - \boldsymbol{a}(\boldsymbol{b} \cdot \boldsymbol{c}) \tag{1.64}$$

and Eq. (1.31), Eq. (1.63) turns into

$$\boldsymbol{A} = \frac{\mu_0}{4\pi} \frac{\boldsymbol{\mu} \times \boldsymbol{s}}{s^3} \tag{1.65}$$

Through Eq. (1.14), the magnetic field intensity generated by a magnetic dipole at a point in space is given by:

$$\boldsymbol{B} = \nabla \times \boldsymbol{A} = \frac{\mu_0}{4\pi}\left(\frac{3\boldsymbol{s}(\boldsymbol{\mu} \cdot \boldsymbol{s})}{s^5} - \frac{\boldsymbol{\mu}}{s^3}\right) \tag{1.66}$$

The Ampere force $\boldsymbol{F}_{e2}$ on coil 2 due to the magnetic field produced by coil 1 is:

$$\begin{aligned} \boldsymbol{F}_{e2} &= \nabla\left(\boldsymbol{\mu}_2 \cdot \boldsymbol{B}_1\right) \\ &= -\frac{3\mu_0}{4\pi}\left[-\frac{\boldsymbol{\mu}_1 \cdot \boldsymbol{\mu}_2}{d^5}\boldsymbol{d} - \frac{\boldsymbol{\mu}_1 \cdot \boldsymbol{d}}{d^5}\boldsymbol{\mu}_2 - \frac{\boldsymbol{\mu}_2 \cdot \boldsymbol{d}}{d^5}\boldsymbol{\mu}_1 + 5\frac{(\boldsymbol{\mu}_1 \cdot \boldsymbol{d})(\boldsymbol{\mu}_2 \cdot \boldsymbol{d})}{d^7}\boldsymbol{d}\right] \end{aligned}$$

(1.67)

The moment τ_{e2} can be calculated by using the fact that a moment applied over a rotation is equal to the change in potential energy, which can be represented by differentiating both sides of Eq. (1.59)

$$dU = -\boldsymbol{\tau} \cdot d\boldsymbol{\theta} = -d\boldsymbol{\mu}\boldsymbol{B} \tag{1.68}$$

The change in the magnetic dipole due to the rotation is given by

$$d\boldsymbol{\mu} = d\boldsymbol{\theta} \times \boldsymbol{\mu} \tag{1.69}$$

Thus, one has

$$\boldsymbol{\tau}_{e2} = \boldsymbol{\mu}_2 \times \boldsymbol{B}_1 \tag{1.70}$$

The moment $\boldsymbol{\tau}_{e2}$ on coil 2 due to the magnetic field produced by coil 1 is:

$$\boldsymbol{\tau}_{e2} = \frac{\mu_0}{4\pi}\left(\boldsymbol{\mu}_2 \times \left(\frac{3\boldsymbol{s}(\boldsymbol{\mu} \cdot \boldsymbol{s})}{s^5} - \frac{\boldsymbol{\mu}}{s^3}\right)\right) \tag{1.71}$$

The far field model is the most widely used model in the controller design of spacecraft electromagnetic docking (Zhang et al., 2013; Zhang et al., 2020) and electromagnetic formation flight (Huang et al., 2016; Wang et al., 2024). However, as the electromagnetic far-field model completely neglects the geometry of the coil, this leads to a rapid increase in the error of the near-field model at close distances. Consider two coils, both with a radius of $R = 1$m, turns $N = 200$, and current $I = 5$A, which are illustrated in Fig. 1.5. Fig. 1.6 shows the error characteristics of the far-field model at different distances d when the two coils are aligned. The near field force results are calculated by Eq. (1.54). It can be observed that when the coil separation exceeds 6 to 8 times the coil radius, the relative amplitude force error of the far-field model is less than 10%. As the distance decreases, the relative error increases with a power-law-like dependence, which arises from the distance raised to the fourth power in the denominator of the far-field model. Fig. 1.7 shows the force error characteristics of the far-field model for two coils at different angles α between their magnetic moments. It can be observed that at large distances, the far-field model can be supposed to have the same relative amplitude error in all directions. At closer distances, however, the relative amplitude error is minimized when the angle between the magnetic moments is $\pm 90°$. Meanwhile, the directional error, defined as the angle between the far-field results and near-field results, consistently

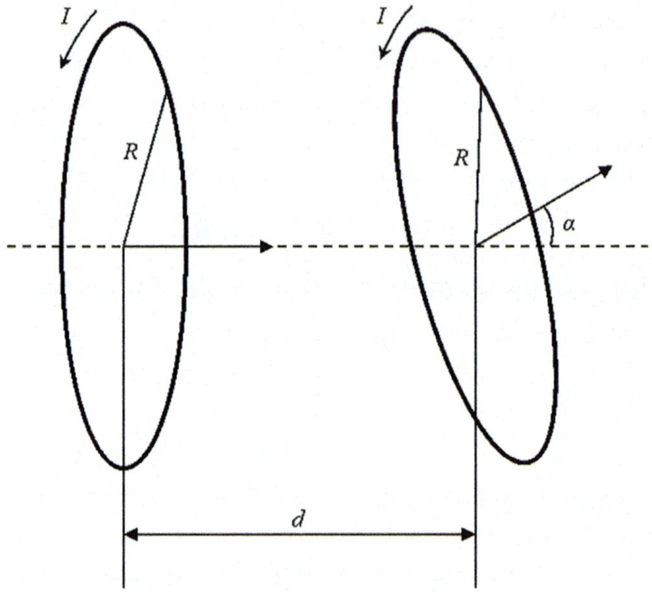

Figure 1.5 Relative position of coils. The following error analyses are based on this relative configuration.

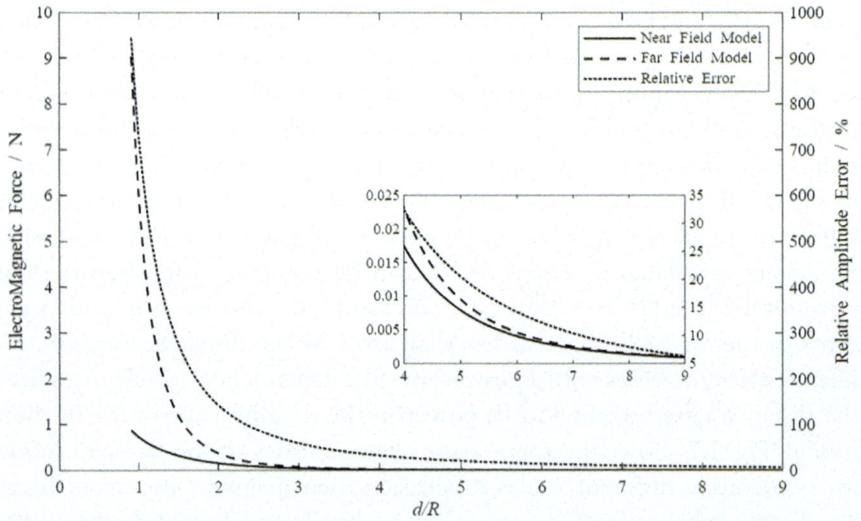

Figure 1.6 Error characteristics of the far-field model at different distances. The far-field model (dashed lines) shows good agreement with the near-field model (solid lines) at distances greater than 6~8 coil radii.

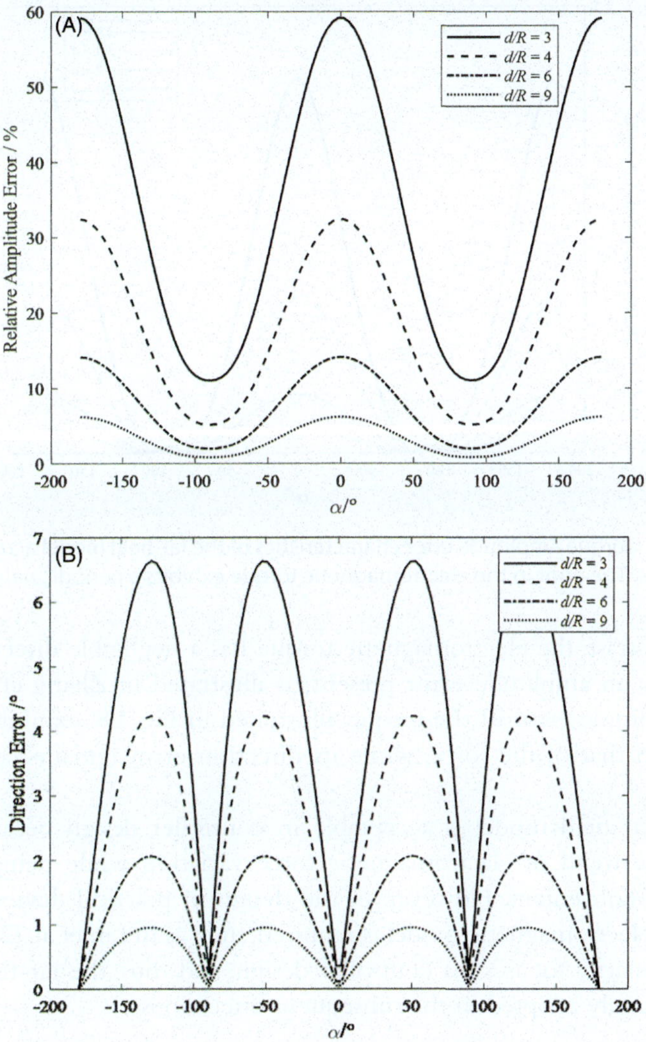

Figure 1.7 Force error characteristics of the far-field model across different axial angles. The error in the electromagnetic force exhibits a periodic nature due to its rotational symmetry. Specifically, when the coils are aligned, the magnitude error is relatively large, while the direction error remains small.

remains within the acceptable range (<1°). The directional error reaches its minimum at multiples of 90° and gradually increases as the relative distance decreases.

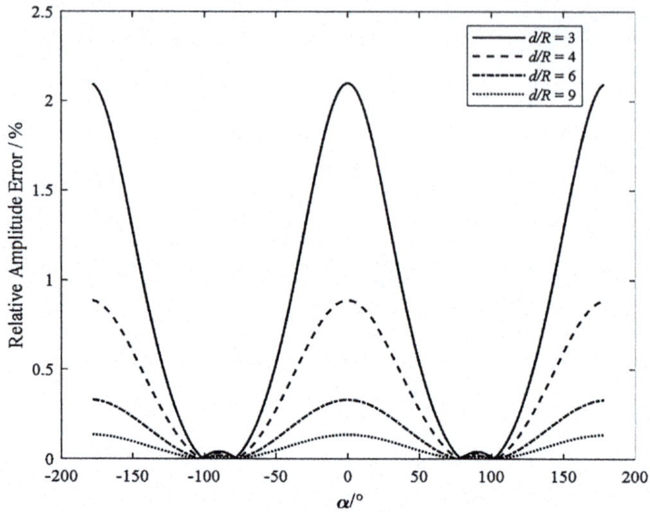

Figure 1.8 Torque amplitude error characteristics of the far-field model across different axial angles. The error in the electromagnetic torque exhibits a periodic nature as well.

In contrast, the electromagnetic torque has a negligible direction error, with only an amplitude error present, as illustrated in Zhang et al. (2011). The amplitude error of the torque, illustrated in Fig. 1.8, exhibits a similar pattern to that of the force, while its direction error remains consistently small.

The far-field model is acceptable in controller design because it can reflect the trend of electromagnetic force variation while being computationally convenient. However, in the design of practical devices, a more accurate electromagnetic model is required, such as in Lin et al. (2025). This book primarily focuses on controller design, and thus, the far-field model will be mainly adopted in the subsequent discussions.

1.2 Dynamic model of spacecraft electromagnetic docking and separation

1.2.1 Relative motion dynamics of spacecraft

To describe the associated dynamics and facilitate the following derivation, "CM" is defined first to denote the center of mass of the spacecraft electromagnetic docking system. To describe the relative motion of the spacecraft, the following coordinate systems are formed.

(1) Earth centered inertial frame (EI)

The Earth-centered inertial frame is a global coordinate system whose origin lies at the Earth's center. Its Z-axis coincides with the Earth's rotation axis, and its X-axis points toward the vernal equinox.

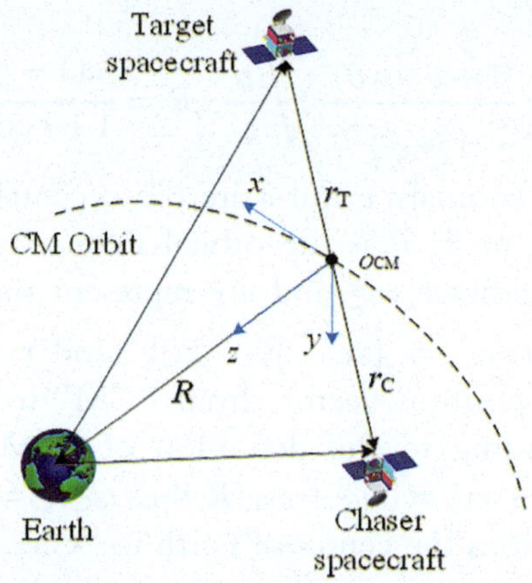

Figure 1.9 CM orbital frame. The CM orbital frame refers to the orbital coordinate system located at the center of mass of the docking system.

(2) CM orbital frame (CM)

The CM orbital frame is an orbital frame where the origin is attached to "CM" with z-axis pointing radially toward the Earth's center, x-axis perpendicular to z-axis in the nominal direction of flight, and y-axis completing the right-handed coordinate system. The motion of frame $O_{CM}xyz$ is not affected by electromagnetic force, and thus it could be hypothesized as a reference frame, in which the dynamic models are developed. The relative motions satisfy constraint conditions based on Tschauner–Hempel (TH) equations, and extending them to an eccentric orbit by using the true anomaly of "CM" gives the dynamic models of chaser and target spacecraft relative to "CM" expressed by Eqs. (1.73) and (1.73) respectively.

The relevant coordinate system adopted for spacecraft electromagnetic docking missions is the orbital reference frame shown in Fig. 1.9,

$$\begin{cases} \ddot{x}_C - \omega^2 x_C - 2\omega \dot{z}_C - \dot{\omega} z_C = -\dfrac{\mu x_C}{\|\boldsymbol{R}+\boldsymbol{r}_C\|^3} + \dfrac{F_{eCx}}{m_C} + \dfrac{w_{Cx}}{m_C} \\ \ddot{y}_C = -\dfrac{\mu y_C}{\|\boldsymbol{R}+\boldsymbol{r}_C\|^3} + \dfrac{F_{eCy}}{m_C} + \dfrac{w_{Cy}}{m_C} \\ \ddot{z}_C - \omega^2 z_C + 2\omega \dot{x}_C + \dot{\omega} x_C = \dfrac{2\mu z_C}{\|\boldsymbol{R}+\boldsymbol{r}_C\|^3} + \dfrac{F_{eCz}}{m_C} + \dfrac{w_{Cz}}{m_C} \end{cases} \tag{1.72}$$

$$\begin{cases} \ddot{x}_T - \omega^2 x_T - 2\omega \dot{z}_T - \dot{\omega} z_T = -\dfrac{\mu x_T}{\|\boldsymbol{R}+\boldsymbol{r}_T\|^3} + \dfrac{F_{eTx}}{m_T} + \dfrac{w_{Tx}}{m_T} \\ \ddot{y}_T = -\dfrac{\mu y_T}{\|\boldsymbol{R}+\boldsymbol{r}_T\|^3} + \dfrac{F_{eTy}}{m_T} + \dfrac{w_{Ty}}{m_T} \\ \ddot{z}_T - \omega^2 z_T + 2\omega \dot{x}_T + \dot{\omega} x_T = \dfrac{2\mu z_T}{\|\boldsymbol{R}+\boldsymbol{r}_T\|^3} + \dfrac{F_{eTz}}{m_T} + \dfrac{w_{Tz}}{m_T} \end{cases} \tag{1.73}$$

with

$$\omega = \frac{(1 + e\cos\theta)^2}{(1 - e^2)^{\frac{3}{2}}}\sqrt{\frac{\mu}{a^3}},\ R = \frac{a(1 - e^2)}{1 + e\cos\theta} \tag{1.74}$$

where θ is the true anomaly, e and a are the eccentricity and semi-major axis of "CM" orbit; $\omega = \dot{\theta}$ is the orbital frequency of "CM" and μ is the gravitational parameter; m_C and m_T represent the mass of the chaser and target spacecraft; $\boldsymbol{r}_C = [x_C \quad y_C \quad z_C]^T$ and $\boldsymbol{r}_T = [x_T \quad y_T \quad z_T]^T$ denote the relative position vectors from "CM" to the chaser and target spacecraft, according to the definition of "CM" point, one knows $m_C(\boldsymbol{R} + \boldsymbol{r}_C) + m_T(\boldsymbol{R} + \boldsymbol{r}_T) = (m_C + m_T)\boldsymbol{R}$, thus $m_C\boldsymbol{r}_C = -m_T\boldsymbol{r}_T$; $\boldsymbol{R}$ represents the position vector from the center of Earth to "CM," with $R = ||\boldsymbol{R}||$ as "CM" orbit radius and $||\bullet||$ denotes 2-norm; F_{eCi} and F_{eTi} ($i = x, y, z$) denote the ith component of electromagnetic forces acting on the two spacecraft with $F_{eTi} = -F_{eCi}$; w_{Ci} and w_{Ti} ($i = x, y, z$) are the i-th component of external disturbances acting on the chaser and target spacecraft.

The external disturbances acting on the chaser and target spacecraft are often generated due to many factors. For example, the gradient of the geomagnetic field strength gives the force on the two spacecraft due to the geomagnetic field, which is due to the core of the planet and can be described as a combination of main field. In addition, an external component (that can be as much as 10% of the main field) often exists, which is due to sources such as the ionosphere and solar wind effects on the Earth's magnetosphere.

Denoting $\tilde{\boldsymbol{\eta}} = [x \quad y \quad z]^T$ as the relative position vector from target to chaser spacecraft, that is, $\tilde{\boldsymbol{\eta}} = r_C - r_T$ and $||\tilde{\boldsymbol{\eta}}|| = ||\boldsymbol{r}_C|| + ||\boldsymbol{r}_T||$, with Eqs. (1.72) and (1.73), one has

$$\begin{cases} \ddot{x} - \omega^2 x - 2\omega\dot{z} - \dot{\omega}z = -\frac{\mu x_C}{||\boldsymbol{R}+\boldsymbol{r}_C||^3} + \frac{\mu x_T}{||\boldsymbol{R}+\boldsymbol{r}_T||^3} + \frac{F_{eCx}}{m_C} + \frac{F_{eCx}}{m_T} + \frac{w_{Cx}}{m_C} - \frac{w_{Tx}}{m_T} \\ \ddot{y} = -\frac{\mu y_C}{||\boldsymbol{R}+\boldsymbol{r}_C||^3} + \frac{\mu y_T}{||\boldsymbol{R}+\boldsymbol{r}_T||^3} + \frac{F_{eCy}}{m_C} + \frac{F_{eCy}}{m_T} + \frac{w_{Cy}}{m_C} - \frac{w_{Ty}}{m_T} \\ \ddot{z} - \omega^2 z + 2\omega\dot{x} + \dot{\omega}x = \frac{2\mu z_C}{||\boldsymbol{R}+\boldsymbol{r}_C||^3} - \frac{2\mu z_T}{||\boldsymbol{R}+\boldsymbol{r}_T||^3} + \frac{F_{eCz}}{m_C} + \frac{F_{eCz}}{m_T} + \frac{w_{Cz}}{m_C} - \frac{w_{Tz}}{m_T} \end{cases} \tag{1.75}$$

For small relative distance between the chaser and target spacecraft in comparison with "CM" orbit radius, that is, $||\tilde{\boldsymbol{\eta}}|| \ll R$, Eq. (1.75) is converted into

$$\begin{cases} \ddot{x} - \omega^2 x - 2\omega\dot{z} - \dot{\omega}z = -\frac{\mu x}{R^3} + \frac{1}{m_{CT}}(F_{eCx} + d_x) \\ \ddot{y} = -\frac{\mu y}{R^3} + \frac{1}{m_{CT}}(F_{eCy} + d_y) \\ \ddot{z} - \omega^2 z + 2\omega\dot{x} + \dot{\omega}x = \frac{2\mu z}{R^3} + \frac{1}{m_{CT}}(F_{eCz} + d_z) \end{cases} \tag{1.76}$$

where, $m_{CT} = \frac{m_C m_T}{m_C + m_T}$, $d_{i=x,y,z} = \frac{m_T w_{Ci} - m_C w_{Ti}}{m_C + m_T}$.

Let

$$\boldsymbol{\eta} = \begin{bmatrix} x & y & z & \dot{x} & \dot{y} & \dot{z} \end{bmatrix}^{\mathrm{T}}, \ \boldsymbol{u} = \begin{bmatrix} F_{eCx} & F_{eCy} & F_{eCz} \end{bmatrix}^{\mathrm{T}}, \ \boldsymbol{w}_0 = \begin{bmatrix} d_x & d_y & d_z \end{bmatrix}^{\mathrm{T}}$$

Due to the parameter perturbations, modeling errors, and environmental changes, the mass of chaser and target spacecraft may not be exactly known; thus, the representative of a valid mass can be used in controller design, and then the mass uncertainty resulting from the difference between m_0 and m_{CT} will appear. In particular, the mass of chaser and target spacecraft may be unknown, which leads to unknown m_{CT}. To study analytically the electromagnetic docking of spacecraft, the admissible parameter uncertainties are assumed to have a norm-bounded form in order to have a sufficient condition for exponential stability, which is the most adopted form in robust stability analysis. With Eq. (1.74), Eq. (1.76) can be written as

$$\dot{\boldsymbol{\eta}}(t) = (\boldsymbol{A} + \Delta\boldsymbol{A})\boldsymbol{\eta}(t) + (\boldsymbol{B} + \Delta\boldsymbol{B})(\boldsymbol{u}(t) + \boldsymbol{w}_0(t)) \tag{1.77}$$

where

$$\boldsymbol{A} = \begin{bmatrix} 0 & 0 & 0 & 1 & 0 & 0 \\ 0 & 0 & 0 & 0 & 1 & 0 \\ 0 & 0 & 0 & 0 & 0 & 1 \\ 0 & 0 & 0 & 0 & 0 & 2\omega_0 \\ 0 & -\omega_0^2 & 0 & 0 & 0 & 0 \\ 0 & 0 & 3\omega_0^2 & -2\omega_0 & 0 & 0 \end{bmatrix}$$

$$\Delta\boldsymbol{A} = \begin{bmatrix} 0 & 0 & 0 & 0 & 0 & 0 \\ 0 & 0 & 0 & 0 & 0 & 0 \\ 0 & 0 & 0 & 0 & 0 & 0 \\ \frac{e\cos\theta}{1+e\cos\theta}\omega^2 & 0 & \dot{\omega} & 0 & 0 & 2(\omega - \omega_0) \\ 0 & \omega_0^2 - \frac{\omega^2}{1+e\cos\theta} & 0 & 0 & 0 & 0 \\ -\dot{\omega} & 0 & \frac{3+e\cos\theta}{1+e\cos\theta}\omega^2 - 3\omega_0^2 & 2(\omega_0 - \omega) & 0 & 0 \end{bmatrix}$$

$$\boldsymbol{B} = \begin{bmatrix} \mathbf{0}_{3\times3} & \mathrm{diag}\left(\begin{bmatrix} m_0^{-1} & m_0^{-1} & m_0^{-1} \end{bmatrix}\right) \end{bmatrix}^{\mathrm{T}}$$

$$\Delta\boldsymbol{B} = \begin{bmatrix} \mathbf{0}_{3\times3} & \mathrm{diag}\left(\begin{bmatrix} m_{CT}^{-1} - m_0^{-1} & m_{CT}^{-1} - m_0^{-1} & m_{CT}^{-1} - m_0^{-1} \end{bmatrix}\right) \end{bmatrix}^{\mathrm{T}}$$

where $\omega_0 = \sqrt{\frac{\mu}{a^3}}$ and m_0 denotes a valid mass.

The control input $\boldsymbol{u}$ can be calculated by Eq. (1.67), which is equivalent to

$$\begin{aligned}\boldsymbol{F}_{\mathrm{eC}} &= \frac{3\mu_0}{4\pi}\left[\frac{1}{\|\tilde{\boldsymbol{\eta}}\|^5}\left(\tilde{\boldsymbol{\eta}}\left(\boldsymbol{\mu}_{\mathrm{T}}^{\mathrm{T}}\boldsymbol{\mu}_{\mathrm{C}}\right)+\left(\boldsymbol{\mu}_{T}^{\mathrm{T}}\tilde{\boldsymbol{\eta}}\right)\boldsymbol{\mu}_{\mathrm{C}}+\boldsymbol{\mu}_{\mathrm{T}}\left(\tilde{\boldsymbol{\eta}}^{\mathrm{T}}\boldsymbol{\mu}_{\mathrm{C}}\right)\right)\right.\\ &\quad\left.-\frac{5}{\|\tilde{\boldsymbol{\eta}}\|^7}\tilde{\boldsymbol{\eta}}\left(\boldsymbol{\mu}_{\mathrm{T}}^{\mathrm{T}}\tilde{\boldsymbol{\eta}}\right)\left(\tilde{\boldsymbol{\eta}}^{\mathrm{T}}\boldsymbol{\mu}_{\mathrm{C}}\right)\right]\\ &= \frac{3\mu_0}{4\pi}\left[\frac{1}{\|\tilde{\boldsymbol{\eta}}\|^5}\left(\tilde{\boldsymbol{\eta}}\,\boldsymbol{\mu}_{\mathrm{T}}^{\mathrm{T}}+\boldsymbol{\mu}_{\mathrm{T}}^{\mathrm{T}}\,\tilde{\boldsymbol{\eta}}\,\boldsymbol{I}_3+\boldsymbol{\mu}_{\mathrm{T}}\tilde{\boldsymbol{\eta}}^{\mathrm{T}}\right)-\frac{5}{\|\tilde{\boldsymbol{\eta}}\|^7}\,\tilde{\boldsymbol{\eta}}\,\boldsymbol{\mu}_{\mathrm{T}}^{\mathrm{T}}\,\tilde{\boldsymbol{\eta}}\,\tilde{\boldsymbol{\eta}}^{\mathrm{T}}\right]\boldsymbol{\mu}_{\mathrm{C}}\\ &= \frac{3\mu_0}{4\pi}\Xi\left(\boldsymbol{\mu}_{\mathrm{T}},\tilde{\boldsymbol{\eta}}\right)\boldsymbol{\mu}_{\mathrm{C}}\end{aligned} \tag{1.78}$$

where

$$\Xi\left(\boldsymbol{\mu}_{\mathrm{T}},\tilde{\boldsymbol{\eta}}\right)=\frac{1}{\left|\tilde{\boldsymbol{\eta}}\right|^5}\left(\tilde{\boldsymbol{\eta}}\,\boldsymbol{\mu}_{\mathrm{T}}^{\mathrm{T}}+\boldsymbol{\mu}_{\mathrm{T}}^{\mathrm{T}}\,\tilde{\boldsymbol{\eta}}\,\boldsymbol{I}_3+\boldsymbol{\mu}_{\mathrm{T}}\tilde{\boldsymbol{\eta}}^{\mathrm{T}}\right)-\frac{5}{\left|\tilde{\boldsymbol{\eta}}\right|^7}\,\tilde{\boldsymbol{\eta}}\,\boldsymbol{\mu}_{\mathrm{T}}^{\mathrm{T}}\,\tilde{\boldsymbol{\eta}}\,\tilde{\boldsymbol{\eta}}^{\mathrm{T}}$$

where $\boldsymbol{F}_{\mathrm{eC}}$ and $\boldsymbol{\tau}_{\mathrm{eC}}$ denote the electromagnetic force and torque generated by the dipole on target spacecraft affecting the dipole on chaser spacecraft; $\tilde{\boldsymbol{\eta}}$ is the position vector from the target spacecraft to the chaser spacecraft; $\boldsymbol{\mu}_0$ is the permeability of free space; $\boldsymbol{\mu}_{\mathrm{C}}$ and $\boldsymbol{\mu}_{\mathrm{T}}$ denote magnetic moment vectors of the chaser and target spacecraft, respectively.

It is worth noting that the magnetic moment vectors are defined as

$$\boldsymbol{\mu}_{\mathrm{C,T}} = NiS\hat{\boldsymbol{n}} \tag{1.79}$$

1.2.2 Relative attitude dynamics of spacecraft

To describe the relative attitude dynamics of spacecraft electromagnetic docking, the following coordinate systems are added.

(1) Electromagnetic Frame (EM)

The electromagnetic frame is defined with its origin at the CM point. Its x-axis points from the mass center of target to the mass center of chaser. Its z-axis is placed in the orbital plane of the CM point and is normal to x-axis. Its y-axis completes the right-handed system. It is utilized to simplify the formulas of electromagnetic force and torque, and is the reference frame of relative attitude. In the EM frame, we define the following relative attitude angles.

α: The angle between the X-axis of EM frame and the x-axis of body frame.

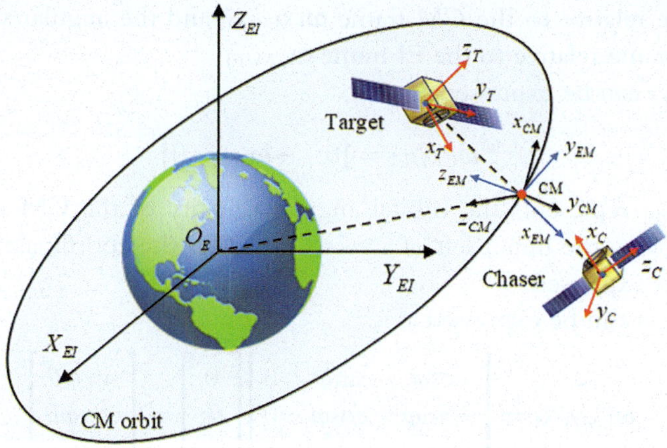

Figure 1.10 Coordinate systems of electromagnetic docking. The EM frame and the CM frame share the same origin and can be transformed into each other through two Euler rotations. The body frames of both spacecraft are located at their respective centers of mass and are aligned with their principal axes of inertia.

χ: The angle between the projection of the x-axis of the body frame onto the yz-plane of the EM frame and the y-axis of the EM frame.

(2) Body frames (BC and BT)

The body frames of chaser and target are attached to the center of mass of each spacecraft. Their axes are chosen to coincide with the principal axes of inertia.

All four coordinate systems in this section are illustrated in Fig. 1.10.

The Euler equation is widely used to describe the attitude dynamics of a rigid spacecraft, which is illustrated as follows:

$$\boldsymbol{I}\dot{\boldsymbol{\omega}} + \boldsymbol{\omega} \times (\boldsymbol{I\omega}) = \boldsymbol{M} + \boldsymbol{\tau} \tag{1.80}$$

where $\boldsymbol{I} = \mathrm{diag}([I_x\ I_y\ I_z])$ is the inertia matrix of the spacecraft; $\boldsymbol{\omega}$ is the spacecraft angular velocity relative to the Earth-centered inertial frame; $\boldsymbol{M} = [M_x\ M_y\ M_z]^{\mathrm{T}}$ is the control torque, while $\boldsymbol{\tau}$ is the disturbance torque including electromagnetic torque, gravity gradient torque, aerodynamic torque, etc.

To derive the Euler equations expressed in terms of relative attitude angles α and χ, the kinematic equations of attitude motion must first be established. The relative attitude angles describe the orientation of the body frame with respect to the EM frame. Therefore, the absolute angular velocity of the spacecraft can be decomposed into the sum of: the angular velocity of the spacecraft relative to the EM frame $\omega_{\mathrm{B/EM}}$, the angular velocity of the

EM frame relative to the CM frame $\boldsymbol{\omega}_{\mathrm{EM/CM}}$, and the angular velocity of the CM frame relative to the EI frame $\boldsymbol{\omega}_{\mathrm{CM/EI}}$.

$\boldsymbol{\omega}_{\mathrm{CM/EI}}$ can be expressed as

$$^{\mathrm{CM}}\boldsymbol{\omega}_{\mathrm{CM/EI}} = \begin{bmatrix} 0 & -\omega_{\mathrm{orb}} & 0 \end{bmatrix}^{\mathrm{T}}$$

where ω_{orb} represents the orbital angular velocity of the CM orbit. The superscript on the upper left of a vector indicates the coordinate system in which it is expressed.

$\boldsymbol{\omega}_{\mathrm{EM/CM}}$ can be expressed as

$$\boldsymbol{\omega}_{\mathrm{EM/CM}} = \begin{bmatrix} \cos\varphi & \sin\varphi & 0 \\ -\sin\varphi & \cos\varphi & 0 \\ 0 & 0 & 1 \end{bmatrix} \begin{bmatrix} 0 \\ \dot{\theta}_2 \\ \dot{\varphi}_2 \end{bmatrix} = \begin{bmatrix} \sin\varphi\dot{\theta} \\ \cos\varphi\dot{\theta} \\ \dot{\varphi} \end{bmatrix}$$

where φ and θ are the relative position angles defined as $\varphi = \arcsin(y/r)$ and $\theta = \arctan2(z/x)$.

$\boldsymbol{\omega}_{\mathrm{B/EM}}$ can be expressed as

$$\boldsymbol{\omega}_{\mathrm{B/EM}} = \begin{bmatrix} \cos\alpha & \sin\alpha & 0 \\ -\sin\alpha & \cos\alpha & 0 \\ 0 & 0 & 1 \end{bmatrix} \begin{bmatrix} \dot{\chi} \\ 0 \\ \dot{\alpha} \end{bmatrix} = \begin{bmatrix} \cos\alpha\dot{\chi} \\ -\sin\alpha\dot{\chi} \\ \dot{\alpha} \end{bmatrix}$$

Thus, $\boldsymbol{\omega}$ in the body frame is equal to

$$^{\mathrm{B}}\boldsymbol{\omega} = \begin{bmatrix} ^{\mathrm{B}}\omega_x \\ ^{\mathrm{B}}\omega_y \\ ^{\mathrm{B}}\omega_z \end{bmatrix} = \begin{bmatrix} \cos\alpha\dot{\chi} \\ -\sin\alpha\dot{\chi} \\ \alpha \end{bmatrix} + \boldsymbol{T}_{\mathrm{B}}^{\mathrm{EM}} \begin{bmatrix} \sin\varphi(\dot{\theta} - \omega_{\mathrm{orb}}) \\ \cos\varphi(\dot{\theta} - \omega_{\mathrm{orb}}) \\ \dot{\varphi} \end{bmatrix}$$

where $\boldsymbol{T}_{\mathrm{B}}^{\mathrm{EM}}$ represents the coordinate rotation matrix from the EM frame to the body frame, which can be expressed as

$$\boldsymbol{T}_{\mathrm{B}}^{\mathrm{EM}} = \begin{bmatrix} \cos\alpha & \sin\alpha\cos\chi & \sin\alpha\sin\chi \\ -\sin\alpha & \cos\alpha\cos\chi & \cos\alpha\sin\chi \\ 0 & -\sin\chi & \cos\chi \end{bmatrix} \tag{1.81}$$

Define the transport angular velocity as $\boldsymbol{\omega}_e = \boldsymbol{\omega}_{\mathrm{EM/EI}}$. Substituting Eq. (1.81) into Eq. (1.80), one can get the relative attitude dynamics model:

$$\begin{cases} \ddot{\alpha} = \frac{1}{I_z}\left[M - I_z\dot{\omega}_{ez} - (I_y - I_x)(\cos\alpha\dot{\chi} + {}^{\mathrm{B}}\omega_{ex})(-\sin\alpha\dot{\chi} + {}^{\mathrm{B}}\omega_{ey})\right] \\ \ddot{\chi} = \frac{1}{I_x\cos\alpha}\left[M_x - I_x(-\dot{\alpha}\dot{\chi}\sin\alpha + \dot{\omega}_{ex}) - (I_z - I_y)(-\sin\alpha\dot{\chi} + {}^{\mathrm{B}}\omega_{ey})(\dot{\alpha} + {}^{\mathrm{B}}\omega_{ez})\right] \end{cases} \tag{1.82}$$

Table 1.1 PD controller parameters.

Variable/control parameters	K_p	K_d
α_C	−6	−10
χ_C	−5	−2
α_T	−30	−100
χ_T	−5	−2

It is easy to design a PD controller for the dynamics system denoted by Eq. (1.82) to stabilize the relative attitude angles during the docking procedure. The following is a simple closed-loop simulation for electromagnetic docking. The inertial matrices are chosen as $\boldsymbol{I}_C = \text{diag}([0.36\ 3.39\ 3.63])\text{kg}\bullet\text{m}^2$, $\boldsymbol{I}_T = \text{diag}\ ([1.68\ 9.26\ 9.08])\text{kg}\bullet\text{m}^2$. The masses of the spacecraft are $m_C = 40\text{kg}$ and $m_T = 100\text{kg}$. The coils are mounted along the x-axis of the body frame, with $||\boldsymbol{\mu}_T||_2 = 1, 0000\text{Am}^2$ and radius $R_{\text{coil}} = 0.1\text{m}$. The orbit elements are as follows. The eccentricity is 0.15, the semimajor axis is 10,000km, the inclination is 30°, the longitude of the ascending node is 30°, the argument of periapsis is 30°, and the initial true anomaly is 0°. The initial state is chosen as $\boldsymbol{\eta}(0) = [0.6\text{m}\ 0.3\text{m}\ 0.4\text{m}\ 0\ 0\ 0]^T$, $\alpha_C(0) = \pi/8$, $\alpha_C(0) = \pi/12$, $\chi_C(0) = -\pi/12$, $\chi_T(0) = \pi/12$ and $\dot{\alpha}_C(0) = \dot{\alpha}_T(0) = \dot{\chi}_C(0) = \dot{\chi}_T(0) = 0$. Considering the coil configuration and the spacecraft volume, the target state of relative motion is set to ${}^{EM}\boldsymbol{\eta}(0) = [0.4\text{m}\quad 0\text{m}\quad 0\text{m}\quad 0\quad 0\quad 0]^T$. The controller parameters are illustrated in Table 1.1. The disturbance torque is the geomagnetic disturbance torque calculated based on the 14th International Geomagnetic Reference Field (IGRF) model (Alken et al., 2021).

The desired docking configuration should be such that it maximizes the component of the relative electromagnetic force along the x-axis of the EM frame, while keeping any additional electromagnetic torque between the coils relatively small. Since the discrepancy between the far-field and near-field models of the electromagnetic torque lies primarily in magnitude, with the direction error being negligible (Zhang et al., 2011), an analysis based on the far-field model can be used to identify configurations that minimize the relative electromagnetic torque.

The electromagnetic torques in EM can be expressed as

$$\begin{cases} {}^{EM}\tau_{Cx} = \frac{\mu_0\mu_C\mu_T}{4\pi r^3}\sin\alpha_C\sin\alpha_T\sin(\chi_C - \chi_T) \\ {}^{EM}\tau_{Cy} = \frac{\mu_0\mu_C\mu_T}{4\pi r^3}(2\sin\alpha_C\sin\chi_C\cos\alpha_T + \cos\alpha_C\sin\alpha_T\sin\chi_T) \\ {}^{EM}\tau_{Cz} = -\frac{\mu_0\mu_C\mu_T}{4\pi r^3}(\cos\alpha_C\sin\alpha_T\cos\chi_T + 2\sin\alpha_C\cos\chi_C\cos\alpha_T) \end{cases} \tag{1.83}$$

Table 1.2 Comparisons of the two docking configurations.

	${}^{B}F_{ecx}$	Ratio of α-angle Restoring Torque to Disturbance Torque	χ-angle Static Stability
Config. 1	$-\frac{3\mu_0\mu_T\mu_C}{2\pi r^4}$	$2\frac{\delta\alpha_C}{\delta\alpha_T}$ (Chaser), $2\frac{\delta\alpha_T}{\delta\alpha_C}$ (Target)	Neutrally stable
Config. 2	$-\frac{3\mu_0\mu_T\mu_C}{4\pi r^4}$	$\frac{1}{2}\frac{\delta\alpha_C}{\delta\alpha_T}$ (Chaser), $\frac{1}{2}\frac{\delta\alpha_T}{\delta\alpha_C}$ (Target)	Static stable

$$\begin{cases} {}^{EM}\tau_{Tx} = \frac{\mu_0\mu_C\mu_T}{4\pi r^3}\sin\alpha_C\sin\alpha_T\sin(\chi_T-\chi_C) \\ {}^{EM}\tau_{Ty} = \frac{\mu_0\mu_C\mu_T}{4\pi r^3}(2\cos\alpha_C\sin\alpha_T\sin\chi_T+\sin\alpha_C\sin\chi_C\cos\alpha_T) \\ {}^{EM}\tau_{Tz} = -\frac{\mu_0\mu_C\mu_T}{4\pi r^3}(\sin\alpha_C\cos\chi_C\cos\alpha_T+2\cos\alpha_C\sin\alpha_T\cos\chi_T) \end{cases} \tag{1.84}$$

By observing Eqs. (1.83) and (1.84), and assuming that the current direction follows the right-hand rule with respect to the x-axis of the body frame, it can be seen that the electromagnetic torque exhibits the following four balanced configurations. Among them, the first and second are docking configurations, where the electromagnetic force between the spacecraft is attractive, while the third and fourth are separation configurations, where the electromagnetic force between the spacecraft is repulsive.

(1) $\alpha_C = \alpha_T = 0,\ \pi, \forall\chi_T, \chi_C$

(2) $\chi_T = \chi_C,\ \alpha_C = -\alpha_T = \pm\frac{\pi}{2}$

(3) $\alpha_C = -\alpha_T = 0, \pi, \forall\chi_T, \chi_C$

(4) $\chi_T = \chi_C,\ \alpha_C = \alpha_T = \pm\frac{\pi}{2}$

Actually, the difference between configurations (1), (2), and (3), (4) is only the direction of force and torque. Focusing on docking configurations (1) and (2), their advantages and disadvantages are summarized in Table 1.2, where $\delta\alpha$ denotes a small perturbation in the α-angle.

It is noted that the electromagnetic force between magnetic dipoles is conservative, and consequently, the electromagnetic torque is also conservative. Therefore, its potential energy function can be analyzed for this purpose.

$$\begin{aligned} U &= -\frac{\mu_0}{4\pi r^3}\left[3(\boldsymbol{\mu}_C\cdot\hat{\boldsymbol{r}})(\boldsymbol{\mu}_T\cdot\hat{\boldsymbol{r}}) - \boldsymbol{\mu}_C\cdot\boldsymbol{\mu}_T\right] \\ &= -\frac{\mu_0\mu_C\mu_T}{4\pi r^3}[2\cos\alpha_C\cos\alpha_T - \sin\alpha_C\sin\alpha_T\cos(\chi_C-\chi_T)] \end{aligned} \tag{1.85}$$

where $\hat{\boldsymbol{r}}$ represents the unit vector of $\boldsymbol{r}$. By analyzing the gradient and Hessian matrix of the potential energy function Eq. (1.85), similar conclusions can be drawn.

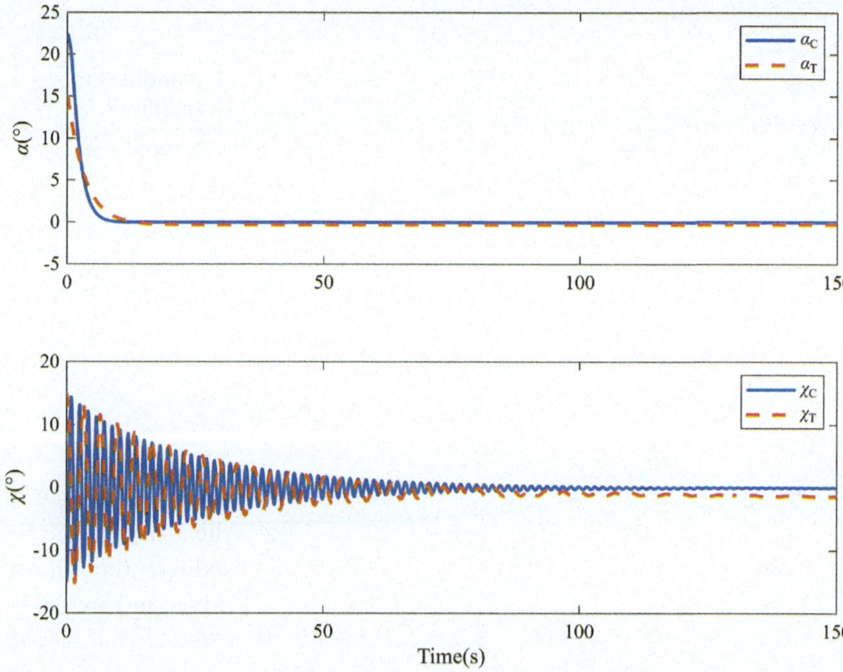

Figure 1.11 Time response of relative angles. The α converges rapidly to near zero; however, α_T exhibits a small steady-state error, while the χ shows oscillatory convergence.

Through Table 1.2, the docking configuration is chosen as $\alpha_C = \alpha_T = 0$ and $\chi_C = \chi_T = 0$, which is a stable docking and separation configuration. The simulation procedure is as follows:

(1) Initially, the relative attitude is controlled to the desired configuration by utilizing the attitude actuators.

(2) Once both α_C and α_T are less than 5 degrees, the electromagnetic coils are activated to provide the necessary electromagnetic force for docking/separation.

Simulation results are illustrated in Figs. 1.11 to 1.14. Fig. 1.11 is the time response of relative angles. It is shown that α_C and α_T converge to the neighborhood of zero within 15s. χ_C and χ_T exhibit oscillatory convergence to near zero over approximately 80s. Fig. 1.12 is the time response of relative position in CM frame and EM frame, where it can be observed that the relative distance between the spacecraft smoothly converges to the target position. Figs. 1.13 and 1.14 are the required inputs of the system, which are acceptable for a microsatellite.

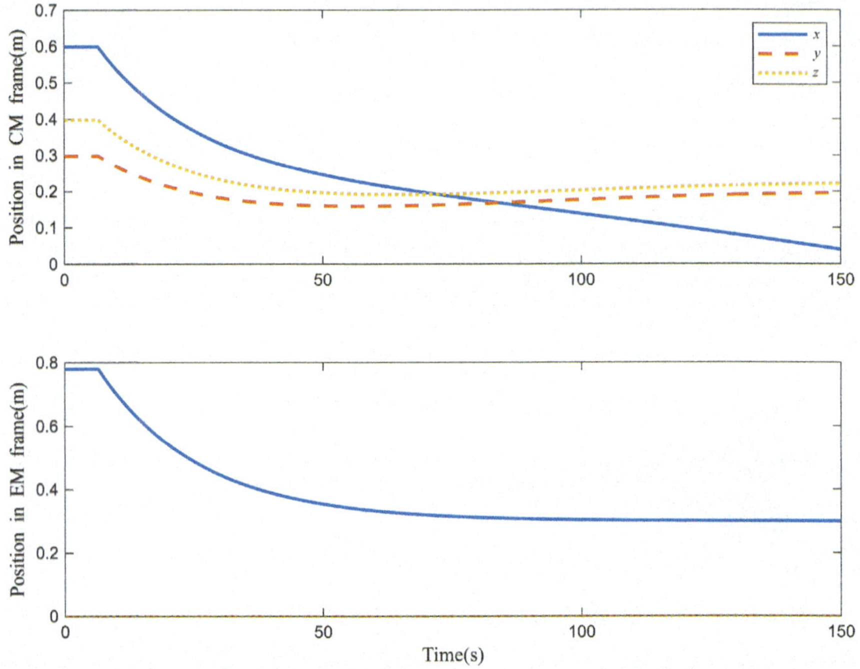

Figure 1.12 Time response of relative position. Compared with the CM frame, when the spacecraft is aligned, the relative position in the EM frame has only an *x*-axis component.

The purpose of this simulation is to demonstrate that the relative attitude for electromagnetic docking can be achieved through simple PD control. Therefore, in most subsequent chapters, it is assumed that the attitude has been stabilized in advance. For further content on coupled orbit-attitude control in electromagnetic docking, please refer to Chapter 2.

1.3 Useful lemmas

In this section, some useful lemmas will be introduced.

Lemma 1.1 (Liu et al., 2022). Let $\overline{\boldsymbol{H}}$, $\overline{\boldsymbol{E}}$ and $\overline{\boldsymbol{F}}(t)$ be real matrices of appropriate dimensions with $\|\overline{\boldsymbol{F}}(t)\| \leq 1$, then for any scalar $\overline{\xi} > 0$, one has

$$\overline{\boldsymbol{H}}\overline{\boldsymbol{F}}(t)\overline{\boldsymbol{E}} + \overline{\boldsymbol{E}}^{\mathrm{T}}\overline{\boldsymbol{F}}(t)^{\mathrm{T}}\overline{\boldsymbol{H}}^{\mathrm{T}} \leq \overline{\xi}^{-1}\overline{\boldsymbol{H}}\overline{\boldsymbol{H}}^{\mathrm{T}} + \overline{\xi}\overline{\boldsymbol{E}}^{\mathrm{T}}\overline{\boldsymbol{E}}$$

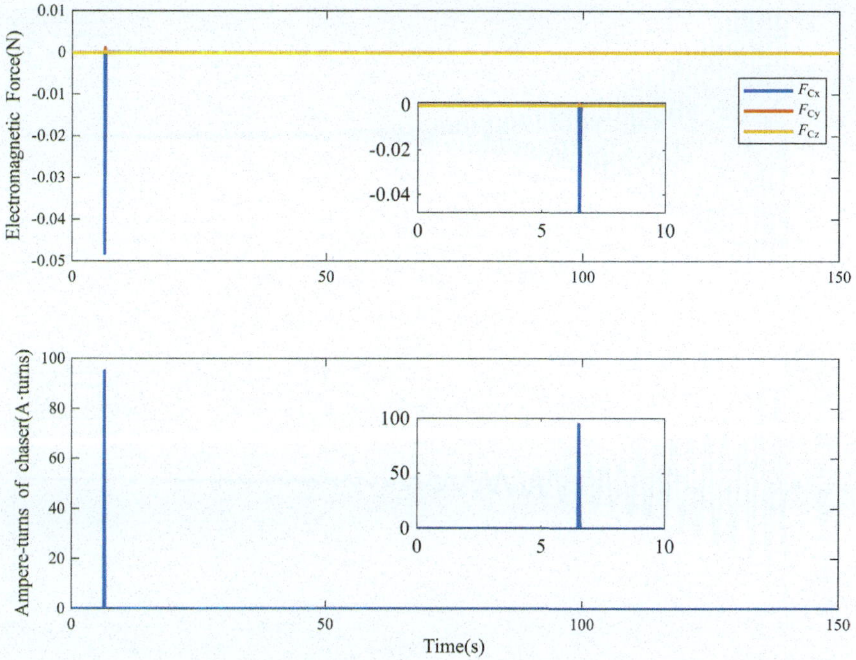

Figure 1.13 Time response of electromagnetic force and ampere-turns. Upon alignment, the required current peaks briefly before quickly decaying to near zero.

Lemma 1.2 (Ge and Wang, 2004). Let function $V(t) \geq 0$ be a continuous function defined $\forall t \geq 0$ and $V(0)$ bounded, if the following inequality holds:

$$\dot{V}(t) \leq -c_1 V(t) + c_2$$

where $c_1 > 0$, c_2 are constants, then one can conclude that $V(t)$ is bounded.

Lemma 1.3 (Schur complement lemma) Let the partitioned matrix

$$\boldsymbol{A} = \begin{bmatrix} \boldsymbol{A}_{11} & \boldsymbol{A}_{12} \\ * & \boldsymbol{A}_{22} \end{bmatrix}$$

be symmetric. Then

$$\boldsymbol{A} < 0 \Leftrightarrow \boldsymbol{A}_{11} < 0,\ \boldsymbol{A}_{22} - \boldsymbol{A}_{12}^{\mathrm{T}} \boldsymbol{A}_{11}^{-1} \boldsymbol{A}_{12} < 0 \Leftrightarrow \boldsymbol{A}_{22} < 0,\ \boldsymbol{A}_{11} - \boldsymbol{A}_{12} \boldsymbol{A}_{22}^{-1} \boldsymbol{A}_{12}^{\mathrm{T}} < 0$$

1.4 Summary

This section introduces the fundamental theoretical groundwork for electromagnetic spacecraft docking. It covers essential concepts and models,

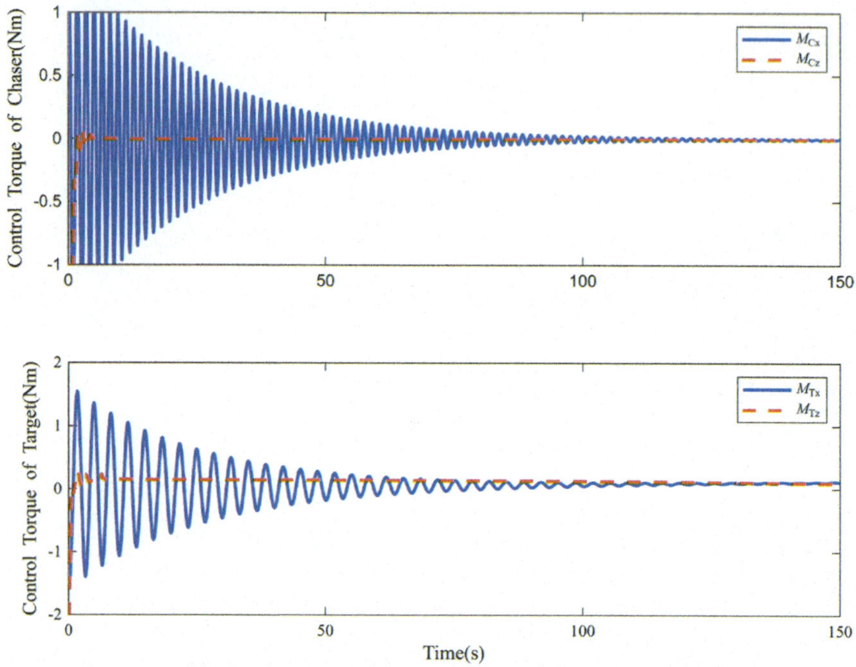

Figure 1.14 Time response of control torque. Oscillatory convergence is observed in the *x*-axis torque, in contrast to the smooth convergence of the *z*-axis torque.

including the magnetic field, the electromagnetic force/torque model and its simplified far-field representation, the relative translational dynamics based on the TH equation under various disturbances, and the relative attitude dynamics derived from Euler's equations combined with a systematic analysis and comparison of candidate docking configurations. Several useful mathematical lemmas are also provided. These fundamentals serve as the basis for the controller design throughout the subsequent chapters.

References

Alken, T., Beggan, A., Aubert, B., Bondar, B., Califf, C., Chulliat, C., & ..., Zhou, B. (2021). International Geomagnetic Reference Field: The thirteenth generation. *Earth, Planets and Space, 73*(1). https://doi.org/10.1186/s40623-020-01288-x.

Babic, S., & Akyel, C. (2012). Magnetic force between inclined circular loops (Lorentz approach). *Progress in Electromagnetics Research B, 38*(38), 333–349. https://doi.org/10.2528/pierb12011501.

Buck, A. J. (2013). *Path planning and position control of an underacted electromagnetic formation flight satellite system in the near field* (pp. 25–32). Massachusetts Institute of Technology.

Fukushima, T. (2009). Fast computation of complete elliptic integrals and Jacobian elliptic functions. *Celestial Mechanics and Dynamical Astronomy, 105*(4), 305–328. https://doi.org/10.1007/s10569-009-9228-z.

Ge, S. S., & Wang, C. (2004). Adaptive neural control of uncertain MIMO nonlinear systems. *IEEE Transactions on Neural Networks, 15*(3), 674–692. https://doi.org/10.1109/TNN.2004.826130.

Huang, X., Zhang, C., Lu, H., & Li, M. (2016). Adaptive reaching law based sliding mode control for electromagnetic formation flight with input saturation. *Journal of the Franklin Institute, 353*(11), 2398–2417. https://doi.org/10.1016/j.jfranklin.2016.04.004.

Liu, C., Yue, X., Zhang, J., & Shi, K. (2022). Active disturbance rejection control for delayed electromagnetic docking of spacecraft in elliptical orbits. *IEEE Transactions on Aerospace and Electronic Systems, 58*(3), 2257–2268. https://doi.org/10.1109/TAES.2021.3130830.

Lin, C. W., Zhang, Y. W., Wei, D. D., Li, A., & Huang, H. (2025). Design and capability analysis of a generalized electromagnetic capture device for microsatellites. In *Proceedings of the 37th Chinese control and decision conference, CCDC 2025* (pp. 6568–6573). IEEE. https://doi.org/10.1109/CCDC65474.2025.11090669.

Notaros, B. (2010). *Electromagnetics* (1st ed.). London: Pearson.

Schweighart, S. (2006). *Electromagnetic formation flight dipole solution planning [Thesis]*. Massachusetts Institute of Technology.

Wang, J., Zhou, Q., Zheng, W., & Shao, J. (2024). Passivity-based control with disturbance observer of electromagnetic formation flight spacecraft in the Port-Hamiltonian framework. *Applied Sciences, 14*(10). https://doi.org/10.3390/app14104248.

Zhang, J., Zeng, G., Gao, Y., & Wang, J. (2020). Underactuated microsatellite electromagnetic docking control using disturbance observer-based sliding mode controller. *IEEE Access, 8*, 124476–124485. https://doi.org/10.1109/ACCESS.2020.3003272.

Zhang, Y. W., Yang, L. P., Zhu, Y. W., Huang, H., & Cai, W. W. (2013). Angular-momentum management of spacecraft electromagnetic docking considering the earth's magnetic field. *Journal of Guidance, Control, and Dynamics, 36*(3), 860–869. https://doi.org/10.2514/1.57723.

Zhang, Y. W., Yang, L. P., Zhu, Y. W., Ren, X. H., & Huang, H. (2011). Self-docking capability and control strategy of electromagnetic docking technology. *Acta Astronautica, 69*(11-12), 1073–1081. https://doi.org/10.1016/j.actaastro.2011.07.001.

CHAPTER 2

Coupled orbit-attitude tracking control for spacecraft electromagnetic docking

2.1 Introduction

Compared with traditional thruster-based docking mechanisms requiring continuous expenditure of limited fuel that can contaminate the sensitive sensors on board, the benefits of spacecraft docking technology using intercraft noncontact forces are obvious, such as no propellant consumption, no plume contamination, and continuous, reversible, and synchronous controllability. In addition, one topic on electromagnetic docking is applications to in-orbit assembly (Foust et al., 2018). Thus more and more researchers have resorted to intercraft noncontact forces produced by the interaction of magnetic or electric fields generated by spacecraft, including Coulomb force, electromagnetic force, and flux-pinning force (Takahashi et al., 2021; Yu et al., 2021). According to Takahashi et al. (2021) and Zhu et al. (2018)), the electromagnetic force is incapable of 6-DOF control, while flux-pinned space modules are passively stable only when the separation distance is of the order of 10 cm. Therefore this chapter focuses on the intercraft electromagnetic force/torque produced by the target and chaser spacecraft equipped with four energized solenoids, respectively.

In the field of electromagnetic force/torque application in space missions, many researchers have investigated the dynamics and control problem of electromagnetic formation flight or electromagnetic docking. Regarding the issue of electromagnetic formation flight, the accurate trajectory tracking control has been investigated in much literature, for example, finite time control (Zeng & Hu, 2012), active disturbance rejection control (Huang et al., 2020), robust suboptimal control (Qi et al., 2019), etc. However, the electromagnetic docking issue studied in this chapter differs from electromagnetic formation flight since it faces a more serious orbit-attitude coupling problem when both position and attitude are considered. According to Liu et al. (2020), the coupling effect on the spacecraft system makes it a challenging

Spacecraft Electromagnetic Docking and Separation.
DOI: https://doi.org/10.1016/B978-0-443-49334-8.00007-0

problem in controller design, and a dynamic output feedback controller was constructed, but it did not incorporate input saturation. It is worth noting that, affected by external disturbances during the electromagnetic docking process, an attitude deflection of the spacecraft will occur. In this case, it is necessary to develop an attitude controller that achieves orbit control to stabilize the coupled orbit-attitude of the spacecraft with high precision. Therefore the coupled orbit-attitude dynamics and tracking control problem should be a top priority in electromagnetic docking missions. In Shi et al. (2017), the dynamics modeling method of CubeSat electromagnetic docking was proposed, but it was limited to existing project issues, which loses generality and cannot be applied directly in practice. In Shi et al. (2019), a disturbance observer-based controller was developed to deal with the orbit control of spacecraft electromagnetic docking, where external disturbances, fault signals, elliptical eccentricity of target spacecraft, and input magnitude and rate constraints are taken into account. In Zhang et al. (2015), the position control problem of spacecraft electromagnetic release was investigated; however, similar to (Shi et al., 2019), the attitude control problem was unsolved. In Huang et al. (2020), a 6-DOF formation keeping control scheme of spacecraft electromagnetic formation was proposed, where the robust centralized translational control and coordinated attitude control problem was investigated. In Zhang et al. (2020a), the far-field model of the magnetic dipole was used for electromagnetic force/torque, and a disturbance observer-based controller was developed; however, the final required magnetic moment and current are relatively large, which is easy to cause damage during the application process. In the aforementioned works, each spacecraft is equipped with three large orthogonal coils that can generate electromagnetic forces. When only the translational motion is considered, except for the relative position determined by electromagnetic forces, the relative orientation of the spacecraft is often controlled by control moment gyro or reaction wheels to generate counteracting torques (Kong, 2002; Shi et al., 2020). Thus the coupled orbit-attitude control problem was often dealt with by hybrid actuators, which inevitably consume limited fuel. In this case, the advantages of electromagnetic docking technology will no longer be prominent. Future space missions require a high-precision and robust controller to deal with model uncertainties due to nonlinearities and external disturbances, for example, magnetic force, solar radiation pressure, and aerodynamic force. One way to overcome this problem is to use a sliding mode controller, and much literature has addressed the sliding mode control approach. For example, a fuzzy-logic-based adaptive event-triggered sliding mode controller has been proposed for satellite attitude maneuver in Xing et al. (2021), and a robust adaptive variable structure controller was proposed in Liu et al. (2018b) to solve the chaotic attitude tracking control problem. However, they cannot be directly applied to spacecraft electromagnetic docking, which is of strong coupling and high nonlinearity. In addition, in Zhang et al. (2011), a control strategy based on an artificial potential function and an extended state observer was developed to solve the inherent control problem involved in self-docking. As can be seen in the control objective, the relative distance and velocity would gradually decrease to zero with no overshooting, which suggests that the size of the docking spacecraft is ignored. As stated in Elias et al. (2007), the maximum available current density inevitably imposes an input constraint on actuators, and an adaptive sliding mode control approach was proposed in Huang et al. (2016) to address this problem. In addition, the saturation function has been widely adopted in literature to solve the problem of input saturation, such as Liu et al. (2018c) and Zhang et al. (2017), etc.

It is worth noting that in close electromagnetic docking, the actual size and input current saturation should be taken into account, which will make the use of electromagnetic docking closer to realization compared with the state of the art (Zhang et al., 2020b). This is also a focus of this work, where the electromagnetic docking devices on the docking of two spacecraft are both equipped with four energized solenoids. The electromagnetic force/torque models for the docking spacecraft can change in the case of attitude deflection, which will make the dynamics and control problem much more complicated. To facilitate the explicit solution of input current, the simplification, error analysis, and other works of the electromagnetic force/torque models are inevitable.

The objective of this chapter is to derive the simplified forms of electromagnetic force/torque models, to establish the coupled orbit-attitude dynamics, and develop a tracking controller with predefined motion time and input saturation for spacecraft electromagnetic docking. Considering attitude deflection and different currents in the electrified solenoids, the one-dimensional electromagnetic docking strategy is investigated in this work. Here, the electromagnetic force/torque models are derived based on the far-field model of the magnetic dipole, and the error characteristics are analyzed compared with the far-field electromagnetic force/torque models to demonstrate the simplification rationality. Then, the dynamics model of coupled orbit-attitude in the case of small-angle electromagnetic docking is established, and a soft docking trajectory is designed, based on which an improved sliding mode control strategy is developed to achieve effective tracking performance. The main contributions are as follows: First, compared with the far-field electromagnetic force/torque model, the high-precision performance of the derived simplified model is illustrated via numerical analysis. Second, the soft docking process is divided into three successive maneuvers, and an improved sliding mode controller is developed to achieve the soft docking process with high precision, and the time of each maneuver phase can be predefined to satisfy the maximum input current constraint. Third, an improved reaching law is developed to drive the tracking errors to the sliding surface in finite time. The significant differences between this work and previous work are as follows: (1) The electromagnetic docking issue studied in this chapter differs from electromagnetic formation flight since it faces a more serious orbit-attitude coupling problem when both position and attitude are considered. (2) Compared with the far-field electromagnetic model, the derived simplified model in this chapter makes the control approach easier to implement in practice, as the nonlinearity and couplings are effectively reduced.

The remainder of this chapter is organized as follows. The next section focuses on the coupled orbit-attitude dynamics modeling of the electromagnetic docking system, where the electromagnetic force/torque produced by four iron core-containing energized solenoids equipped on the docking of the two spacecraft are simplified to facilitate the following control scheme design. Then, based on the trajectory planned in this chapter, an improved sliding mode controller is presented, where two virtual inputs are introduced to solve the problem of saturated input current. It is followed by numerical simulations to demonstrate the simplification, rationality, and tracking control performance at predefined times.

2.2 Problem formulation

Suppose that the electromagnetic docking device is composed of four energized solenoids with iron cores, and the number of coil turns is N for each energized solenoid. With respect to the iron core, the diameter is d_c, and the length is l_c. The current is provided by an external solar panel with a device mass of m_p. The masses of the chaser and target spacecraft are both m, respectively. The four energized solenoids on the target spacecraft are denoted by a_1, a_2, a_3, and a_4, arranged in two layers up and down, and the energized solenoids are symmetric about the center of mass of the target

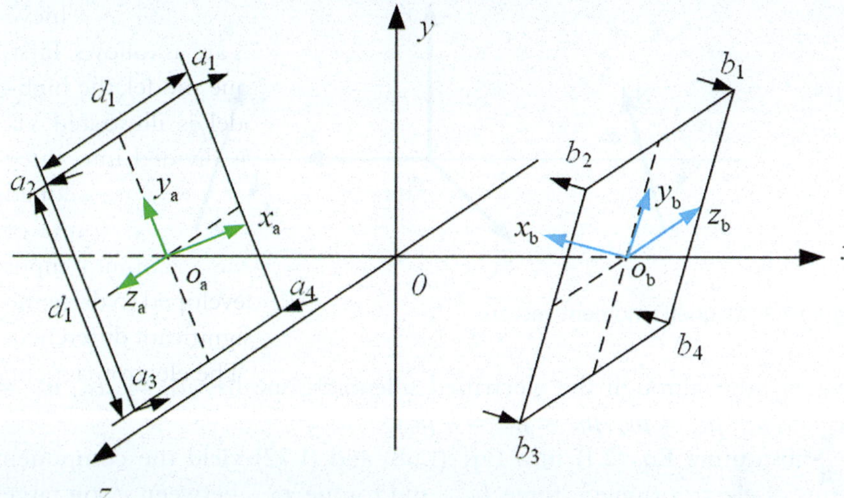

Figure 2.1 The geometric model of the docking system.

spacecraft. The distance between two neighboring energized solenoids is d_1. The four energized solenoids on the chaser spacecraft are numbered with b_1, b_2, b_3, and b_4, and the arrangement is the same as that of the target spacecraft. The geometric model of the docking system is shown in Fig. 2.1.

By approximating the energized coils on each spacecraft to a completely steerable dipole, the electromagnetic force/torque produced by the two spacecraft can be calculated using Eqs. (1.68) and (1.72).

Taking the two energized solenoids on target and chaser spacecraft, that is, a_1, b_2. For an example, suppose that α, β_0 are the angles between the magnetic moment vectors of a_1, b_2 and the x_1-axis, and χ, δ represent the angles between the projection of the magnetic moment vectors of a_1, b_2 on the $O_{CM}y_1z_1$ plane and the y_1-axis, respectively, which is shown in Fig. 2.2.

The coordinates of the magnetic moment vectors of a_1 and b_2 are expressed as

$$\begin{cases} \boldsymbol{\mu}_{a_1} = \mu_T[\cos\alpha \quad \sin\alpha\cos\chi \quad \sin\alpha\sin\chi]^T \\ \boldsymbol{\mu}_{b_2} = \mu_C[\cos\beta_0 \quad \sin\beta_0\cos\delta \quad \sin\beta_0\sin\delta]^T \end{cases} \tag{2.1}$$

On this basis, if the same current flows through the electrified solenoids, one has $\mu_{a_1} = \mu_{a_2} = \mu_{a_3} = \mu_{a_4}$, $\mu_{b_1} = \mu_{b_2} = \mu_{b_3} = \mu_{b_4}$; if the different

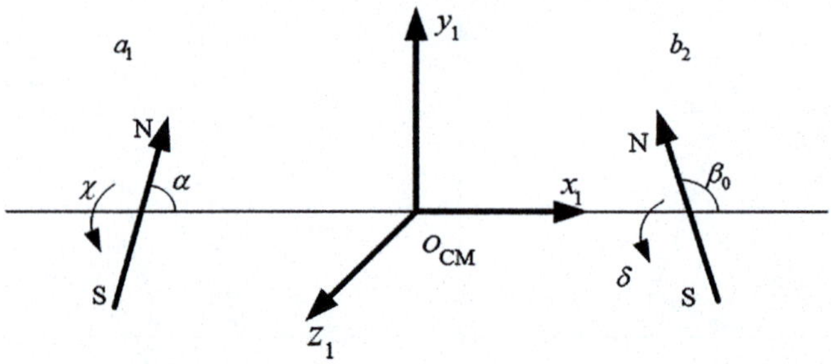

Figure 2.2 Magnetic moment vectors.

current flows through the electrified solenoids, one has $\mu_{a_1} = \mu_{a_3}$, $\mu_{a_2} = \mu_{a_4} = -\mu_{a_1}$, $\mu_{b_2} = \mu_{b_4}$, $\mu_{b_1} = \mu_{b_3} = -\mu_{b_2}$.

Substituting Eq. (2.1) into Eqs. (1.68) and (1.72) yield the component forms of electromagnetic force $F_{b_2a_1}$ and torque $\tau_{b_2a_1}$ between a_1 on target spacecraft and b_2 on the chaser spacecraft as follows:

$$
\begin{aligned}
F_{b_2a_1x} &= \frac{3\mu_0\mu_T\mu_C}{4\pi d_{b_2a_1}^4}(2\sin\alpha\sin\beta_0 + \cos(\delta-\chi)\cos\alpha\cos\beta_0) \\
F_{b_2a_1y} &= -\frac{3\mu_0\mu_T\mu_C}{4\pi d_{b_2a_1}^4}(\sin\alpha\cos\beta_0\cos\delta - \cos\alpha\sin\beta_0\cos\chi) \\
F_{b_2a_1z} &= -\frac{3\mu_0\mu_T\mu_C}{4\pi d_{b_2a_1}^4}(\sin\alpha\cos\beta_0\sin\delta - \cos\alpha\sin\beta_0\sin\chi)
\end{aligned}
\tag{2.2}
$$

$$
\begin{aligned}
\tau_{b_2a_1x} &= -\frac{\mu_0\mu_T\mu_C}{4\pi d_{b_2a_1}^4}(-\cos\alpha\cos\beta_0\sin(\delta-\chi)) \\
\tau_{b_2a_1y} &= \frac{\mu_0\mu_T\mu_C}{4\pi d_{b_2a_1}^4}(\sin\alpha\cos\beta_0\sin\delta - 2\cos\alpha\sin\beta_0\sin\chi) \\
\tau_{b_2a_1z} &= -\frac{\mu_0\mu_T\mu_C}{4\pi d_{b_2a_1}^4}(\sin\alpha\cos\beta_0\cos\delta - 2\cos\alpha\sin\beta_0\cos\chi)
\end{aligned}
\tag{2.3}
$$

Then, the electromagnetic force and torque on target spacecraft are

$$
\boldsymbol{F}_T = \sum_{i=1}^{4}\sum_{j=1}^{4}\boldsymbol{F}_{b_ja_i} \tag{2.4}
$$

$$
\boldsymbol{\tau}_T = \sum_{i=1}^{4}\sum_{j=1}^{4}\boldsymbol{\tau}_{b_ja_i} \tag{2.5}
$$

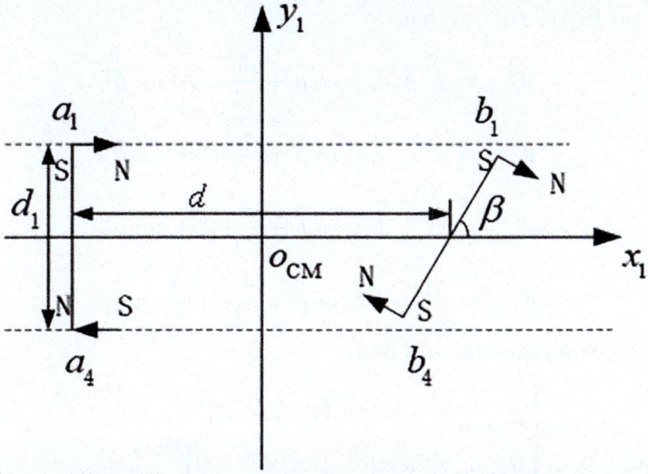

Figure 2.3 The schematic diagram of the electromagnetic docking system expressed in $O_{CM}x_1y_1z_1$.

where $F_{b_ja_i}$ represents the electromagnetic force produced by the j-th coil of chaser spacecraft on the i-th coil of target spacecraft, and $\tau_{b_ja_i}$ denotes the electromagnetic torque produced by the j-th coil of the chaser spacecraft on the i-th coil of the target spacecraft.

For the docking system is left-right symmetrical, only one side needs to be considered in the case of small-angle electromagnetic docking. Then, the schematic diagram of the electromagnetic docking system is expressed in $O_{CM}x_1y_1z_1$ is shown in Fig. 2.3.

As shown above, the center coordinates of the four electrified solenoids on the target and chaser spacecraft are $a_1\left(-\frac{d}{2}, \frac{d_1}{2}, -\frac{d_1}{2}\right)$, $a_4\left(-\frac{d}{2}, -\frac{d_1}{2}, -\frac{d_1}{2}\right)$, $b_1\left(\frac{d}{2}+\frac{d_1}{2}\cos\beta, \frac{d_1}{2}\sin\beta, -\frac{d_1}{2}\right)$, $b_4\left(\frac{d}{2}-\frac{d_1}{2}\cos\beta, -\frac{d_1}{2}\sin\beta, -\frac{d_1}{2}\right)$, then, the distance vectors between them can be expressed as

$$
\begin{aligned}
\mathbf{d}_{b_1a_1} &= \left[-\tfrac{d_1}{2}\cos\beta - d \quad -\tfrac{d_1}{2}\sin\beta + \tfrac{d_1}{2} \quad 0\right]^{\mathrm{T}} \\
\mathbf{d}_{b_4a_1} &= \left[\tfrac{d_1}{2}\cos\beta - d \quad \tfrac{d_1}{2}\sin\beta + \tfrac{d_1}{2} \quad 0\right]^{\mathrm{T}} \\
\mathbf{d}_{b_1a_4} &= \left[-\tfrac{d_1}{2}\cos\beta - d \quad -\tfrac{d_1}{2}\sin\beta - \tfrac{d_1}{2} \quad 0\right]^{\mathrm{T}} \\
\mathbf{d}_{b_4a_4} &= \left[\tfrac{d_1}{2}\cos\beta - d \quad \tfrac{d_1}{2}\sin\beta - \tfrac{d_1}{2} \quad 0\right]^{\mathrm{T}}
\end{aligned} \tag{2.6}
$$

Based on Eq. (2.6), one has

$$\begin{aligned}
\boldsymbol{d}_{b_1a_1}^2 &= d^2 + d_1 d\cos\beta + \frac{{d_1}^2}{2}(1-\sin\beta)\\
\boldsymbol{d}_{b_4a_1}^2 &= d^2 - d_1 d\cos\beta + \frac{{d_1}^2}{2}(1+\sin\beta)\\
\boldsymbol{d}_{b_1a_4}^2 &= d^2 + d_1 d\cos\beta + \frac{{d_1}^2}{2}(1+\sin\beta)\\
\boldsymbol{d}_{b_4a_4}^2 &= d^2 - d_1 d\cos\beta + \frac{{d_1}^2}{2}(1-\sin\beta)
\end{aligned} \tag{2.7}$$

In the $O_{CM}x_1y_1$ plane, one has

$$\begin{cases}
\boldsymbol{\mu}_{a_1\mathrm{CM}} = \pi NR^2 i_a\begin{bmatrix}1 & 0 & 0\end{bmatrix}^{\mathrm{T}}\\
\boldsymbol{\mu}_{a_4\mathrm{CM}} = \pi NR^2 i_a\begin{bmatrix}-1 & 0 & 0\end{bmatrix}^{\mathrm{T}}
\end{cases}$$

$$\begin{cases}
\boldsymbol{\mu}_{b_1\mathrm{CM}} = \pi NR^2 i_{b_1}\begin{bmatrix}\sin\beta & -\cos\beta & 0\end{bmatrix}^{\mathrm{T}}\\
\boldsymbol{\mu}_{b_4\mathrm{CM}} = \pi NR^2 i_{b_4}\begin{bmatrix}-\sin\beta & \cos\beta & 0\end{bmatrix}^{\mathrm{T}}
\end{cases} \tag{2.8}$$

2.2.1 Electromagnetic force model

With Fig. 2.3, substituting Eqs. (2.6) and (2.8) into Eq. (1.68) yields the x-direction electromagnetic force as follows

$$\begin{aligned}
F_{b_1a_1x} = {}& \frac{3\pi N^2R^4 i_a i_{b_1}\mu_0}{4d_{b_1a_1}^5}\left(-3d\sin\beta - d_1\sin\beta\cos\beta - \frac{d_1}{2}\cos\beta - \frac{5}{d_{b_1a_1}^2}(-d^3\sin\beta\right.\\
&\left.\left. - d^2\left(d_1\sin\beta\cos\beta + \frac{d_1}{2}\cos\beta\right) - d\left(\frac{d_1^2}{2}\cos^2\beta + \frac{d_1^2}{4}\sin\beta\cos^2\beta\right) - \frac{d_1^3}{8}\cos^3\beta\right)\right)\\
F_{b_4a_1x} = {}& \frac{3\pi N^2R^4 i_a i_{b_4}\mu_0}{4d_{b_4a_1}^5}\left(3d\sin\beta - d_1\sin\beta\cos\beta + \frac{d_1}{2}\cos\beta - \frac{5}{d_{b_4a_1}^2}(d^3\sin\beta\right.\\
&\left.\left. - d^2\left(d_1\sin\beta\cos\beta - \frac{d_1}{2}\cos\beta\right) + d\left(-\frac{d_1^2}{2}\cos^2\beta + \frac{d_1^2}{4}\sin\beta\cos^2\beta\right) + \frac{d_1^3}{8}\cos^3\beta\right)\right)\\
F_{b_1a_4x} = {}& \frac{3\pi N^2R^4 i_a i_{b_1}\mu_0}{4d_{b_1a_4}^5}\left(3d\sin\beta + d_1\sin\beta\cos\beta - \frac{d_1}{2}\cos\beta - \frac{5}{d_{b_1a_4}^2}(d^3\sin\beta\right.\\
&\left.\left. + d^2\left(d_1\sin\beta\cos\beta - \frac{d_1}{2}\cos\beta\right) + d\left(-\frac{d_1^2}{2}\cos^2\beta + \frac{d_1^2}{4}\sin\beta\cos^2\beta\right) - \frac{d_1^3}{8}\cos^3\beta\right)\right)\\
F_{b_4a_4x} = {}& \frac{3\pi N^2R^4 i_a i_{b_4}\mu_0}{4d_{b_4a_4}^5}\left(-3d\sin\beta + d_1\sin\beta\cos\beta + \frac{d_1}{2}\cos\beta - \frac{5}{d_{b_4a_4}^2}(-d^3\sin\beta\right.\\
&\left.\left. + d^2\left(d_1\sin\beta\cos\beta + \frac{d_1}{2}\cos\beta\right) - d\left(\frac{d_1^2}{2}\cos^2\beta + \frac{d_1^2}{4}\sin\beta\cos^2\beta\right) + \frac{d_1^3}{8}\cos^3\beta\right)\right)
\end{aligned} \tag{2.9}$$

For small-angle docking, defining $\beta_1=\pi/2-\beta$, with approximate equalities $\sin\beta\approx 1$ and $\cos\beta\approx\pi/2-\beta$, Eqs. (2.7) and (2.9) reduce to the following Eqs. (2.10) and (2.11), respectively.

$$\begin{aligned}
d_{b_1a_1}^2 &= d^2 + dd_1\beta_1\\
d_{b_4a_1}^2 &= d^2 - dd_1\beta_1 + {d_1}^2\\
d_{b_1a_4}^2 &= d^2 + dd_1\beta_1 + {d_1}^2\\
d_{b_4a_4}^2 &= d^2 - dd_1\beta_1
\end{aligned}\tag{2.10}$$

$$\begin{aligned}
F_{b_1a_1x} &= \frac{3\pi N^2R^4 i_a i_{b_1}\mu_0}{4d_{b_1a_1}^5}\left(-3d - \frac{3d_1\beta_1}{2} - \frac{5}{d_{b_1a_1}^2}\left(-d^3 - \frac{3d^2d_1\beta_1}{2} - \frac{3dd_1^2\beta_1^2}{4} - \frac{d_1^3\beta_1^3}{8}\right)\right)\\
F_{b_4a_1x} &= \frac{3\pi N^2R^4 i_a i_{b_4}\mu_0}{4d_{b_4a_1}^5}\left(3d - \frac{d_1\beta_1}{2} - \frac{5}{d_{b_4a_1}^2}\left(d^3 - \frac{d^2d_1\beta_1}{2} - \frac{dd_1^2\beta_1^2}{4} + \frac{d_1^3\beta_1^3}{8}\right)\right)\\
F_{b_1a_4x} &= \frac{3\pi N^2R^4 i_a i_{b_1}\mu_0}{4d_{b_1a_4}^5}\left(3d + \frac{d_1\beta_1}{2} - \frac{5}{d_{b_1a_4}^2}\left(d^3 + \frac{d^2d_1\beta_1}{2} - \frac{dd_1^2\beta_1^2}{4} - \frac{d_1^3\beta_1^3}{8}\right)\right)\\
F_{b_4a_4x} &= \frac{3\pi N^2R^4 i_a i_{b_4}\mu_0}{4d_{b_4a_4}^5}\left(-3d + \frac{3d_1\beta_1}{2} - \frac{5}{d_{b_4a_4}^2}\left(-d^3 + \frac{3d^2d_1\beta_1}{2} - \frac{3dd_1^2\beta_1^2}{4} + \frac{d_1^3\beta_1^3}{8}\right)\right)
\end{aligned}\tag{2.11}$$

The first-order Taylor expansions of $1/d_{b_1a_1}^7$, $1/d_{b_4a_1}^7$, $1/d_{b_4a_4}^7$ and $1/d_{b_1a_4}^7$ are as follows:

$$\begin{aligned}
\frac{1}{d_{b_1a_1}^7} &= \frac{1}{d^7}\left(1 - \frac{7d_1\beta_1}{2d}\right)\\
\frac{1}{d_{b_4a_1}^7} &= \left(d^2 + d_1^2\right)^{-\frac{7}{2}} + \frac{7}{2}\left(d^2 + d_1^2\right)^{-\frac{9}{2}} dd_1\beta_1\\
\frac{1}{d_{b_4a_4}^7} &= \frac{1}{d^7}\left(1 + \frac{7d_1\beta_1}{2d}\right)\\
\frac{1}{d_{b_1a_4}^7} &= \left(d^2 + d_1^2\right)^{-\frac{7}{2}} - \frac{7}{2}\left(d^2 + d_1^2\right)^{-\frac{9}{2}} dd_1\beta_1
\end{aligned}\tag{2.12}$$

Substituting Eq. (2.10) into (2.11) and simplifying it yields

$$\begin{aligned}
F_{b_1a_1x} &= \frac{3\pi N^2R^4 i_a i_{b_1}\mu_0}{4d_{b_1a_1}^7}\left(2d^3 + 3d^2d_1\beta + \frac{9dd_1^2\beta_1^2}{4} + \frac{5d_1^3\beta_1^3}{8}\right)\\
F_{b_1a_4x} &= \frac{3\pi N^2R^4 i_a i_{b_1}\mu_0}{4d_{b_1a_4}^7}\left(-2d^3 + d^2d_1\beta_1 + 3d{d_1}^2 + \frac{d_1^3\beta_1}{2} + \frac{7dd_1^2\beta_1^2}{4} + \frac{5d_1^3\beta_1^3}{8}\right)
\end{aligned}$$

$$F_{b_4a_1x} = \frac{3\pi N^2R^4 i_a i_{b_4}\mu_0}{4d_{b_4a_1}^7}\left(-2d^3 - d^2d_1\beta_1 + 3dd_1{}^2 - \frac{d_1^3\beta_1}{2} + \frac{7dd_1^2\beta_1^2}{4} - \frac{5d_1^3\beta_1^3}{8}\right)$$

$$F_{b_4a_4x} = \frac{3\pi N^2R^4 i_a i_{b_4}\mu_0}{4d_{b_4a_4}^7}\left(2d^3 - 3d^2d_1\beta_1 + \frac{9dd_1{}^2\beta_1^2}{4} - \frac{5d_1{}^3\beta_1^3}{8}\right) \tag{2.13}$$

Substituting Eq. (2.12) into Eq. (2.13) yields

$$F_{b_1a_1x} = \frac{3\pi N^2R^4 i_a i_{b_1}\mu_0}{4d^7}\left(1 - \frac{7d_1\beta_1}{2d}\right)\left(2d^3 + 3d^2d_1\beta_1 + \frac{9dd_1^2\beta_1^2}{4} + \frac{5d_1^3\beta_1^3}{8}\right)$$

$$F_{b_1a_4x} = \frac{3\pi N^2R^4 i_a i_{b_1}\mu_0}{4}\left(\left(d^2 + d_1^2\right)^{-\frac{7}{2}} - \frac{7}{2}\left(d^2 + d_1^2\right)^{-\frac{9}{2}}dd_1\beta_1\right)$$
$$\cdot\left(3dd_1{}^2 + \frac{d_1^3\beta_1}{2} - 2d^3 + d^2d_1\beta_1 + \frac{7dd_1^2\beta_1^2}{4} + \frac{5d_1^3\beta_1^3}{8}\right)\Bigg)$$

$$F_{b_4a_1x} = \frac{3\pi N^2R^4 i_a i_{b_4}\mu_0}{4}\left(d^2 + d_1^2\right)^{-\frac{9}{2}}\left(-4d^5 - 16d^4d_1\beta_1 + 2d^3d_1^2 - \frac{7d^3d_1^2\beta_1^2}{2}\right.$$
$$\left. + 18d^2d_1^3\beta_1 + 11d^2d_1^3\beta_1^3 + 6dd_1^4 - \frac{35dd_1^4\beta_1^4}{8} - d_1^5\beta_1 - \frac{5d_1^5\beta_1^3}{4}\right)$$

$$F_{b_4a_4x} = \frac{3\pi N^2R^4 i_a i_{b_4}\mu_0}{8d^8}\left(4d^4 + 8d^3d_1\beta_1 - \frac{33d^2d_1^2\beta_1^2}{2} + \frac{29dd_1^3\beta_1^3}{2} - \frac{35d_1^4\beta_1^4}{8}\right) \tag{2.14}$$

Combining similar items containing i_{b_1} and i_{b_4} leads to

$$F_{b_1a_1x} + F_{b_1a_4x} = \frac{3\pi N^2R^4 i_a i_{b_1}\mu_0}{4}\left(\left(\frac{2}{d^4} - \frac{4d_1\beta_1}{d^5} - \frac{33d_1^2\beta_1^2}{4d^6} - \frac{29d_1^3\beta_1^3}{4d^7} - \frac{35d_1^4\beta_1^4}{16d^8}\right)\right.$$
$$+\left(d^2\left(-4d^5 + 16d^4d_1\beta_1 + 2d^3d_1^2 - \frac{7d^3d_1^2\beta_1^2}{2} - 18d^2d_1^3\beta_1\right.\right.$$
$$\left.\left.\left. - 11d^2d_1^3\beta_1^3 + 6dd_1^4 - \frac{35dd_1^4\beta_1^4}{8} + d_1^5\beta_1 + \frac{5d_1^5\beta_1^3}{4}\right)\right)\right) \tag{2.15}$$

$$F_{b_4a_1x} + F_{b_4a_4x} = \frac{3\pi N^2R^4 i_a i_{b_4}\mu_0}{4}\left(\left(\frac{2}{d^4} + \frac{4d_1\beta_1}{d^5} - \frac{33d_1^2\beta_1^2}{4d^6} + \frac{29d_1^3\beta_1^3}{4d^7} - \frac{35d_1^4\beta_1^4}{16d^8}\right)\right.$$
$$+\left(d^2\left(-4d^5 - 16d^4d_1\beta_1 + 2d^3d_1^2 - \frac{7d^3d_1^2\beta_1^2}{2} + 18d^2d_1^3\beta_1\right.\right.$$
$$\left.\left.\left. + 11d^2d_1^3\beta_1^3 + 6dd_1^4 - \frac{35dd_1^4\beta_1^4}{8} - d_1^5\beta_1 - \frac{5d_1^5\beta_1^3}{4}\right)\right)\right) \tag{2.16}$$

With small-angle assumption, $35d_1^4\beta_1^4/16d^8$, $35dd_1^4\beta_1^4/8$, $5d_1^5\beta_1^3/4$, $29d_1^3\beta_1^3/4d^7$, $11d^2d_1^3\beta_1^3$, $7d^3d_1^2\beta_1^2/2$, and $d_1^5\beta_1$ are regarded as infinitesimals of higher order, so that the simplified expressions of electromagnetic forces given in Eqs. (2.15) and (2.16) are rewritten as

$$F_{b_1a_1x} + F_{b_1a_4x} = \frac{3\pi N^2R^4 i_a i_{b_1}\mu_0}{4}\left(\left(\frac{2}{d^4} - \frac{4d_1\beta_1}{d^5} - \frac{33d_1^2\beta_1^2}{4d^6}\right) + \left(d^2\left(-4d^5 + 16d^4d_1\beta_1 + 2d^3{d_1}^2 - 18d^2d_1^3\beta_1 + 6d{d_1}^4\right)\right)\right) \quad (2.17)$$

$$F_{b_4a_1x} + F_{b_4a_4x} = \frac{3\pi N^2R^4 i_a i_{b_4}\mu_0}{4}\left(\left(\frac{2}{d^4} + \frac{4d_1\beta_1}{d^5} - \frac{33d_1^2\beta_1^2}{4d^6}\right) + \left(d^2\left(-4d^5 - 16d^4d_1\beta_1 + 2d^3d_1^2 + 18d^2d_1^3\beta_1 + 6d{d_1}^4\right)\right)\right) \quad (2.18)$$

As can be seen from Eqs. (2.17) and (2.18), the simplified x-direction electromagnetic force in $O_{CM}x_1y_1z_1$ coordinate frame is

$$\overline{F}_x = \frac{3\pi N^2R^4 i_a\mu_0}{4}\left(\left(i_{b_1} + i_{b_4}\right)\left(\frac{2}{d^4} - \frac{33d_1^2\beta_1^2}{4d^6} - \frac{4d^5 - 2d^3d_1^2 - 6d{d_1}^4}{\left(d^2 + d_1^2\right)^{\frac{9}{2}}}\right) + \left(i_{b_1} - i_{b_4}\right)\left(-\frac{4d_1\beta_1}{d^5} + \frac{16d^4d_1\beta_1 - 18d^2d_1^3\beta_1}{\left(d^2 + d_1^2\right)^{\frac{9}{2}}}\right)\right) \quad (2.19)$$

In this case, the expression of electromagnetic force in "CM" calculation coordinate system $O_{CM}x_1y_1z_1$ is the same as that in the "CM" orbital reference frame $O_{CM}xyz$, that is,

$$F_x = \overline{F}_x = \frac{3\pi N^2R^4 i_a\mu_0}{4}\left(\left(i_{b_1} + i_{b_4}\right)\left(\frac{2}{d^4} - \frac{33d_1^2\beta_1^2}{4d^6} - \frac{4d^5 - 2d^3d_1^2 - 6d{d_1}^4}{\left(d^2 + d_1^2\right)^{\frac{9}{2}}}\right) + \left(i_{b_1} - i_{b_4}\right)\left(-\frac{4d_1\beta_1}{d^5} + \frac{16d^4d_1\beta_1 - 18d^2d_1^3\beta_1}{\left(d^2 + d_1^2\right)^{\frac{9}{2}}}\right)\right) \quad (2.20)$$

2.2.2 Electromagnetic torque model

With Fig. 2.3, substituting Eqs. (2.6) and (2.8) into Eq. (1.72), yields the x-direction electromagnetic torque as follows

$$\tau_{b_1a_1x} = \tau_{b_1a_1y} = \tau_{b_4a_1x} = \tau_{b_4a_1y} = \tau_{b_1a_4x} = \tau_{b_1a_4y} = \tau_{b_4a_4x} = \tau_{b_4a_4y} = 0$$

$$\tau_{b_1a_1z} = \frac{\pi N^2R^4 i_a i_{b_1}\mu_0}{4d_{b_1a_1}^5}\left(-\frac{3d_1d}{2}\sin\beta - \frac{d_1^2}{4}\cos\beta + \frac{3d_1d}{2}\sin^2\beta\right.$$

$$+\frac{d_1^2}{4}\sin\beta\cos\beta + d^2\cos\beta + d_1 d\cos^2\beta\Bigg)$$

$$\tau_{b_1a_4z} = \frac{\pi N^2R^4 i_a i_{b_1}\mu_0}{4d_{b_1a_4}^5}\Bigg(-\frac{3d_1d}{2}\sin\beta + \frac{d_1^2}{4}\cos\beta$$

$$-\frac{3d_1d}{2}\sin^2\beta + \frac{d_1^2}{4}\sin\beta\cos\beta - d^2\cos\beta - d_1d\cos^2\beta\Bigg)$$

$$\tau_{b_4a_1z} = \frac{\pi N^2R^4 i_a i_{b_4}\mu_0}{4d_{b_4a_1}^5}\Bigg(\frac{3d_1d}{2}\sin\beta + \frac{d_1^2}{4}\cos\beta + \frac{3d_1d}{2}\sin^2\beta$$

$$+\frac{d_1^2}{4}\sin\beta\cos\beta - d^2\cos\beta + d_1d\cos^2\beta\Bigg)$$

$$\tau_{b_4a_4z} = \frac{\pi N^2R^4 i_a i_{b_4}\mu_0}{4d_{b_4a_4}^5}\Bigg(\frac{3d_1d}{2}\sin\beta - \frac{d_1^2}{4}\cos\beta - \frac{3d_1d}{2}\sin^2\beta$$

$$+\frac{d_1^2}{4}\sin\beta\cos\beta + d^2\cos\beta - d_1d\cos^2\beta\Bigg) \tag{2.21}$$

For small-angle docking, defining $\beta_1 = \pi/2-\beta$, with approximate equalities $\sin\beta\approx 1$ and $\cos\beta \approx \pi/2-\beta$, Eq. (2.21) can be simplified as follows:

$$\tau_{b_1a_1x} = \tau_{b_1a_1y} = \tau_{b_4a_1x} = \tau_{b_4a_1y} = \tau_{b_1a_4x} = \tau_{b_1a_4y} = \tau_{b_4a_4x} = \tau_{b_4a_4y} = 0$$

$$\tau_{b_1a_1z} = \frac{\pi N^2R^4 i_a i_{b_1}\mu_0}{4d_{b_1a_1}^5}\left(d^2\beta_1 + d_1d\beta_1^2\right)$$

$$\tau_{b_1a_4z} = \frac{\pi N^2R^4 i_a i_{b_1}\mu_0}{4d_{b_1a_4}^5}\left(-3d_1d + \frac{d_1^2\beta_1}{2} - d^2\beta_1 - d_1d\beta_1^2\right)$$

$$\tau_{b_4a_1z} = \frac{\pi N^2R^4 i_a i_{b_4}\mu_0}{4d_{b_4a_1}^5}\left(3d_1d + \frac{d_1^2\beta_1}{2} - d^2\beta_1 + d_1d\beta_1^2\right)$$

$$\tau_{b_4a_4z} = \frac{\pi N^2R^4 i_a i_{b_4}\mu_0}{4d_{b_4a_4}^5}\left(d^2\beta_1 - d_1d\beta_1^2\right) \tag{2.22}$$

The first-order Taylor expansions of $1/d_{b_1a_1}^5$, $1/d_{b_4a_1}^5$, $1/d_{b_4a_4}^5$ and $1/d_{b_1a_4}^5$ are as follows:

$$\frac{1}{d_{b_1a_1}^5} = \frac{1}{d^5}\left(1 - \frac{5d_1\beta_1}{2d}\right)$$

$$\frac{1}{d_{b_4a_1}^5} = \left(d^2 + d_1^2\right)^{-\frac{5}{2}} + \frac{5}{2}\left(d^2 + d_1^2\right)^{-\frac{7}{2}} dd_1\beta_1$$
$$\frac{1}{d_{b_4a_4}^5} = \frac{1}{d^5}\left(1 + \frac{5d_1\beta_1}{2d}\right)$$
$$\frac{1}{d_{b_1a_4}^5} = \left(d^2 + d_1^2\right)^{-\frac{5}{2}} - \frac{5}{2}\left(d^2 + d_1^2\right)^{-\frac{7}{2}} dd_1\beta_1 \tag{2.23}$$

Substituting Eq. (2.23) into Eq. (2.22) yields

$$\tau_{b_1a_1z} = \frac{\pi N^2 R^4 i_a i_{b_1}\mu_0}{4}\left(\frac{\beta_1}{d^3} - \frac{3d_1\beta_1^2}{2d^4} - \frac{5d_1^2\beta_1^3}{2d^5}\right)$$
$$\tau_{b_1a_4z} = \frac{\pi N^2 R^4 i_a i_{b_1}\mu_0}{4}\left(d^2 + d_1^2\right)^{-\frac{7}{2}} \times\left(-3d^3d_1 + 7d^2d_1^2\beta_1 - d^4\beta_1 + \frac{3d^3d_1\beta_1^2}{2} - 3dd_1^3 + \frac{d_1^4\beta_1}{2} - \frac{9dd_1^3\beta_1^2}{4} + \frac{5d^2d_1^2\beta_1^3}{2}\right)$$
$$\tau_{b_4a_1z} = \frac{\pi N^2 R^4 i_a i_{b_4}\mu_0}{4}\left(d^2 + d_1^2\right)^{-\frac{7}{2}} \times\left(3d^3d_1 + 7d^2d_1^2\beta_1 - d^4\beta_1 - \frac{3d^3d_1\beta_1^2}{2} + 3dd_1^3 + \frac{d_1^4\beta_1}{2} + \frac{9dd_1^3\beta_1^2}{4} + \frac{5d^2d_1^2\beta_1^3}{2}\right)$$
$$\tau_{b_4a_4z} = \frac{\pi N^2 R^4 i_a i_{b_4}\mu_0}{4}\left(\frac{\beta_1}{d^3} + \frac{3d_1\beta_1^2}{2d^4} - \frac{5d_1^2\beta_1^3}{2d^5}\right) \tag{2.24}$$

Combining similar items containing i_{b_1} and i_{b_4} leads to

$$\tau_{b_1a_1z} + \tau_{b_1a_4z} = \frac{\pi N^2 R^4 i_a i_{b_1}\mu_0}{4}\left(\left(\frac{\beta_1}{d^3} - \frac{3d_1\beta_1^2}{2d^4} - \frac{5d_1^2\beta_1^3}{2d^5}\right) + \left(d^2\left(-3d^3d_1 + 7d^2d_1^2\beta_1 - d^4\beta_1 + \frac{3d^3d_1\beta_1^2}{2} - 3dd_1^3 + \frac{d_1^4\beta_1}{2} - \frac{9dd_1^3\beta_1^2}{4} + \frac{5d^2d_1^2\beta_1^3}{2}\right)\right)\right) \tag{2.25}$$

$$\tau_{b_4a_1z} + \tau_{b_4a_4z} = \frac{\pi N^2 R^4 i_a i_{b_4}\mu_0}{4}\left(\left(d^2\left(3d^3d_1 + 7d^2d_1^2\beta_1 - d^4\beta_1 - \frac{3d^3d_1\beta_1^2}{2} + 3dd_1^3 + \frac{d_1^4\beta_1}{2} + \frac{9dd_1^3\beta_1^2}{4} + \frac{5d^2d_1^2\beta_1^3}{2}\right) + \left(\frac{\beta_1}{d^3} + \frac{3d_1\beta_1^2}{2d^4} - \frac{5d_1^2\beta_1^3}{2d^5}\right)\right)\right) \tag{2.26}$$

With the small-angle assumption, $5dd_1^2\beta_1^3$, $3d^2d_1\beta_1^2$, $5d^2d_1^2\beta_1^3/2$, $9dd_1^3\beta_1^2/4$ and $3d^3d_1\beta_1^2/2$ are regarded as infinitesimals of higher order, so that the simplified expressions of electromagnetic torques given in Eqs. (2.25) and (2.26) are rewritten as

$$\tau_{b_1a_1z}+\tau_{b_1a_4z}=\frac{\pi N^2R^4 i_a i_{b_1}\mu_0}{4}\left(\frac{\beta_1}{d^3}+\left(d^2+d_1^2\right)^{-\frac{7}{2}}\left(-3d^3d_1+7d^2d_1^2\beta_1-d^4\beta_1-3dd_1^3+\frac{d_1^4\beta_1}{2}\right)\right) \tag{2.27}$$

$$\tau_{b_4a_1z}+\tau_{b_4a_4z}=\frac{\pi N^2R^4 i_a i_{b_4}\mu_0}{4}\left(\left(d^2+d_1^2\right)^{-\frac{7}{2}}\left(3d^3d_1+7d^2d_1^2\beta_1-d^4\beta_1+3dd_1^3+\frac{d_1^4\beta_1}{2}\right)+\frac{\beta_1}{d^3}\right) \tag{2.28}$$

As can be seen from Eqs. (2.27) and (2.28), the simplified x-direction electromagnetic torque in $O_{CM}x_1y_1z_1$ coordinate frame is

$$\overline{\tau}_z=\frac{\pi N^2R^4 i_a\mu_0}{4}\left(\left(i_{b_1}+i_{b_4}\right)\left(\frac{\beta_1}{d^3}+\frac{7d^2d_1^2\beta_1-d^4\beta_1+\frac{d_1^4\beta_1}{2}}{\left(d^2+d_1^2\right)^{\frac{7}{2}}}\right)+\left(i_{b_1}-i_{b_4}\right)\frac{-3d^3d_1-3dd_1^3}{\left(d^2+d_1^2\right)^{\frac{7}{2}}}\right) \tag{2.29}$$

In this case, the expression of electromagnetic torque in "CM" calculation coordinate system $O_{CM}x_1y_1z_1$ is the same as that in "CM" orbital reference frame $O_{CM}xyz$, that is,

$$\tau_z=\overline{\tau}_z=\frac{\pi N^2R^4 i_a\mu_0}{4}\left(\left(i_{b_1}+i_{b_4}\right)\left(\frac{\beta_1}{d^3}+\frac{7d^2d_1^2\beta_1-d^4\beta_1+\frac{d_1^4\beta_1}{2}}{\left(d^2+d_1^2\right)^{\frac{7}{2}}}\right)+\left(i_{b_1}-i_{b_4}\right)\frac{-3d^3d_1-3dd_1^3}{\left(d^2+d_1^2\right)^{\frac{7}{2}}}\right) \tag{2.30}$$

Defining $\lambda=3\pi N^2R^4 i_a\mu_0/4$, $x_1=d$, $x_2=\beta_1$, $u_1=i_{b_1}$, $u_2=i_{b_4}$, the electromagnetic force can be simplified as

$$F_x=\frac{3\pi N^2R^4 i_a\mu_0}{4}\left(\left(i_{b_1}+i_{b_4}\right)\left(\frac{2}{d^4}-\frac{33d_1^2\beta_1^2}{4d^6}-\frac{4d^5-2d^3d_1^2-6dd_1{}^4}{\left(d^2+d_1^2\right)^{\frac{9}{2}}}\right)\right.$$

$$+\left(i_{b_1}-i_{b_4}\right)\left(-\frac{4d_1\beta_1}{d^5}+\frac{16d^4d_1\beta_1-18d^2d_1^3\beta_1}{\left(d^2+d_1^2\right)^{\frac{9}{2}}}\right)\Bigg)$$

$$=\lambda\big((u_1+u_2)g_1(x)+(u_1-u_2)g_2(x)\big)$$

$$=\lambda\begin{bmatrix}g_1(x)+g_2(x) & g_1(x)-g_2(x)\end{bmatrix}\begin{bmatrix}u_1\\u_2\end{bmatrix} \tag{2.31}$$

where

$$g_1(x)=\frac{2}{x_1^4}-\frac{33d_1^2x_2^2}{4x_1^6}-\frac{4x_1^5-2d_1^2x_1^3-6{d_1}^4x_1}{\left(x_1^2+d_1^2\right)^{\frac{9}{2}}} \tag{2.32}$$

$$g_2(x)=-\frac{4d_1x_2}{x_1^5}+\frac{16d_1x_1^4x_2-18d_1^3x_1^2x_2}{\left(x_1^2+d_1^2\right)^{\frac{9}{2}}} \tag{2.33}$$

Similarly, the electromagnetic torque can be simplified as

$$\tau_z=\frac{\pi N^2R^4i_a\mu_0}{4}\left(\left(i_{b_1}+i_{b_4}\right)\left(\frac{\beta_1}{d^3}+\frac{7d^2d_1^2\beta_1-d^4\beta_1+\frac{d_1^4\beta_1}{2}}{\left(d^2+d_1^2\right)^{\frac{7}{2}}}\right)\right.$$

$$\left.+\left(i_{b_1}-i_{b_4}\right)\frac{-3d^3d_1-3dd_1^3}{\left(d^2+d_1^2\right)^{\frac{7}{2}}}\right)$$

$$=\frac{\lambda}{3}\big((u_1+u_2)g_3(x)+(u_1-u_2)g_4(x)\big)$$

$$=\frac{\lambda}{3}\begin{bmatrix}g_3(x)+g_4(x) & g_3(x)-g_4(x)\end{bmatrix}\begin{bmatrix}u_1\\u_2\end{bmatrix} \tag{2.34}$$

where

$$g_3(x)=\frac{x_2}{x_1^3}+\frac{7d_1^2x_1^2x_2-x_1^4x_2+\frac{d_1^4x_2}{2}}{\left(x_1^2+d_1^2\right)^{\frac{7}{2}}} \tag{2.35}$$

$$g_4(x)=\frac{-3d_1x_1^3-3d_1^3x_1}{\left(x_1^2+d_1^2\right)^{\frac{7}{2}}} \tag{2.36}$$

Suppose that the spacecraft's moment of inertia about the z-axis is I, taking the external disturbance force d_f and torque d_τ into account, the coupled orbit-attitude dynamics of the electromagnetic docking system are given as follows

$$\begin{cases} \ddot{x}_1 - 3n^2 x_1 = \frac{2F_x}{m} + \frac{d_f}{m} \\ -I\ddot{x}_2 = \tau_z + d_\tau \end{cases} \tag{2.37}$$

Combining Eqs. (2.31), (2.34), and (2.37), one has

$$\begin{cases} \ddot{x}_1 - 3n^2 x_1 = \frac{2\lambda}{m}\begin{bmatrix} g_1(x) + g_2(x) & g_1(x) - g_2(x) \end{bmatrix}\begin{bmatrix} u_1 \\ u_2 \end{bmatrix} + \frac{d_f}{m} \\ \ddot{x}_2 = -\frac{\lambda}{3I}\begin{bmatrix} g_3(x) + g_4(x) & g_3(x) - g_4(x) \end{bmatrix}\begin{bmatrix} u_1 \\ u_2 \end{bmatrix} - \frac{d_\tau}{I} \end{cases} \tag{2.38}$$

Eq. (2.38) is expressed in vector form as follows

$$\ddot{\mathbf{x}} = \boldsymbol{f}(\boldsymbol{x}) + \boldsymbol{g}(\boldsymbol{x})\boldsymbol{u} + \boldsymbol{d} \tag{2.39}$$

where $\boldsymbol{x} = [x_1 \quad x_2]^{\mathrm{T}}$, $\boldsymbol{f}(\boldsymbol{x}) = [3n^2 x_1 \quad 0]^{\mathrm{T}}$, $\boldsymbol{u} = [u_1 \quad u_2]^{\mathrm{T}}$, $\boldsymbol{d} = [d_f/m \quad -d_\tau/I]^{\mathrm{T}}$ and

$$\boldsymbol{g}(x) = \begin{bmatrix} \frac{2\lambda}{m}\left(g_1(x) + g_2(x)\right) & \frac{2\lambda}{m}\left(g_1(x) - g_2(x)\right) \\ -\frac{\lambda}{3I}\left(g_3(x) + g_4(x)\right) & -\frac{\lambda}{3I}\left(g_3(x) - g_4(x)\right) \end{bmatrix}$$

2.3 Orbit-attitude tracking sliding mode controller design

The reference trajectories of the relative distance and attitude angle of the chaser spacecraft are assumed to be $x_{1d} = d_d$ and $x_{2d} = \beta_{1d}$. Similar to Liu et al. (2018) and You and Dong (2019), the maneuvering process is divided into three successive parts, that is, the acceleration phase, the uniform phase, and the deceleration phase, where the reference trajectories of relative distance and attitude angle are assumed to have the same characteristic time instants, that is, t_1, t_2, and t_3. To satisfy the requirement of control input constraint, the relative velocity and angular velocity of the docking spacecraft should track the given reference trajectory. Taking the attitude motion for example, the reference trajectory of angular velocity is shown in Fig. 2.4, one can obtain the mathematical expressions of reference attitude angle and angular velocity used for attitude control, see details below.

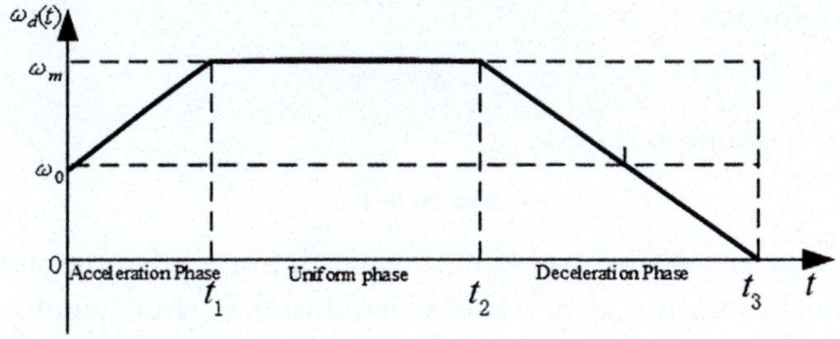

Figure 2.4 Reference trajectory of angular velocity.

According to Fig. 2.4, one can calculate the mathematical expression of angular velocity as

$$\dot{x}_{2d}(t) = \begin{cases} -\omega_0 - \frac{\omega_m - \omega_0}{t_1} t & 0 \leq t \leq t_1 \\ -\omega_m & t_1 < t \leq t_2 \\ -\omega_m + \frac{t - t_2}{t_3 - t_2} \omega_m & t_2 < t \leq t_3 \\ 0 & t > t_3 \end{cases} \tag{2.40}$$

where $\dot{x}_{2d}(t)$ represents the reference angular velocity, ω_0 denotes the initial angular velocity, ω_m is the maximum angular velocity.

Based on Eq. (2.40), one has

$$x_{2d}(t) = \begin{cases} \beta_0 - \omega_0 t - \frac{\omega_m - \omega_0}{2t_1} t^2 & 0 \leq t \leq t_1 \\ \beta_0 - \frac{\omega_m + \omega_0}{2} t_1 - \omega_m (t - t_1) & t_1 < t \leq t_2 \\ \beta_0 - \frac{\omega_m + \omega_0}{2} t_1 - \omega_m (t_2 - t_1) - \omega_m (t - t_2) + \frac{(tt_2)^2}{2(t_3 - t_2)} \omega_m & t_2 < t \leq t_3 \\ 0 & t > t_3 \end{cases} \tag{2.41}$$

where $x_{2d}(t)$ represents the reference attitude angle, β_0 denotes the initial attitude angle.

The maximum value of the reference angular velocity is calculated by

$$\omega_m = \frac{2\beta_0 - \omega_0 t_1}{(t_2 + t_3 - t_1)} \tag{2.42}$$

Thus the mathematical expressions of reference velocity, reference distance, and maximum reference velocity are similar to Eqs. (2.40)–(2.42), then the desired state is $\boldsymbol{x}_d = [x_{1d} \quad x_{2d}]^{\mathrm{T}}$ and the actual state is $\boldsymbol{x} = [x_1 \quad x_2]^{\mathrm{T}}$. Furthermore, the tracking errors of relative distance and attitude angle are defined as $\boldsymbol{e} = \boldsymbol{x}_d - \boldsymbol{x}$, respectively, then the derivative of tracking errors with respect to time is

$$\dot{e} = \dot{x}_d - \dot{x} \tag{2.43}$$

Choose a sliding surface as

$$s = ce + \dot{e} \tag{2.44}$$

where c is a given positive constant. It is worth noting that the sliding mode control law is constituted of equivalent control u_{eq} and switching control $\boldsymbol{u}_{sw}$. With $\dot{\boldsymbol{s}} = 0$, the equivalent control term can be obtained, then let $\boldsymbol{u}=\boldsymbol{u}_{eq}+\boldsymbol{u}_{sw}$, by analyzing $\dot{\boldsymbol{s}}$ and substitute $\boldsymbol{u}$, such that $\boldsymbol{s}\dot{\boldsymbol{s}} < 0$ satisfies, and the switching control term can be obtained. The equivalent control term ensures that the state of the error system remains on the sliding surface, and the switching control term ensures that the state does not leave the sliding surface.

Next, the first derivative of the sliding surface is derived as

$$\dot{\boldsymbol{s}} = c(\dot{\boldsymbol{x}}_d - \dot{\boldsymbol{x}}) + \left(\ddot{\boldsymbol{x}}_{\mathrm{d}} - \boldsymbol{f}(\boldsymbol{x}) - \boldsymbol{g}(\boldsymbol{x})\boldsymbol{u} - \boldsymbol{d}\right) \tag{2.45}$$

Let $\dot{\boldsymbol{s}} = 0$, the equivalent control term

$$\boldsymbol{u}_{eq} = \boldsymbol{g}(\boldsymbol{x})^{-1}\left[c(\dot{\boldsymbol{x}}_d - \dot{\boldsymbol{x}}) + \left(\ddot{\boldsymbol{x}}_d - \boldsymbol{f}(\boldsymbol{x}) - \boldsymbol{d}\right)\right] \tag{2.46}$$

Lemma 2.1. Choose the relevant law as

$$\dot{\boldsymbol{s}} = -k\boldsymbol{s} - \varepsilon|\boldsymbol{s}|^{\tilde{\alpha}}\mathrm{sgn}(\boldsymbol{s}) \tag{2.47}$$

Then the error $\boldsymbol{e}$ will reach the sliding surface within a finite time t_f, which satisfies

$$t_f = \frac{1}{k(\tilde{\alpha} - 1)}\ln\frac{\varepsilon}{k|\boldsymbol{s}(0)|^{1-\tilde{\alpha}} + \varepsilon} \tag{2.48}$$

where k is a given positive constant, ε is a small positive constant, and $\tilde{\alpha}$ satisfies $0 < \tilde{\alpha} < 1$.

Proof. With Eq. (2.47), one has

$$\begin{aligned}\boldsymbol{s}^{\mathrm{T}}\dot{\boldsymbol{s}} &= -k\boldsymbol{s}^{\mathrm{T}}\boldsymbol{s} - \varepsilon|\boldsymbol{s}|^{\tilde{\alpha}}\boldsymbol{s}^{\mathrm{T}}\mathrm{sgn}(\boldsymbol{s}) = -k|\boldsymbol{s}|^2 - \varepsilon|\boldsymbol{s}|^{\tilde{\alpha}-1}\boldsymbol{s}^{\mathrm{T}}\left(|\boldsymbol{s}|\mathrm{sgn}(\boldsymbol{s})\right)\\ &= -\left(k + \varepsilon|\boldsymbol{s}|^{\tilde{\alpha}-1}\right)|\boldsymbol{s}|^2 = \frac{1}{2}\frac{d\left(\boldsymbol{s}^{\mathrm{T}}\boldsymbol{s}\right)}{dt} = \frac{1}{2}\frac{d|\boldsymbol{s}|^2}{dt} = |\boldsymbol{s}|\frac{d|\boldsymbol{s}|}{dt}\end{aligned}$$

Then, one has

$$\frac{d|\boldsymbol{s}|}{dt} = -\left(k + \varepsilon|\boldsymbol{s}|^{\tilde{\alpha}-1}\right)|\boldsymbol{s}|$$

$$dt = -\frac{d|\boldsymbol{s}|}{\left(k + \varepsilon|\boldsymbol{s}|^{\tilde{\alpha}-1}\right)|\boldsymbol{s}|} = -\frac{|\boldsymbol{s}|^{-\tilde{\alpha}}d|\boldsymbol{s}|}{k|\boldsymbol{s}|^{1-\tilde{\alpha}} + \varepsilon} = \frac{1}{\tilde{\alpha} - 1}\frac{d|\boldsymbol{s}|^{1-\tilde{\alpha}}}{k|\boldsymbol{s}|^{1-\tilde{\alpha}} + \varepsilon} \tag{2.49}$$

Integrating both sides of Eq. (2.49) yields

$$t_f = \frac{1}{\tilde{\alpha}-1}\int_{|s(0)|}^{|s(t_f)|} \frac{d|s|^{1-\tilde{\alpha}}}{k|s|^{1-\tilde{\alpha}}+\varepsilon} = \frac{1}{k(\tilde{\alpha}-1)}\ln\left(k|s|^{1-\tilde{\alpha}}+\varepsilon\right)\Big|_{|s(0)|}^{|s(t_f)|}$$

$$= \frac{1}{k(\tilde{\alpha}-1)}\ln\frac{\varepsilon}{k|s(0)|^{1-\tilde{\alpha}}+\varepsilon}$$

When $t > t_f$, one has $\boldsymbol{s} = 0, \dot{\boldsymbol{s}} = 0$.

This completes the proof.

Remark 2.1. Compared with the exponential reaching law $\dot{\boldsymbol{s}} = -k\boldsymbol{s} - \varepsilon\,\mathrm{sgn}(\boldsymbol{s})$ and the power reaching law $\dot{\boldsymbol{s}} = -\varepsilon|\boldsymbol{s}|^{\tilde{\alpha}}\mathrm{sgn}(\boldsymbol{s})$, it has the advantages of both reaching laws and has better performance. As one can see, when the error $\boldsymbol{e}$ is far from the sliding surface, for $0 < \tilde{\alpha} < 1$, the linear part $-k\boldsymbol{s}$ in the fast finite-time reaching law, a main role is played, and the fast finite-time reaching law has a similar structure to the exponential reaching law and satisfies the exponential convergence condition, which allows the error $\boldsymbol{e}$ to quickly approach the sliding surface; when the error $\boldsymbol{e}$ is near the sliding surface, for example, $|\boldsymbol{s}| < 1$, the fractional power term $-\varepsilon|\boldsymbol{s}|^{\tilde{\alpha}}\mathrm{sgn}(\boldsymbol{s})$ in the finite-time reaching law has a much greater effect than the linear part $-k\boldsymbol{s}$, and the fast finite-time reaching law has a similar structure to the power reaching law and can drive the error $\boldsymbol{e}$ to reach the sliding surface within finite time. Thus it has the advantages of both reaching laws and has better performance.

Theorem 2.1. Design a sliding mode control law as

$$\boldsymbol{u} = \boldsymbol{g}(\boldsymbol{x})^{-1}\left[c(\dot{\boldsymbol{x}}_d - \dot{\boldsymbol{x}}) + \left(\ddot{\boldsymbol{x}}_d - \boldsymbol{f}(\boldsymbol{x}) - \boldsymbol{d}\right) + k\boldsymbol{s} + \varepsilon|\boldsymbol{s}|^{\tilde{\alpha}}\mathrm{sgn}(\boldsymbol{s})\right] \tag{2.50}$$

Then, the error state starting from any initial point will reach the sliding surface in finite time. Consequently, the closed-loop system generated by Eqs. (2.43) and (2.50) are globally asymptotically stable.

Proof. Consider a Lyapunov function candidate as

$$V = \frac{1}{2}\boldsymbol{s}^{\mathrm{T}}\boldsymbol{s} \tag{2.51}$$

Differentiating the Lyapunov function V defined by Eq. (2.51) with respect to time yields

$$\begin{aligned}\dot{V} &= \boldsymbol{s}^{\mathrm{T}}\dot{\boldsymbol{s}} = \boldsymbol{s}^{\mathrm{T}}\left[c(\dot{\boldsymbol{x}}_d - \dot{\boldsymbol{x}}) + \ddot{\boldsymbol{x}}_d - \boldsymbol{f}(\boldsymbol{x}) - \boldsymbol{g}(\boldsymbol{x})\boldsymbol{u} - \boldsymbol{d}\right] \\ &= \boldsymbol{s}^{\mathrm{T}}\Big[c(\dot{\boldsymbol{x}}_d - \dot{\boldsymbol{x}}) + \ddot{\boldsymbol{x}}_d - \boldsymbol{f}(\boldsymbol{x}) - \Big(c(\dot{\boldsymbol{x}}_d - \dot{\boldsymbol{x}}) + \left(\ddot{\boldsymbol{x}}_d - \boldsymbol{f}(\boldsymbol{x}) - \boldsymbol{d}\right) \\ &\quad + k\boldsymbol{s} + \varepsilon|\boldsymbol{s}|^{\tilde{\alpha}}\mathrm{sgn}(\boldsymbol{s})\Big) - \boldsymbol{d}\Big]\end{aligned}$$

$$= s^{\mathrm{T}}\left[-ks - \varepsilon|s|^{\tilde{\alpha}}\operatorname{sgn}(s)\right] = -k|s|^2 - \varepsilon|s|^{\tilde{\alpha}+1} < 0$$

Letting the virtual input be $\boldsymbol{v} = c(\dot{\boldsymbol{x}}_d - \dot{\boldsymbol{x}}) + (\ddot{\boldsymbol{x}}_d - \boldsymbol{f}(\boldsymbol{x}) - \boldsymbol{d}) + k\boldsymbol{s} + \varepsilon|\boldsymbol{s}|^{\tilde{\alpha}}\operatorname{sgn}(\boldsymbol{s})$, one has

$$\boldsymbol{g}(\boldsymbol{x})\boldsymbol{u} = \boldsymbol{v} \tag{2.52}$$

which leads to $\boldsymbol{u} = \boldsymbol{g}(\boldsymbol{x})^{-1}\boldsymbol{v}$.

To facilitate the derivation of subsequent control input, the components of $\boldsymbol{f}(\boldsymbol{x})$, $\boldsymbol{g}(\boldsymbol{x})$ and $\boldsymbol{d}$ are utilized to make the control input more concise and understandable. In this case, define $\boldsymbol{v} = [v_1 \quad v_2]^{\mathrm{T}}$, $\boldsymbol{g}(\boldsymbol{x}) = \begin{bmatrix} g_{11} & g_{12} \\ g_{21} & g_{22} \end{bmatrix}$, one has

$$\begin{cases} u_1 = \dfrac{g_{22}v_1 - g_{12}v_2}{g_{11}g_{22} - g_{21}g_{12}} \\ u_2 = \dfrac{g_{21}v_1 - g_{11}v_2}{g_{21}g_{12} - g_{11}g_{22}} \end{cases} \tag{2.53}$$

Therefore the control law can ensure that the error states reach the sliding surface in finite time. Once the error states reach the sliding surface, it will remain on the sliding surface. For the sliding surface is $\boldsymbol{s} = c\boldsymbol{e} + \dot{\boldsymbol{e}}$, when $t \geq t_f$, one has $\boldsymbol{s} = c\boldsymbol{e} + \dot{\boldsymbol{e}} = 0$, that is, $\dot{\boldsymbol{e}} = -c\boldsymbol{e}$. Then, defining a Lyapunov function candidate $V = \frac{1}{2}\boldsymbol{e}^{\mathrm{T}}\boldsymbol{e}$and differentiating it yields $\dot{V} = \boldsymbol{e}^{\mathrm{T}}\dot{\boldsymbol{e}} = -c\boldsymbol{e}^{\mathrm{T}}\boldsymbol{e} \leq 0$, so $\lim_{t\to\infty}\boldsymbol{e} = 0$, that is, the error system will be globally asymptotically stable. Therefore the error system denoted by Eq. (2.43) is globally asymptotically stable with the controller (2.50), that is, (2.53), such that $\lim_{t\to\infty}\boldsymbol{e} = 0$, and this completes the proof.

As the disturbance $\boldsymbol{d} = [\tilde{d}_1 \quad \tilde{d}_2]^{\mathrm{T}}$ is unknown, the control law in Eq. (2.53) cannot be applied. Hence, the boundary of disturbance, that is $\boldsymbol{D} = [D_1 \quad D_2]^{\mathrm{T}}$ satisfying $D_1 \geq \tilde{d}_1, D_2 \geq \tilde{d}_2$, is used such that the improved control law is

$$\begin{cases} u_1 = \dfrac{g_{22}\kappa_1 - g_{12}\kappa_2}{g_{11}g_{22} - g_{21}g_{12}} \\ u_2 = \dfrac{g_{21}\kappa_1 - g_{11}\kappa_2}{g_{21}g_{12} - g_{11}g_{22}} \end{cases} \tag{2.54}$$

where$\boldsymbol{\kappa} = c(\dot{\boldsymbol{x}}_d - \dot{\boldsymbol{x}}) + (\ddot{x}_d - \boldsymbol{f}(\boldsymbol{x}) - \boldsymbol{D}) + k\boldsymbol{s} + \varepsilon|\boldsymbol{s}|^{\tilde{\alpha}}\operatorname{sgn}(\boldsymbol{s})$.

Substituting Eq. (2.54) into Eq. (2.45) yields

$$\dot{s} = -k\boldsymbol{s} - \varepsilon|\boldsymbol{s}|^{\tilde{\alpha}}\operatorname{sgn}(s) + \boldsymbol{D} - \boldsymbol{d} \tag{2.55}$$

With an appropriate selection of $\boldsymbol{D}$, the sliding surface arrival condition can be satisfied, and the system represented by Eq. (2.39) can be stabilized.

For i=1, 2, suppose $\tilde{d}_{iL} \leq \tilde{d}_i \leq \tilde{d}_{iU}$, where $\tilde{d}_{iL}$ and $\tilde{d}_{iU}$ denote the lower and upper boundaries of disturbance $\tilde{d}_i$, respectively. The value of D_i is chosen as follows

When $s_i(t) > 0$, $\dot{s}_i = -ks_i - \varepsilon|s_i|^{\tilde{\alpha}} + D_i - \tilde{d}_i$, to guarantee $\dot{s}_i(t) < 0$, one has $D_i = d_{iL}$.

When $s_i(t) < 0$, $\dot{s}_i = -ks_i + \varepsilon|s_i|^{\tilde{\alpha}} + D_i - \tilde{d}_i$, to guarantee $\dot{s}_i(t) > 0$, one has $D_i = d_{iU}$.

If $\tilde{d}_1 = \frac{d_{iU}-d_{iL}}{2}$, $\tilde{d}_2 = \frac{d_{iU}+d_{iL}}{2}$, the value of D_i is obtained as follows

$$D_i = \tilde{d}_2 - \tilde{d}_1 \operatorname{sgn}(s_i) \tag{2.56}$$

If the control input current calculated by Eq. (2.54) exceeds the maximum current that can be provided, it should be saturated to get the actual control input current, that is, $\operatorname{sat}(\boldsymbol{u}) = [\operatorname{sat}(\boldsymbol{u}_1) \quad \operatorname{sat}(\boldsymbol{u}_2)]^{\mathrm{T}}$, which satisfies

$$\operatorname{sat}(u_i) = \begin{cases} u_{im} & u_i > u_{im} \\ u_i & -u_{im} \leq u_i \leq u_{im} \\ -u_{im} & u_i < -u_{im} \end{cases} \tag{2.57}$$

where u_{im}, $i = 1, 2$ are the maximum available currents that the spacecraft can provide.

2.4 Simulation test

In this section, the effectiveness of the electromagnetic force/torque models and the coupled orbit-attitude controller is illustrated with simulations of a CubeSat electromagnetic docking system. In this example, for "CM" orbit, the eccentricity $e = 0$, the semimajor axis $a = 6978.14$km, and the initial true anomaly $\theta(0) = 0$. The masses of the chaser and target spacecraft are assumed to be $m_C = 6$kg and m_T=6kg, respectively; the inertia matrix of both spacecraft is $\boldsymbol{I} = \operatorname{diag}([0.01 \quad 0.01 \quad 0.01])\mathrm{kg}\cdot\mathrm{m}^2$; the initial attitude angle and angular velocity are assumed to be $\beta_0 = \pi/18$rad and $\omega_0 = -0.0002$rad/s, respectively. In addition, the initial relative distance is $x_{10} = 0.3$m, and the initial relative velocity is $v_0 = -0.0003$m/s. It is worth mentioning that the final relative distance should be $x_{1f} = 0.1$m. Once the docking task is accomplished, to avoid collision. The characteristic time instants are chosen as $t_1 = 180$s, $t_2 = 200$s, $t_3 = 250$s, and according to Eq. (2.42), the maximum values of the reference relative velocity and reference angular

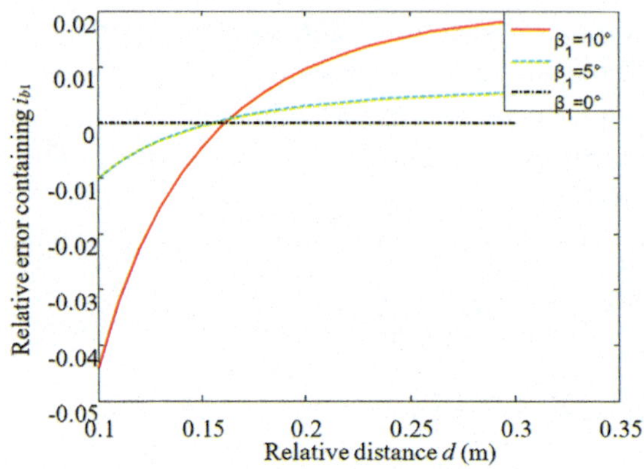

Figure 2.5 Relative error of the electromagnetic force with i_{b1} along the x-direction.

velocity are calculated as $v_m = 0.0013\text{m/s}$ and $\omega_m = 0.0012\text{rad/s}$. In addition, the environmental disturbance has been analyzed in Shi et al. (2020), Liu et al. (2019), and Yang & Sun (2002)) in detail, which includes gravity-gradient force, aerodynamic force, and the force of the Earth magnetic field. As pointed out in this work, the sinusoid is a prototype of periodic disturbances, and the environmental disturbance force, such as magnetic force and gravity-gradient force, are essentially cyclic, which can be represented by sinusoids. Thus the external disturbance force and torque are assumed to be $d_1 = 6 \times 10^{-6}\sin(10^{-3}\pi t)\text{N}$ and $d_2 = 1 \times 10^{-8}\sin(10^{-3}\pi t)\text{Nm}$, and the maximum current is $u_{im} = 2\text{A}, i = 1, 2$. Furthermore, the number of coil turns $N = 1440$ for each energized solenoid. Regarding the iron core, the diameter $d_c = 10\text{mm}$ and the length $l_c = 100\text{mm}$. The current is provided by an external solar panel with the device mass of $m_p = 0.5\text{kg}$. The distance between two neighbored energized solenoids $d_1 = 0.05\text{m}$. The parameters involved in the control law for the simulation are $c = 10, k = 5, \varepsilon = 10^{-6}, \tilde{\alpha} = 0.4$.

With Eqs. (1.68) and (2.20), one can obtain the relative errors between the electromagnetic force with i_{b1}, i_{b4}, and the far-field electromagnetic force along the x-direction. As can be seen from Figs. 2.5 and 2.6, the relative errors between the electromagnetic force containing the current terms i_{b1}, i_{b4}, and the far-field electromagnetic force are within 5%, which shows that it is reasonable to ignore the higher-order small items to simplify the electromagnetic force that satisfies the error requirement. Furthermore, for the small-angle docking task, it can be observed from Fig. 2.7 that the

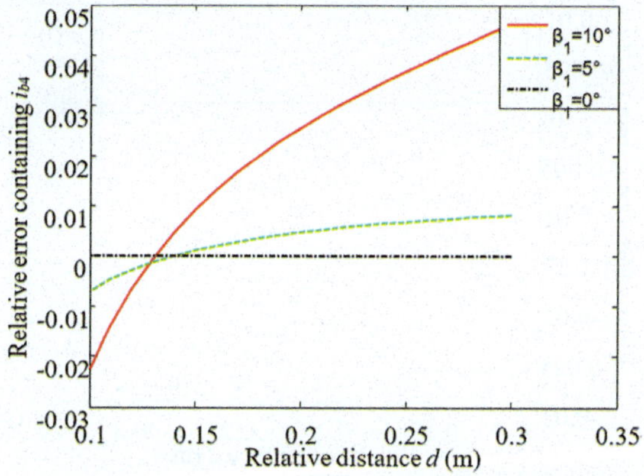

Figure 2.6 Relative error of the electromagnetic force with i_{b4} along the *x*-direction.

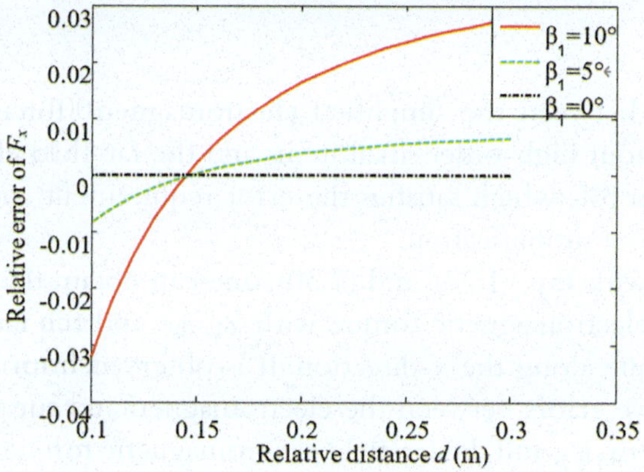

Figure 2.7 Relative error of the electromagnetic force along the *x*-direction.

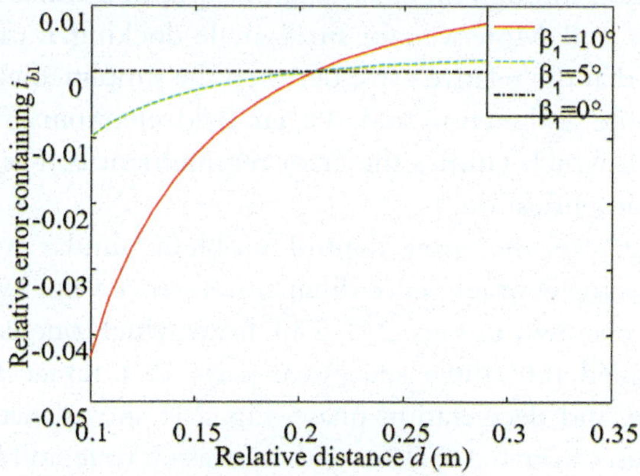

Figure 2.8 Relative error of the electromagnetic torque with i_{b1} along the *x*-direction.

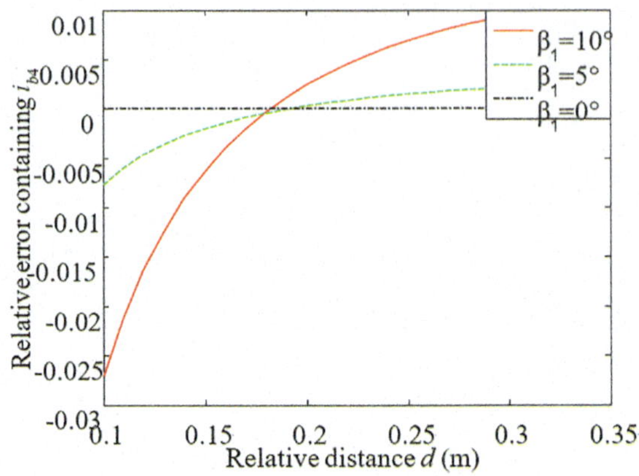

Figure 2.9 Relative error of the electromagnetic torque with i_{b4} along the x-direction.

relative error between the simplified electromagnetic force along the x-direction without high-order small items and the far-field electromagnetic force is within 3%, which satisfies the error requirement and accounts for the rationality of simplification.

Likewise, with Eqs. (1.72). and (2.30), one can obtain the relative errors between the electromagnetic torque with i_{b1}, i_{b4}, and the far-field electromagnetic torque along the x-direction. It is observed from Fig. 2.8 to 2.9 that the relative errors between the electromagnetic torque containing the current terms i_{b1}, i_{b4}, and the far-field electromagnetic torque are within 5%, which demonstrates the reasonability of ignoring the high-order small items and simplification through the small-angle assumption and satisfies the error requirement as well. Moreover, for small-angle docking, it can be observed from Fig. 2.10 that the relative error between the simplified electromagnetic torque along the x-direction and the far-field electromagnetic torque is also within 3%, which satisfies the error requirement and accounts for the rationality of simplification.

With respect to the orbit control problem, similar to Eqs. (2.40)–(2.42), the trajectories of reference distance, reference velocity, and reference acceleration are shown in Figs. 2.11–2.13, from which one can see that the process is divided into three successive parts, that is, acceleration phase, uniform phase, and deceleration phase. Fig. 2.11 shows that the reference distance decreases from 0.3 to 0.1m with increasing time, and Fig. 2.12 shows that the reference velocity increases from 0.0003 to 0.0013m/s with a fixed acceleration for 180s, then maintains a constant value for 20s, after which

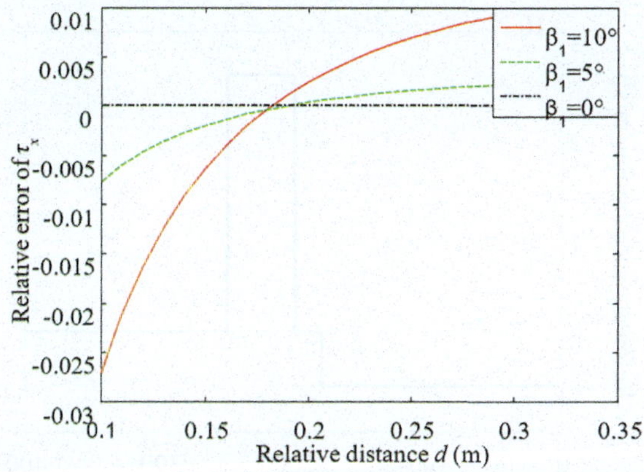

Figure 2.10 Relative error of the electromagnetic torque along the x-direction.

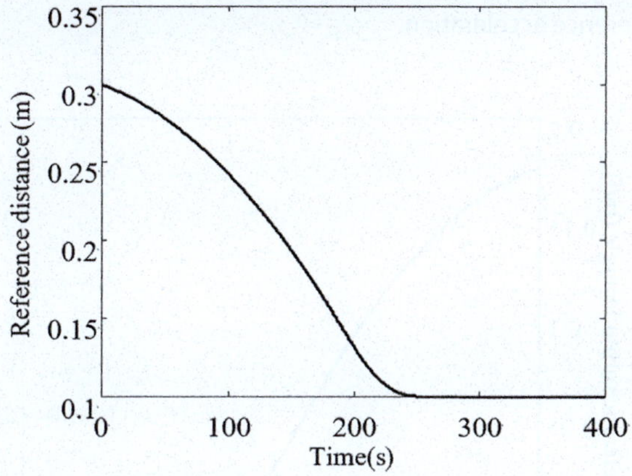

Figure 2.11 Reference distance.

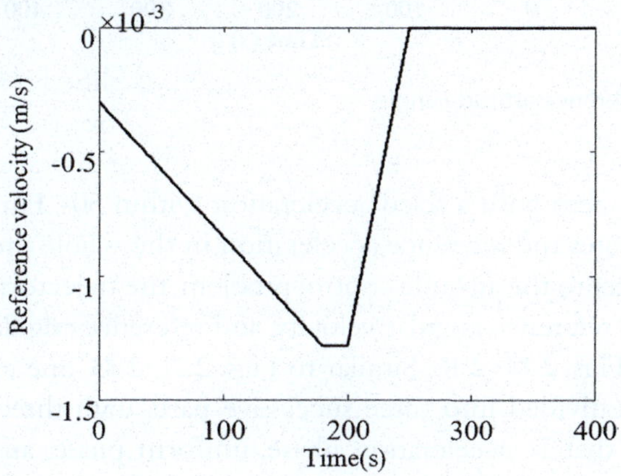

Figure 2.12 Reference velocity.

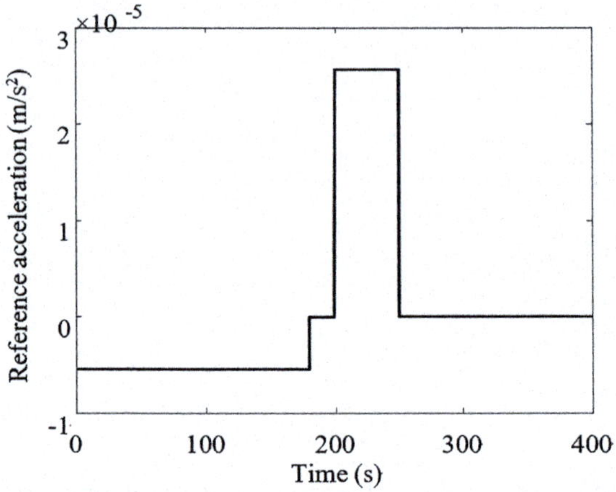

Figure 2.13 Reference acceleration.

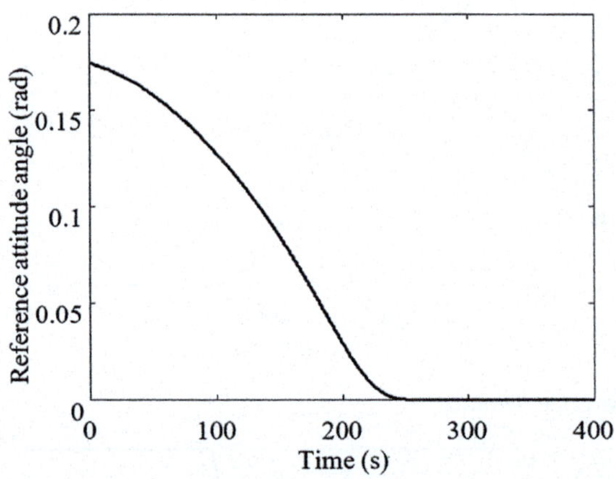

Figure 2.14 Reference attitude angle.

it decreases to zero with a fixed acceleration within 50s. Furthermore, Fig. 2.13 demonstrates the reference acceleration in the whole process.

With respect to the attitude control problem, the trajectories of reference attitude angle, reference angular velocity, and reference angular acceleration are shown in Figs. 2.14–2.16. Similar to Figs. 2.11–2.13, one can see that the process is also divided into three successive parts with three characteristic time instants, that is, acceleration phase, uniform phase, and deceleration phase. Fig. 2.14 shows that the reference attitude angle decreases from $\pi/18$rad to zero with increasing time, and Fig. 2.15 shows that the reference

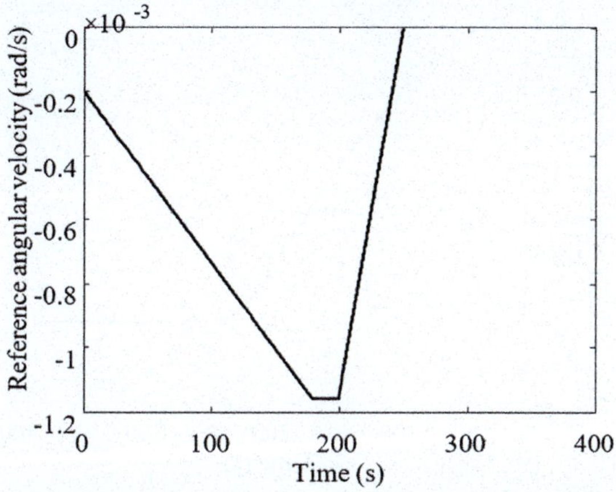

Figure 2.15 Reference angular velocity.

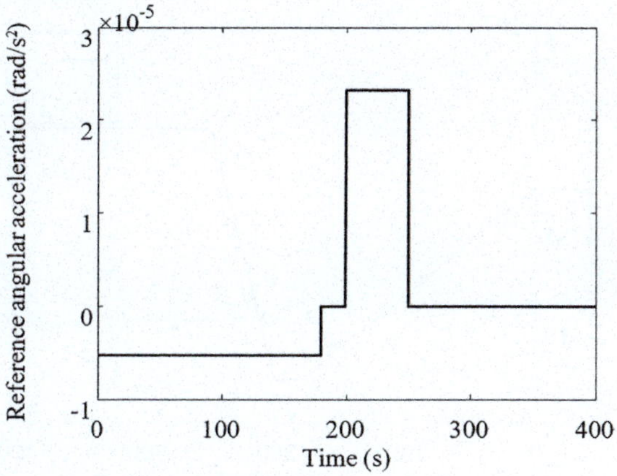

Figure 2.16 Reference angular acceleration.

angular velocity increases from 0.0002 to 0.0012rad/s with a fixed acceleration before 180s, then maintains a constant value for 20s, after which it decreases to zero with a fixed acceleration within 50s. Furthermore, Fig. 2.16 demonstrates the reference angular acceleration in the whole process.

Applying controller (2.54) to the coupling orbit-attitude dynamics system represented by Eq. (2.39) gives the time responses of position/velocity tracking and attitude angle/angular velocity tracking results as shown in Figs. 2.17–2.20. It is observed that the tracking errors of the system states all converge to within a small bound. In terms of coupling orbit-attitude control, the final relative distance decreases to 0.1m, and the attitude angle

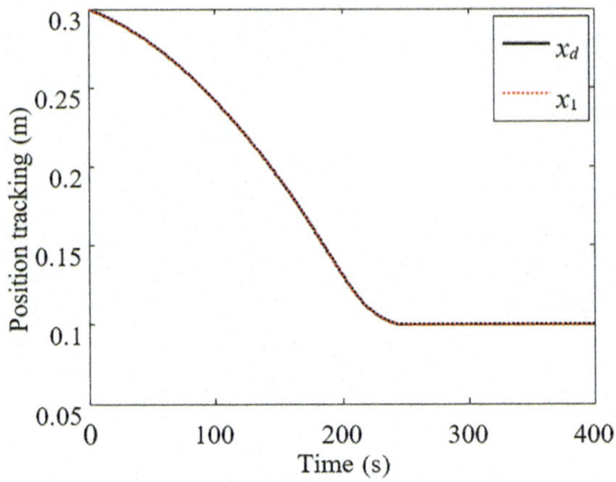

Figure 2.17 Time response of position tracking.

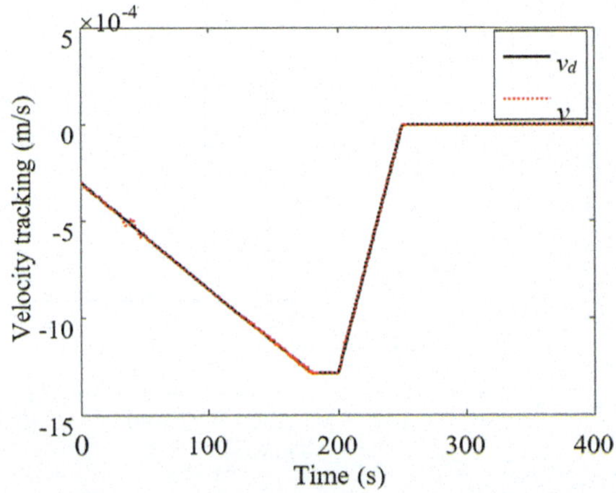

Figure 2.18 Time response of velocity tracking.

decreases to zero, respectively, both within the characteristic time instant t_3 −250s. It can be seen from Figs. 2.18-2.20 that both velocity and angular velocity track the reference trajectories effectively, and the chattering phenomenon from a theoretical viewpoint indeed exists for both velocity tracking and angular velocity tracking.

In addition, the plots of the corresponding control input currents applied on the chaser spacecraft during the application of the electromagnetic force/torque are shown in Fig. 2.21, from which one can see that the control input is no more than 2A, that is, the input saturation is satisfied. However,

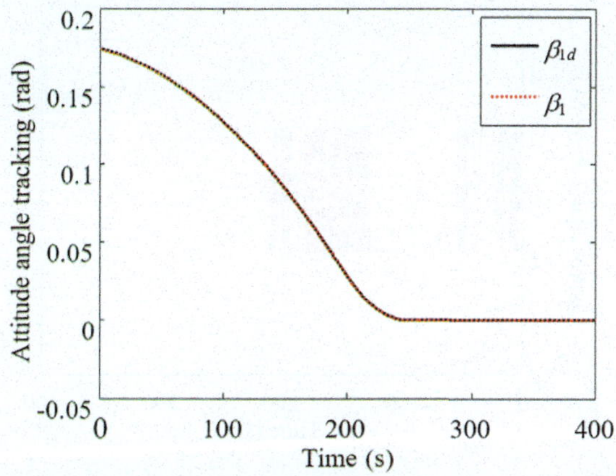

Figure 2.19 Time response of attitude angle tracking.

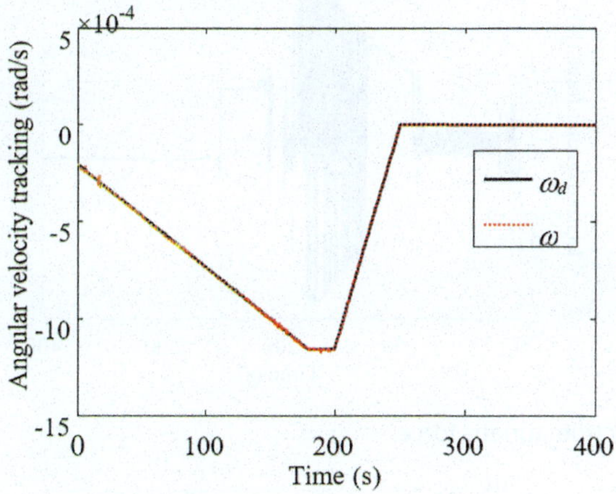

Figure 2.20 Time response of angular velocity tracking.

chattering cannot be ignored due to the discontinuous control law, that is, the implementation of the sign function. The chattering phenomenon can be weakened or even avoided by combining the second-order sliding mode controller or adaptive law (Zhang et al., 2018). Furthermore, the traditional control inputs, that is, electromagnetic force and electromagnetic torque, are depicted in Figs. 2.22 and 2.23, respectively. It can be seen that as the distance between the two docking spacecraft decreases, the maximum electromagnetic force is less than 2×10^{-4}N and the maximum electromagnetic torque is less than 2×10^{-6}Nm, which is easy to achieve in reality.

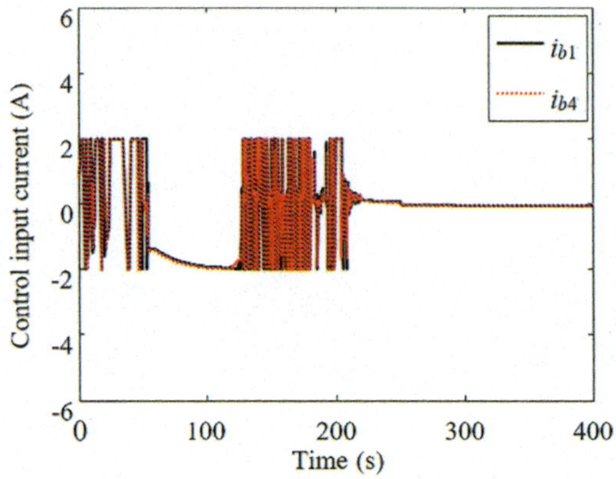

Figure 2.21 Control input current.

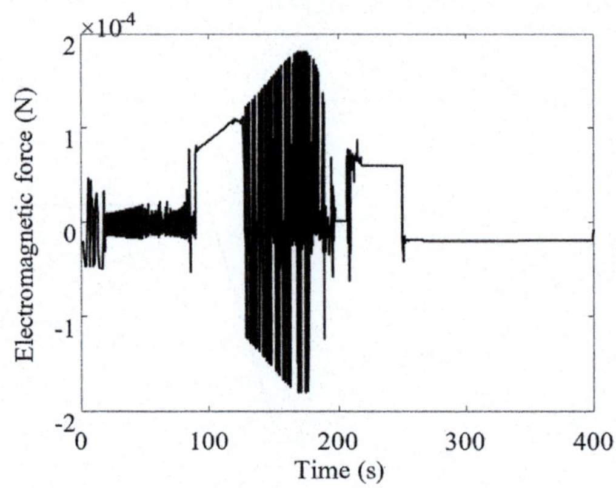

Figure 2.22 Electromagnetic force.

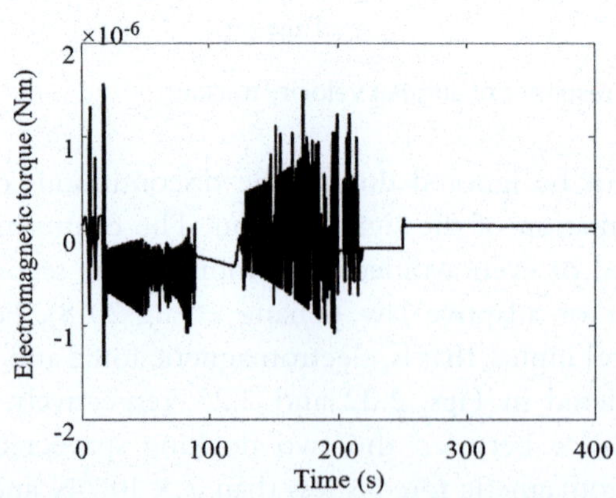

Figure 2.23 Electromagnetic torque.

2.5 Conclusions

In this chapter, a simplified form of the x-direction electromagnetic force/torque is firstly derived for small-angle electromagnetic docking with different currents in the electrified solenoids on spacecraft, based on which the nonlinear equation of one-dimensional coupled orbit-attitude dynamics is obtained. Then, a soft docking trajectory for relative position/attitude is planned, and the tracking errors of relative position/attitude are defined. Based on the tracking errors, an improved sliding mode control approach is proposed to asymptotically stabilize the orbit and attitude simultaneously for spacecraft electromagnetic docking, which is proved via Lyapunov analysis. By defining two virtual inputs and introducing the boundary of disturbance and saturation function, the corresponding control input currents can be solved. Simulation results demonstrate the rationality of model simplification in terms of relative errors of electromagnetic force/torque, and the effectiveness of the developed sliding mode controller in terms of trajectory tracking performance and maximum control input currents.

References

Elias, L. M., Kwon, D. W., Sedwick, R. J., & Miller, D. W. (2007). Electromagnetic formation flight dynamics including reaction wheel gyroscopic stiffening effects. *Journal of Guidance, Control, and Dynamics, 30*(2), 499–511. https://doi.org/10.2514/1.18679.

Foust, R. C., Lupu, E. S., Nakka, Y. K., Chung, S. J., & Fred, Y. H. (2018). Ultra-soft electromagnetic docking with applications to in-orbit assembly. *Proceedings of the 69th International Astronautical Congress (IAC)* (pp. 1–14). Bremen, Germany: International Astronautical Federation (IAF).

Huang, H., Cai, W. W., & Yang, L. P. (2020). *6-DOF formation keeping control for an invariant three-craft triangular electromagnetic formation Advances in Space Research, 65*, 312–325. https://doi.org/10.1016/j.asr.2019.09.033.

Huang, X., Zhang, C., Lu, H., & Li, M. (2016). *Adaptive reaching law based sliding mode control for electromagnetic formation flight with input saturation Journal of the Franklin Institute, 353*, 2398–2417. https://doi.org/10.1016/j.jfranklin.2016.04.004.

Kong, E. M. C. (2002). *Spacecraft formation flight exploiting potential fields* (pp. 145–146). Cambridge: Massachusetts Institute of Technology.

Liu, C., Shi, K., & Sun, Z. (2018a). *Spacecraft chaotic attitude control with certain actuator failure based on integral sliding mode Journal of Engineering Research, 6*, 153–169. https://doi.org/10.1016/S2307-1877(25)00781-3.

Liu, C., Vukovich, G., Sun, Z., & Shi, K. (2018b). *Observer-based fault-tolerant attitude control for spacecraft with input delay Journal of Guidance, Control, and Dynamics, 41*, 2041–2053. https://doi.org/10.2514/1.g003555.

Liu, C., Sun, Z., Ye, D., & Shi, K. (2018c). *Robust adaptive variable structure tracking control for spacecraft chaotic attitude motion IEEE Access, 6*, 3851–3857. https://doi.org/10.1109/access.2017.2788860.

Liu, C., Shi, K., Yue, X., & Sun, Z. (2020). *Inertia-free saturated output feedback attitude stabilization for uncertain spacecraft International Journal of Robust and Nonlinear Control, 30*, 5101–5121. https://doi.org/10.1002/rnc.5044.

Liu, C., Yue, X., Shi, K., & Sun, Z. (2019). *Inertia-free attitude stabilization for flexible spacecraft with active vibration suppression International Journal of Robust and Nonlinear Control, 29*, 6311–6336. https://doi.org/10.1002/rnc.4742.

Qi, D. W., Yang, L. P., Zhang, Y. W., & Cai, W. W. (2019). *Indirect robust suboptimal control of two-satellite electromagnetic formation reconfiguration with geomagnetic effect Advances in Space Research, 64*, 2331–2344. https://doi.org/10.1016/j.asr.2019.08.011.

Shi, K., Liu, C., Biggs, J. D., Sun, Z., & Yue, X. (2020). Observer-based control for spacecraft electromagnetic docking. *Aerospace Science and Technology, 99*, 105759. https://doi.org/10.1016/j.ast.2020.105759.

Shi, K., Liu, C., & Sun, Z. (2019). *Constrained fuel-free control for spacecraft electromagnetic docking in elliptical orbits Acta Astronautica, 162*, 14–24. https://doi.org/10.1016/j.actaastro.2019.05.016.

Shi, K., Sun, Z., Liu, C., & Ye, D. (2017). Dynamics modeling and simulation of space electromagnetic docking for CubeSat. *Proceedings of the 8th International Conference on Mechanical and Aerospace Engineering (ICMAE 2017)* (pp. 603–608). China: Institute of Electrical and Electronics Engineers Inc. 9781538633052. https://doi.org/10.1109/ICMAE.2017.8038716.

Takahashi, Y., Sakamoto, H., & Sakai, S. I. (2021). Simultaneous control of relative position and absolute attitude for electromagnetic spacecraft swarm. *AIAA Scitech 2021 Forum* (pp. 1–23). Japan: American Institute of Aeronautics and Astronautics Inc, AIAA. https://doi.org/10.2514/6.2021-1104.

Xing, L., Zhang, J., Liu, C., & Zhang, X. (2021). Fuzzy-logic-based adaptive event-triggered sliding mode control for spacecraft attitude tracking. *Aerospace Science and Technology, 108*, 106394. https://doi.org/10.1016/j.ast.2020.106394.

Yang, C. D., & Sun, Y. P. (2002). *Mixed H_2/H_∞ state-feedback design for microsatellite attitude control Control Engineering Practice, 10*, 951–970. https://doi.org/10.1016/s0967-0661(02)00049-7.

You, L., & Dong, Y. (2019). *Near time-optimal controller based on analytical trajectory planning algorithm for satellite attitude maneuver Aerospace Science and Technology, 84*, 497–509. https://doi.org/10.1016/j.ast.2018.10.030.

Yu, Y., Yang, F., Yue, H., Lu, Y., Li, S., & Zhao, H. (2021). *Prospects of de-tumbling large space debris using a two-satellite electromagnetic formation Advances in Space Research, 67*, 1816–1829. https://doi.org/10.1016/j.asr.2020.12.039.

Zeng, G., & Hu, M. (2012). *Finite-time control for electromagnetic satellite formations Acta Astronautica, 74*, 120–130. https://doi.org/10.1016/j.actaastro.2012.01.003.

Zhang, J., Ye, D., Liu, M., & Sun, Z. (2017). *Adaptive fuzzy finite-time control for spacecraft formation with communication delays and changing topologies Journal of the Franklin Institute, 354*, 4377–4403. https://doi.org/10.1016/j.jfranklin.2017.04.018.

Zhang, J., Ye, D., Sun, Z., & Liu, C. (2018). *Extended state observer based robust adaptive control on SE(3) for coupled spacecraft tracking maneuver with actuator saturation and misalignment Acta Astronautica, 143*, 221–233. https://doi.org/10.1016/j.actaastro.2017.11.034.

Zhang, J., Zeng, G., Zhang, S., & Biggs, J. D. (2020a). *Disturbance observer-based nonsingular terminal sliding mode control for spacecraft electromagnetic docking International Journal of Aerospace Engineering, 2020*, 1–20. https://doi.org/10.1155/2020/8887699.

Zhang, J., Zeng, G., Gao, Y., & Wang, J. (2020b). *Underactuated microsatellite electromagnetic docking control using disturbance observer-based sliding mode controller IEEE Access, 8*, 124476–124485. https://doi.org/10.1109/access.2020.3003272.

Zhang, Y. W., Yang, L. P., Zhu, Y. W., Huang, H., & Cai, W. W. (2015). *An investigation of on-orbit release with inter-satellite electromagnetic force Aerospace Science and Technology, 45*, 309–315. https://doi.org/10.1016/j.ast.2015.04.021.

Zhang, Y. W., Yang, L. P., Zhu, Y. W., Ren, X. H., & Huang, H. (2011). *Self-docking capability and control strategy of electromagnetic docking technology Acta Astronautica, 69*, 1073–1081. https://doi.org/10.1016/j.actaastro.2011.07.001.

Zhu, Y. W., Cai, W. W., Yang, L. P., & Huang, H. (2018). *Flatness-based trajectory planning for electromagnetic spacecraft proximity operations in elliptical orbits Acta Astronautica, 152*, 342–351. https://doi.org/10.1016/j.actaastro.2018.08.031.

CHAPTER 3

Intermediate observer-based control for spacecraft electromagnetic docking

3.1 Introduction

Spacecraft electromagnetic docking is a novel technology that applies an intercraft electromagnetic force, generated by magnetic coils equipped on the neighboring target and chaser spacecraft, to control the relative motion between them. Electromagnetic docking has distinct advantages over other docking mechanisms; for example, it requires no propellant, produces no plume contamination, and offers continuous, reversible, and noncontact actuation (Ahsun, Miller, & Ramirez, 2010; Zhang, Yang, Zhu, Huang, & Cai, 2013). However, along with these advantages come challenges related to its complex and coupled dynamics, uncertainties that can influence the robust performance of the docking system.

Autonomous rendezvous and docking technologies for small spacecraft are necessary to enable on-orbit assembly schemes. Small spacecraft have the distinct advantages of low cost and weight, which make them particularly attractive to multi-agent missions such as on-orbit assembly. Noncontact docking uses the interaction of magnetic or electric fields generated by the spacecraft, including Coulomb force, electromagnetic force, and flux-pinning force (Norman & Peck, 2010; Zhang, Yang, Zhu, Huang, & Cai, 2012). According to Zhang, Yang, Zhu, Huang, & Cai, 2012 and Zhu, Cai, Yang, & Huang, 2018, the Coulomb force is incapable of 6-dimensions-of-freedom (6-DOF) control, while flux-pinned space modules are passively stable only when the separation distance is on the order of 10 cm. Therefore, this chapter focuses on the intercraft electromagnetic force. In spacecraft electromagnetic docking, each spacecraft is equipped with three large orthogonal coils that can generate electromagnetic forces. The relative positions between spacecraft are then determined by these electromagnetic forces, while the relative orientation of the dipoles on spacecraft is controlled using reaction wheels or control moment gyros to

Spacecraft Electromagnetic Docking and Separation.
DOI: https://doi.org/10.1016/B978-0-443-49334-8.00006-9

generate the counteracting torques (Kong, 2002). The focus of this chapter is on the position control problem of spacecraft electromagnetic docking in the presence of external disturbances, fault signals, unknown mass, elliptical eccentricity, measurement errors, and input constraints.

The translational dynamics of a spacecraft using electromagnetic force are highly coupled and nonlinear, which increases the challenge for precise docking maneuvers. Most of the existing work in the field of electromagnetic docking only considers circular orbits of the spacecraft, which would fail in missions where the target spacecraft is in a large-eccentric orbit. Electromagnetic docking systems are also constrained by the maximum available current density of the electromagnetic coils (Elias et al., 2007; Shi et al., 2019ba). Moreover, for traditional docking mechanisms, the magnitude of the order of the external disturbance is negligible over a short period of time, while electromagnetic forces increase as the relative distance decreases. Consequently, the external disturbance force and the electromagnetic force will be of the same magnitude of order when the relative distance is greater than a certain value. Since spacecraft electromagnetic docking technology has not yet matured, controllers designed for fault-free cases may not be directly applicable to reality. In contrast to this, a fault-tolerant controller can guarantee stability and performance in case of component fault/failure (Liu et al., 2018a; Narula & Biggs, 2018). Thus, the possible fault signals should be taken into account in the controller design.

In addition to the problems mentioned, it is often expensive or even impossible to determine the mass properties of the spacecraft, and this can lead to a loss of control precision. To circumvent this need, the controllers developed in Sanyal et al. (2009) and Liu et al. (2019) were inertia-free in the sense that no prior information on the inertia matrix is required. Analogous to this, for electromagnetic docking missions, the controller developed by Shi et al. (2019a) required no mass information on the chaser or target spacecraft, but the relative motion information should be accurately known. Various navigation sensors have been used to determine the target spacecraft's relative state using light detection and ranging (LIDAR), which inevitably have measurement errors that can lead to poorer control performance (Singla et al., 2006; Sun & Zheng, 2019).

A challenging problem that arises in the domain of electromagnetic docking is that external disturbances, fault signals, mass properties, and measurement errors may be unknown, further increasing the complexity of control design. Future space missions demand that the new generation of spacecraft should be able to perform on-orbit servicing missions with high precision and better robustness to inexact knowledge of spacecraft mass and other uncertainties. One way to overcome this problem is by using disturbance rejection control, where the lumped disturbance is estimated and compensated for in the control (Chen, Yang, Guo, & Li, 2016). Thus, disturbance observer-based control has been widely studied and explored in practical applications (Chen et al., 2018; Sun et al., 2018). However, the state information and mass properties are often needed for control design.

To our knowledge, no study has explicitly focused on the aforementioned issues, i.e., external disturbances, fault signals, unknown mass, elliptical eccentricity, measurement errors, and input constraints, in the case that the relative motion information on electromagnetic docking spacecraft is unavailable. These gaps make spacecraft electromagnetic docking much more difficult, but through the use of well-designed controllers and docking actuators, these difficulties can be overcome. Thus, it is necessary to develop effective control schemes for the spacecraft electromagnetic docking system in the presence of external disturbances, fault signals, unknown mass, elliptical eccentricity, measurement errors, and input constraints. This chapter aims to develop a novel intermediate observer-based control scheme without exactly knowing the external disturbances, fault signals, measurement errors, and mass properties to achieve high-precision control of electromagnetic docking for spacecraft in elliptical orbits. This chapter documents several key contributions made to the field of spacecraft electromagnetic docking. First, compared with Shi et al. (2019a), where the relative position and velocity are required to be measurable for direct feedback control, the control objective of spacecraft electromagnetic docking can be achieved if the relative position and velocity are unavailable. Second, compared with Liu et al. (2019) and Liu et al. (2018b), where the observer and controller gains are independently given in advance according to Hurwitz stability theory, the observer and controller gains are obtained simultaneously by solving linear matrix inequalities (LMIs), such that the given input constraints are satisfied. As the observer and controller gains depend on each other for electromagnetic docking, the control strategy here has made the use of electromagnetic docking closer to realization compared with the state of the art. Third, compared with Sanyal et al. (2009) and Weiss et al. (2013), no parameter identification or neural or fuzzy nonlinear approximations for the mass of the chaser and target spacecraft are needed. Fourth, the developed control scheme is not only applicable to spacecraft electromagnetic docking systems subject to input constraints, but also easily extendable to a more general class of second-order systems with input constraints. Here, the lumped disturbance resulting from external disturbances, fault signals, unknown mass, and elliptical eccentricity is addressed for an anti-disturbance controller design, where an intermediate observer is developed to estimate the relative motion information and the lumped disturbance.

The rest of this chapter is organized as follows. The next section focuses on the mathematical models for electromagnetic docking systems of the chaser and target spacecraft, where external disturbances, fault signals, unknown mass, elliptical eccentricity, and measurement errors are taken into account, and the corresponding control problem is formulated. Then, the main results of this work are presented, where a novel intermediate observer-based control scheme with input constraints is proposed, and the corresponding stability analysis is performed via a Lyapunov approach. It is followed by the numerical simulations to demonstrate the performance and superiority of the proposed control strategy. Finally, conclusions are drawn.

3.2 Problem formulation

3.2.1 Dynamic modeling

The dynamic model is illustrated in Section 1.2, Eq. (1.78). In addition, the following important points should be noted.

3.2.1.1 Measurement errors

The output model in Eq. (1.78) is based on the assumption that true position and velocity states of the spacecraft are known, but in any real case, only the estimates derived from sensor measurements for spacecraft states are available for feedback purposes. Hence, the sensor measurement error $v(t) \in L_2[0,+\infty)$ is present, which represents an aggregation of sensor calibration errors, systematic biases, and stochastic disturbances present in any real sensor measurement.

3.2.1.2 Actuator faults

Since the performance of a spacecraft docking system can be severely impaired by improper actuator actions, actuator faults have been considered to be one of the most critical challenges to be addressed (Chen & Shan, 2020b). Taking this issue into account, let $\boldsymbol{E}$ with full-column rank, which has a similar structure to $\boldsymbol{B}$, denote the distribution matrix of fault signals $f(t)$ appearing in the input. Specifically, it represents the process fault if $\boldsymbol{E} \neq \boldsymbol{B}$, and it also represents the actuator fault if $\boldsymbol{E} = \boldsymbol{B}$.

3.2.1.3 Actuator saturation

Due to the maximum available current density of the electromagnetic coils and other physical limitations, the actuator constraint, if unaccounted for, will often have a negative impact on the stability and performance of an electromagnetic docking system designed for the "ideal" case. In order not to violate the requirements necessary for input constraints, a positive scalar λ is introduced satisfying

$$\|\boldsymbol{u}\| < \lambda \tag{3.1}$$

where $||\boldsymbol{u}||$ denotes the 2-norm of the control input.

In other cases, when the electromagnetic docking system works, if the theoretical control force is larger than the saturation value of the actual control force, the actual control force $\text{sat}(\boldsymbol{u})$ should be described as (Zhang et al., 2018).

$$\text{sat}\big(u_{i=1,2,3}(t)\big) = \text{sign}(u_i(t)) \min\{|u_i(t)|, u_{mi}\} \tag{3.2}$$

where u_{mi} is the upper boundary of control input that the actuator can provide.

Considering the above factors, the new form of spacecraft electromagnetic docking system can be expressed as:

$$\begin{cases} \dot{\boldsymbol{\eta}}(t) = \boldsymbol{A}\boldsymbol{\eta}(t) + \boldsymbol{B}\boldsymbol{u}(t) + \boldsymbol{B}\boldsymbol{w}(t) \\ \boldsymbol{y}(t) = \boldsymbol{C}\boldsymbol{\eta}(t) + \boldsymbol{v}(t) \end{cases} \tag{3.3}$$

where $\boldsymbol{w}(t) = \boldsymbol{B}^{\mathrm{T}}(\boldsymbol{B}\boldsymbol{B}^{\mathrm{T}})^{*}(\Delta\boldsymbol{A}\boldsymbol{\eta}(t)+\Delta\boldsymbol{B}\boldsymbol{u}(t)+(\boldsymbol{B}+\Delta\boldsymbol{B})\boldsymbol{w}_0(t)+\boldsymbol{E}\boldsymbol{f}(t))$ is regarded as the lumped disturbance from external disturbances, fault signals, unknown mass, and elliptical eccentricity, and where the matrix $(\boldsymbol{B}\boldsymbol{B}^{\mathrm{T}})^{*}$ is the pseudoinverse of $\boldsymbol{B}\boldsymbol{B}^{\mathrm{T}}$.

3.2.2 Preliminaries

The subsection presents the following necessary assumptions, remarks, and lemmas, which are needed to provide theoretical support before further analysis.

Assumption 3.1 : The disturbance force $\boldsymbol{w}_0(t)$ and its first-time derivative is assumed to be piecewise continuous and bounded.

Remark 3.1 : Assumption 3.1 is reasonable. The main reason is that the external disturbances are primarily caused by solar radiation, geopotential anomaly, magnetic effects and residual atmosphere, which are all continuous and bounded (Chen & Shan, 2020a; Li & Ye, 2019). In addition, because the total control authority is limited, all three available control input forces should be continuous and bounded, which is reasonable in practice. It is worth noting that the desired translational motion represented by relative position and velocity is also bounded and differentiable in practice. In particular, the derivative of $f(t)$ with respect to time is often assumed to be norm-bounded (Huang & Yang, 2014). They together result in the fact that the lumped disturbance force is bounded and differentiable, and the first derivative of $\boldsymbol{w}(t)$ satisfies $\|\dot{\boldsymbol{w}}(t)\| \leq \kappa_1$, where $\kappa_1 \geq 0$. It should be noted that the norm of the lumped disturbance $w(t)$ and its components could be unknown. Thus it is more general than the fault-tolerant sliding-mode observer approach by Shi et al. (2015), which requires the preliminary knowledge of the bounds of fault signals.

Assumption 3.2 : The measurement error signal $\boldsymbol{v}(t)$ satisfies $||\boldsymbol{v}(t)|| \leq \kappa_2$with $\kappa_2 \geq 0$, which describes the boundary of measurement errors.

Remark 3.2 :Assumption 3.2 is common in the literature (Shi et al.,2019b; Singla et al.,2006). For κ_2 would be unknown when the H_∞ control method is applied, it is more general compared with Shi et al. (2019b), where κ_2 should be known.

Assumption 3.3 : rank$(\boldsymbol{B}, \boldsymbol{E})$ = rank$(\boldsymbol{B})$, and there exists a matrix $\boldsymbol{B}^*$ such that $(\boldsymbol{I} - \boldsymbol{B}\boldsymbol{B}^*)\boldsymbol{E} = 0$.

Remark 3.3 :Assumption 3.3 is general in Tabatabaeipour and Bak (2014), which means that the faults appear in the actuator and can be compensated for by the control input.

3.2.3 Control objectives

Based on the spacecraft electromagnetic docking model governed by Eq. (3.3), one should design an effective control scheme to achieve the following objectives: (1) the states of the closed-loop system are stabilized in the presence of external disturbances, fault signals, unknown mass, elliptical eccentricity, measurement errors, and input constraints; (2) the input constraint, i.e. $||\boldsymbol{u}|| < \lambda$ is satisfied, where λ is a known positive constant; (3) the relative motion information, as well as the lumped disturbance, is estimated with high precision.

3.3 Intermediate observer-based controller design

In this section, the relative motion information as well as the lumped disturbance should be estimated simultaneously, such that the control scheme is designed based on the estimation. These requirements frequently necessitate the use of an intermediate variable for observer design. The new variable is described as

$$\boldsymbol{\xi}(t) = \boldsymbol{w}(t) - \varpi \boldsymbol{B}^{\mathrm{T}} \boldsymbol{\eta}(t) \tag{3.4}$$

$$\dot{\boldsymbol{\xi}}(t) = \dot{\boldsymbol{w}}(t) - \varpi \boldsymbol{B}^{\mathrm{T}} \left[\boldsymbol{A}\boldsymbol{\eta}(t) + \boldsymbol{B}\boldsymbol{u}(t) + \boldsymbol{B}\boldsymbol{\xi}(t) + \varpi \boldsymbol{B}\boldsymbol{B}^{\mathrm{T}} \boldsymbol{\eta}(t) \right] \tag{3.5}$$

For $\boldsymbol{w}(t)$ and $\boldsymbol{\eta}(t)$ are both bounded, the intermediate variable $\boldsymbol{\xi}(t)$ is also bounded, which satisfies $||\boldsymbol{\xi}(t)|| \leq \kappa_3$ with $\kappa_3 > 0$. Then, the intermediate observer is designed based on

$$\dot{\hat{\boldsymbol{\eta}}}(t) = \boldsymbol{A}\hat{\boldsymbol{\eta}}(t) + \boldsymbol{B}\boldsymbol{u}(t) + \boldsymbol{B}\hat{\boldsymbol{w}}(t) + \boldsymbol{L}\big(\boldsymbol{y}(t) - \hat{\boldsymbol{y}}(t)\big) \tag{3.6}$$

$$\dot{\hat{\boldsymbol{\xi}}}(t) = -\varpi \boldsymbol{B}^{\mathrm{T}} \left[\boldsymbol{A}\hat{\boldsymbol{\eta}}(t) + \boldsymbol{B}\boldsymbol{u}(t) + \boldsymbol{B}\hat{\boldsymbol{\xi}}(t) + \varpi \boldsymbol{B}\boldsymbol{B}^{\mathrm{T}} \hat{\boldsymbol{\eta}}(t) \right] \tag{3.7}$$

$$\boldsymbol{y}(t) = \boldsymbol{C}\hat{\boldsymbol{\eta}}(t) \tag{3.8}$$

$$\hat{\boldsymbol{w}}(t) = \boldsymbol{\xi}(t) + \varpi \boldsymbol{B}^{\mathrm{T}} \hat{\boldsymbol{\eta}}(t) \tag{3.9}$$

where $\hat{\boldsymbol{\eta}}(t)$, $\hat{\boldsymbol{\xi}}(t)$, $\boldsymbol{y}(t)$ and $\hat{\boldsymbol{w}}(t)$ are the estimates of $\boldsymbol{\eta}(t)$, $\boldsymbol{\xi}(t)$, $\boldsymbol{y}(t)$ and $\boldsymbol{w}(t)$, respectively.

Based on the above estimations, the antidisturbance controller is designed as

$$\boldsymbol{u}(t) = -\boldsymbol{K}\hat{\boldsymbol{\eta}}(t) - \hat{\boldsymbol{w}}(t) \tag{3.10}$$

where $\boldsymbol{K}$ is the controller gain matrix and $\hat{\boldsymbol{w}}(t)$ is added to the input to compensate for the effect of the lumped disturbance.

Denote $\boldsymbol{e}_\eta(t) = \boldsymbol{\eta}(t) - \hat{\boldsymbol{\eta}}(t)$and $\boldsymbol{e}_\xi(t) = \boldsymbol{\xi}(t) - \hat{\boldsymbol{\xi}}(t)$, substituting Eq. (3.10) into the first of Eq. (3.3) yields

$$\dot{\boldsymbol{\eta}}(t) = (\boldsymbol{A} - \boldsymbol{B}\boldsymbol{K})\boldsymbol{\eta}(t) + \boldsymbol{B}\boldsymbol{K}\boldsymbol{e}_\eta(t) + \boldsymbol{B}\boldsymbol{e}_\xi(t) + \varpi \boldsymbol{B}\boldsymbol{B}^{\mathrm{T}} \boldsymbol{e}_\eta(t) \tag{3.11}$$

Then, it can be concluded that

$$\dot{\mathbf{e}}_\eta(t) = (\boldsymbol{A} - \boldsymbol{L}\boldsymbol{C})\boldsymbol{e}_\eta(t) + \boldsymbol{B}\boldsymbol{e}_\xi(t) + \varpi \boldsymbol{B}\boldsymbol{B}^{\mathrm{T}} \boldsymbol{e}_\eta(t) - \boldsymbol{L}\boldsymbol{v}(t) \tag{3.12}$$

$$\dot{\boldsymbol{e}}_\xi(t) = \dot{\boldsymbol{w}}(t) - \varpi \boldsymbol{B}^{\mathrm{T}} \left[\boldsymbol{A}\boldsymbol{e}_\eta(t) + \boldsymbol{B}\boldsymbol{e}_\xi(t) + \varpi \boldsymbol{B}\boldsymbol{B}^{\mathrm{T}} \boldsymbol{e}_\eta(t) \right] \tag{3.13}$$

Denoting the aggregate state and disturbance as $\chi(t) = \left[\boldsymbol{\eta}(t)^{\mathrm{T}} \quad \boldsymbol{e}_{\eta}(t)^{\mathrm{T}} \quad \boldsymbol{e}_{\xi}(t)^{\mathrm{T}}\right]^{\mathrm{T}}$ and $\boldsymbol{w}_{v}(t) = \left[\boldsymbol{v}(t)^{\mathrm{T}} \quad \dot{\boldsymbol{w}}(t)^{\mathrm{T}}\right]^{\mathrm{T}}$, one has

$$\dot{\chi}(t) = \boldsymbol{A}_{\chi}\chi(t) + \boldsymbol{B}_{\chi}\boldsymbol{w}_{v}(t) \tag{3.14}$$

where

$$\boldsymbol{A}_{\chi} = \begin{bmatrix} \boldsymbol{A}-\boldsymbol{B}\boldsymbol{K} & \boldsymbol{B}\boldsymbol{K}+\varpi\boldsymbol{B}\boldsymbol{B}^{\mathrm{T}} & \boldsymbol{B} \\ 0 & \boldsymbol{A}-\boldsymbol{L}\boldsymbol{C}+\varpi\boldsymbol{B}\boldsymbol{B}^{\mathrm{T}} & \boldsymbol{B} \\ 0 & -\varpi\boldsymbol{B}^{\mathrm{T}}\boldsymbol{A}-\varpi^{2}\boldsymbol{B}^{\mathrm{T}}\boldsymbol{B}\boldsymbol{B}^{\mathrm{T}} & -\varpi\boldsymbol{B}^{\mathrm{T}}\boldsymbol{B} \end{bmatrix},$$

$$\boldsymbol{B}_{\chi} = \begin{bmatrix} 0 & 0 \\ -\boldsymbol{L} & 0 \\ 0 & \boldsymbol{I} \end{bmatrix}$$

Remark 3.4 : As a consequence of the fact that $w(t)$ includes uncertain magnitudes, the initial estimation errors $\boldsymbol{e}_{\eta}(0)$ and $\boldsymbol{e}_{\xi}(0)$ are not easy to be exactly known, and the initial states of spacecraft may be unavailable, but they should be norm-bounded in practice. Then, for any symmetric positive definite matrix $\tilde{\boldsymbol{P}}_{0}$ of appropriate dimensions, there exist three positive scalars $\overline{\kappa}_{k}$, $k = 1, 2, 3$ such that $\chi_{k}(0)^{\mathrm{T}}\tilde{\boldsymbol{P}}_{0}\chi_{k}(0) \le \overline{\kappa}_{k}^{2}\boldsymbol{1}_{1\times n}\tilde{\boldsymbol{P}}_{0}\boldsymbol{1}_{n\times 1}$, where $\chi_{1}(0) = \boldsymbol{\eta}(0)$, $\chi_{2}(0) = \boldsymbol{e}_{\eta}(0)$, $\chi_{3}(0) = \boldsymbol{e}_{\xi}(0)$, $\boldsymbol{1}_{i\times j}$ denotes an i by j matrix (i rows, j columns) whose elements are all ones, and n is the number of elements for $\chi_{k}(0)$, $k = 1, 2, 3$.

Theorem 3.1 now gives the existence conditions of the controller gain matrix $\boldsymbol{K}$ and observer gain matrix $\boldsymbol{L}$.

Theorem 3.1 The closed-loop system represented by Eq. (3.14) is quadratically stable under controller (Eq. (3.10)), and $\boldsymbol{y}(t)$ satisfies an H_{∞} performance constraint if there exists a positive scalar $\gamma_{2} = \gamma^{2}$, for the given positive scalars $\varpi, \kappa_{3}, \overline{\kappa}_{k} > 0$, $k = 1, 2, 3$, the given matrix $\boldsymbol{M}_{0}$, there exist symmetric positive definite matrices $\boldsymbol{X}_{1}$, $\boldsymbol{X}_{3}$ matrices $\boldsymbol{W}$ and $\boldsymbol{H}$, such that the following LMIs hold:

$$\begin{bmatrix} \boldsymbol{\Pi}_{11} & \boldsymbol{\Pi}_{12} & \boldsymbol{B}\boldsymbol{X}_{3} & 0 & \boldsymbol{X}_{1}\boldsymbol{C}^{\mathrm{T}} & \boldsymbol{X}_{1}\boldsymbol{C}^{\mathrm{T}} \\ * & \boldsymbol{\Pi}_{22} & \boldsymbol{\Pi}_{23} & 0 & -\boldsymbol{X}_{1}\boldsymbol{M}_{0} & 0 \\ * & * & \boldsymbol{\Pi}_{33} & \boldsymbol{I} & 0 & 0 \\ * & * & * & -\gamma_{2}\boldsymbol{I} & 0 & 0 \\ * & * & * & * & (1-\gamma_{2})\boldsymbol{I} & 0 \\ * & * & * & * & * & -\boldsymbol{I} \end{bmatrix} < 0 \tag{3.15}$$

$$\begin{bmatrix} -\alpha & \overline{\kappa}_1\mathbf{1}_{1\times6} & \overline{\kappa}_2\mathbf{1}_{1\times6} & \overline{\kappa}_3\mathbf{1}_{1\times3} \\ * & -\boldsymbol{X}_1 & 0 & 0 \\ * & * & -\boldsymbol{X}_2 & 0 \\ * & * & * & -\boldsymbol{X}_3 \end{bmatrix} < 0 \tag{3.16}$$

$$\begin{bmatrix} -k_0\boldsymbol{X}_1 & 0 & 0 & 0 & \boldsymbol{W}^{\mathrm{T}} + \varpi\boldsymbol{X}_1\boldsymbol{B} \\ * & -k_0\boldsymbol{X}_1 & 0 & 0 & -\boldsymbol{W}^{\mathrm{T}} - \varpi\boldsymbol{X}_1\boldsymbol{B} \\ * & * & -k_0\boldsymbol{X}_3 & 0 & -\boldsymbol{X}_3 \\ * & * & * & -k_0\kappa_3^{-2}\boldsymbol{I} & \boldsymbol{I} \\ * & * & * & * & -\boldsymbol{I} \end{bmatrix} < 0 \tag{3.17}$$

where

$$\begin{aligned}
\boldsymbol{\Pi}_{11} &= \boldsymbol{X}_1\boldsymbol{A}^{\mathrm{T}} - \boldsymbol{W}^{\mathrm{T}}\boldsymbol{B}^{\mathrm{T}} + \boldsymbol{A}\boldsymbol{X}_1 - \boldsymbol{B}\boldsymbol{W} \\
\boldsymbol{\Pi}_{12} &= \boldsymbol{B}\boldsymbol{W} + \varpi\boldsymbol{B}\boldsymbol{B}^{\mathrm{T}}\boldsymbol{X}_1 \\
\boldsymbol{\Pi}_{22} &= \boldsymbol{X}_1\boldsymbol{A}^{\mathrm{T}} - \boldsymbol{H}^{\mathrm{T}} + \boldsymbol{A}\boldsymbol{X}_1 - \boldsymbol{H} + \varpi\boldsymbol{B}\boldsymbol{B}^{\mathrm{T}}\boldsymbol{X}_1 + \varpi\boldsymbol{X}_1\boldsymbol{B}\boldsymbol{B}^{\mathrm{T}} \\
\boldsymbol{\Pi}_{23} &= \boldsymbol{B}\boldsymbol{X}_3 - \varpi\boldsymbol{X}_1\boldsymbol{A}^{\mathrm{T}}\boldsymbol{B} - \varpi^2\boldsymbol{X}_1\boldsymbol{B}\boldsymbol{B}^{\mathrm{T}}\boldsymbol{B} \\
\boldsymbol{\Pi}_{33} &= -\varpi\left(\boldsymbol{B}^{\mathrm{T}}\boldsymbol{B}\boldsymbol{X}_3 + \boldsymbol{X}_3\boldsymbol{B}^{\mathrm{T}}\boldsymbol{B}\right)
\end{aligned}$$

and $k_0 = (\alpha+1)^{-1}\lambda^2$.

By minimizing γ_2, the optimal feasible solution of LMIs inequalities Eqs. (3.15)– Eqs. (3.17) is obtained, and one has the controller gain matrix $\boldsymbol{K} = \boldsymbol{W}\boldsymbol{X}_1^{-1}$ and the observer gain matrix $\boldsymbol{L} = \boldsymbol{H}(\boldsymbol{C}\boldsymbol{X}_1)^{-1}$.

Proof: First, it is shown that the closed-loop system (Eq. (3.14)) is quadratically stable while $\boldsymbol{v}(t) = 0$, $\cdot\boldsymbol{w}(t) = 0$, and the definition of quadratic stability has been stated in Liu et al. (2017).

Choose a Lyapunov function candidate as

$$V(t) = \boldsymbol{\eta}(t)^{\mathrm{T}}\boldsymbol{P}_1\boldsymbol{\eta}(t) + \boldsymbol{e}_\eta(t)^{\mathrm{T}}\boldsymbol{P}_2\boldsymbol{e}_\eta(t) + \boldsymbol{e}_\xi(t)^{\mathrm{T}}\boldsymbol{P}_3\boldsymbol{e}_\xi(t) \tag{3.18}$$

satisfying $V(t) \leq \alpha$, where $\boldsymbol{P}_1$, $\boldsymbol{P}_2$ and $\boldsymbol{P}_3$ are Lyapunov variables with $\boldsymbol{P}_1 = \boldsymbol{P}_2 = \boldsymbol{X}_1^{-1}$, $\boldsymbol{P}_3 = \boldsymbol{X}_3^{-1}$.

Before the next step of the derivation, an assumption, i.e., $\boldsymbol{e}_\eta(t)^{\mathrm{T}}\boldsymbol{P}_2\boldsymbol{L}\boldsymbol{v}(t) \geq \boldsymbol{e}_\eta(t)^{\mathrm{T}}\boldsymbol{M}_0\boldsymbol{v}(t)$, is introduced first. As $\boldsymbol{e}_\eta(t)$, $\boldsymbol{v}(t)$ are both bounded and $\boldsymbol{P}_2$, $\boldsymbol{L}$ are both fixed matrices for applications in aerospace engineering, the scalars $\boldsymbol{e}_\eta(t)^{\mathrm{T}}\boldsymbol{P}_2\boldsymbol{L}\boldsymbol{v}(t)$ and $\boldsymbol{e}_\eta(t)^{\mathrm{T}}\boldsymbol{M}_0\boldsymbol{v}(t)$ must be bounded, where $\boldsymbol{M}_0$ is a real matrix of appropriate dimension. As a result, $\boldsymbol{M}_0$ can be chosen appropriately such that the above assumption is established.

The time derivative of $V(t)$ along the system trajectories is given by

$$\begin{aligned}
\dot{V}(t) = {} & \dot{\boldsymbol{\eta}}(t)^{\mathrm{T}}\boldsymbol{P}_1\boldsymbol{\eta}(t) + \boldsymbol{\eta}(t)^{\mathrm{T}}\boldsymbol{P}_1\dot{\boldsymbol{\eta}}(t) + \dot{\boldsymbol{e}}_\eta(t)^{\mathrm{T}}\boldsymbol{P}_2\boldsymbol{e}_\eta(t) + \boldsymbol{e}_\eta(t)^{\mathrm{T}}\boldsymbol{P}_2\dot{\boldsymbol{e}}_\eta(t) \\
& + \dot{\boldsymbol{e}}_\xi(t)^{\mathrm{T}}\boldsymbol{P}_3\boldsymbol{e}_\xi(t) + \boldsymbol{e}_\xi(t)^{\mathrm{T}}\boldsymbol{P}_3\dot{\boldsymbol{e}}_\xi(t)
\end{aligned} \tag{3.19}$$

Now, substitution of Eqs. (3.11)–(3.13) into Eq. (3.18) leads to the following expression for $V(t)$:

$$\begin{aligned}\dot{V}(t) &= \boldsymbol{\eta}(t)^{\mathrm{T}}\left[(\boldsymbol{A}-\boldsymbol{BK})^{\mathrm{T}}\boldsymbol{P}_1+\boldsymbol{P}_1(\boldsymbol{A}-\boldsymbol{BK})\right]\boldsymbol{\eta}(t)+2\boldsymbol{\eta}(t)^{\mathrm{T}}\boldsymbol{P}_1\boldsymbol{BK}\boldsymbol{e}_\eta(t)\\&\quad+2\boldsymbol{\eta}(t)^{\mathrm{T}}\boldsymbol{P}_1\boldsymbol{B}\boldsymbol{e}_\xi(t)+2\varpi\boldsymbol{\eta}(t)^{\mathrm{T}}\boldsymbol{P}_1\boldsymbol{BB}^{\mathrm{T}}\boldsymbol{e}_\eta(t)+\boldsymbol{e}_\eta(t)^{\mathrm{T}}\left[(\boldsymbol{A}-\boldsymbol{LC})^{\mathrm{T}}\boldsymbol{P}_2+\boldsymbol{P}_2(\boldsymbol{A}-\boldsymbol{LC})\right]\boldsymbol{e}_\eta(t)\\&\quad+2\boldsymbol{e}_\eta(t)^{\mathrm{T}}\boldsymbol{P}_2\boldsymbol{B}\boldsymbol{e}_\xi(t)+\boldsymbol{e}_\eta(t)^{\mathrm{T}}\left(\varpi\boldsymbol{P}_2\boldsymbol{BB}^{\mathrm{T}}+\varpi\boldsymbol{BB}^{\mathrm{T}}\boldsymbol{P}_2\right)\boldsymbol{e}_\eta(t)-2\boldsymbol{e}_\eta(t)^{\mathrm{T}}\boldsymbol{P}_2\boldsymbol{L}\boldsymbol{v}(t)+2\boldsymbol{e}_\xi(t)^{\mathrm{T}}\boldsymbol{P}_3\dot{\boldsymbol{w}}(t)\\&\quad-\varpi\boldsymbol{e}_\xi(t)^{\mathrm{T}}\left(\boldsymbol{P}_3\boldsymbol{B}^{\mathrm{T}}\boldsymbol{B}+\boldsymbol{B}^{\mathrm{T}}\boldsymbol{BP}_3\right)\boldsymbol{e}_\xi(t)-2\varpi\boldsymbol{e}_\xi(t)^{\mathrm{T}}\boldsymbol{P}_3\boldsymbol{B}^{\mathrm{T}}\left[\boldsymbol{A}\boldsymbol{e}_\eta(t)+\varpi\boldsymbol{BB}^{\mathrm{T}}\boldsymbol{e}_\eta(t)\right]\end{aligned}\tag{3.20}$$

For $\boldsymbol{v}(t)=0,\ \dot{\boldsymbol{w}}(t)=0$, Eq. (3.20) reduces to

$$\begin{aligned}\dot{V}(t) &= \boldsymbol{\eta}(t)^{\mathrm{T}}\left[(\boldsymbol{A}-\boldsymbol{BK})^{\mathrm{T}}\boldsymbol{P}_1+\boldsymbol{P}_1(\boldsymbol{A}-\boldsymbol{BK})\right]\boldsymbol{\eta}(t)+2\boldsymbol{\eta}(t)^{\mathrm{T}}\boldsymbol{P}_1\boldsymbol{BK}\boldsymbol{e}_\eta(t)+2\boldsymbol{\eta}(t)^{\mathrm{T}}\boldsymbol{P}_1\boldsymbol{B}\boldsymbol{e}_\xi(t)\\&\quad+2\varpi\boldsymbol{\eta}(t)^{\mathrm{T}}\boldsymbol{P}_1\boldsymbol{BB}^{\mathrm{T}}\boldsymbol{e}_\eta(t)+\boldsymbol{e}_\eta(t)^{\mathrm{T}}\left[(\boldsymbol{A}-\boldsymbol{LC})^{\mathrm{T}}\boldsymbol{P}_2+\boldsymbol{P}_2(\boldsymbol{A}-\boldsymbol{LC})\right]\boldsymbol{e}_\eta(t)+2\boldsymbol{e}_\eta(t)^{\mathrm{T}}\boldsymbol{P}_2\boldsymbol{B}\boldsymbol{e}_\xi(t)\\&\quad+\boldsymbol{e}_\eta(t)^{\mathrm{T}}\left(\varpi\boldsymbol{P}_2\boldsymbol{BB}^{\mathrm{T}}+\varpi\boldsymbol{BB}^{\mathrm{T}}\boldsymbol{P}_2\right)\boldsymbol{e}_\eta(t)-\varpi\boldsymbol{e}_\xi(t)^{\mathrm{T}}\left(\boldsymbol{P}_3\boldsymbol{B}^{\mathrm{T}}\boldsymbol{B}+\boldsymbol{B}^{\mathrm{T}}\boldsymbol{BP}_3\right)\boldsymbol{e}_\xi(t)\\&\quad-2\varpi\boldsymbol{e}_\xi(t)^{\mathrm{T}}\boldsymbol{P}_3\boldsymbol{B}^{\mathrm{T}}\boldsymbol{A}\boldsymbol{e}_\eta(t)-2\varpi^2\boldsymbol{e}_\xi(t)\boldsymbol{P}_3\boldsymbol{B}^{\mathrm{T}}\boldsymbol{BB}^{\mathrm{T}}\boldsymbol{e}_\eta(t)\end{aligned}\tag{3.21}$$

Denote

$$\overline{\boldsymbol{\Theta}}_N=\begin{bmatrix}\overline{\boldsymbol{\Pi}}_{11} & \overline{\boldsymbol{\Pi}}_{12} & \boldsymbol{P}_1\boldsymbol{B}\\ * & \overline{\boldsymbol{\Pi}}_{22} & \overline{\boldsymbol{\Pi}}_{23}\\ * & * & \overline{\boldsymbol{\Pi}}_{33}\end{bmatrix}$$

where

$$\begin{aligned}&\overline{\boldsymbol{\Pi}}_{11}=\boldsymbol{A}^{\mathrm{T}}\boldsymbol{P}_1-\boldsymbol{K}^{\mathrm{T}}\boldsymbol{B}^{\mathrm{T}}\boldsymbol{P}_1+\boldsymbol{P}_1\boldsymbol{A}-\boldsymbol{P}_1\boldsymbol{BK}\\&\overline{\boldsymbol{\Pi}}_{12}=\boldsymbol{P}_1\boldsymbol{BK}+\varpi\boldsymbol{P}_1\boldsymbol{BB}^{\mathrm{T}}\\&\overline{\boldsymbol{\Pi}}_{22}=\boldsymbol{A}^{\mathrm{T}}\boldsymbol{P}_2-\boldsymbol{C}^{\mathrm{T}}\boldsymbol{L}^{\mathrm{T}}\boldsymbol{P}_2+\boldsymbol{P}_2\boldsymbol{A}-\boldsymbol{P}_2\boldsymbol{LC}+\varpi\boldsymbol{P}_2\boldsymbol{BB}^{\mathrm{T}}+\varpi\boldsymbol{BB}^{\mathrm{T}}\boldsymbol{P}_2\\&\overline{\boldsymbol{\Pi}}_{23}=\boldsymbol{P}_2\boldsymbol{B}-\varpi\boldsymbol{A}^{\mathrm{T}}\boldsymbol{BP}_3-\varpi^2\boldsymbol{BB}^{\mathrm{T}}\boldsymbol{BP}_3\\&\overline{\boldsymbol{\Pi}}_{33}=-\varpi\left(\boldsymbol{P}_3\boldsymbol{B}^{\mathrm{T}}\boldsymbol{B}+\boldsymbol{B}^{\mathrm{T}}\boldsymbol{BP}_3\right)\end{aligned}$$

The first derivative of the Lyapunov function becomes

$$\dot{V}(t)=\boldsymbol{\chi}(t)^{\mathrm{T}}\overline{\boldsymbol{\Theta}}_N\boldsymbol{\chi}(t)$$

Considering inequality Eqs. (3.15), multiplied by diag ($[\boldsymbol{P}_1\quad \boldsymbol{P}_2\quad \boldsymbol{P}_3\quad \boldsymbol{I}\quad \boldsymbol{I}\quad \boldsymbol{I}]$) at both sides simultaneously, one has

$$\begin{bmatrix}\overline{\boldsymbol{\Pi}}_{11} & \overline{\boldsymbol{\Pi}}_{12} & \boldsymbol{P}_1\boldsymbol{B} & 0 & \boldsymbol{C}^{\mathrm{T}} & \boldsymbol{C}^{\mathrm{T}}\\ * & \overline{\boldsymbol{\Pi}}_{22} & \overline{\boldsymbol{\Pi}}_{23} & 0 & -\boldsymbol{M}_0 & 0\\ * & * & \overline{\boldsymbol{\Pi}}_{33} & \boldsymbol{P}_3 & 0 & 0\\ * & * & * & -\gamma_2\boldsymbol{I} & 0 & 0\\ * & * & * & * & (1-\gamma_2)\boldsymbol{I} & 0\\ * & * & * & * & * & -\boldsymbol{I}\end{bmatrix}<0\tag{3.22}$$

According to inequality (Eqs. 3.22) and Lemma 1.3, one has $\overline{\boldsymbol{\Theta}}_N < 0$, so that

$$\dot{V}(t) = \chi(t)^{\mathrm{T}}\overline{\boldsymbol{\Theta}}_N\chi(t) \le \lambda_{\max}\left(\overline{\boldsymbol{\Theta}}_N\right)||\chi||^2$$

Furthermore, let $\overline{\alpha} = -\lambda_{\max}(\overline{\boldsymbol{\Theta}}_N) > 0$, then,

$$\dot{V}(t) \le -\overline{\alpha}||\chi||^2 \tag{3.23}$$

i.e., the closed-loop system (Eq. (3.14)) is quadratically stable under the controller (Eq. (3.10)).

Next, it is shown that the output $\boldsymbol{y}(t)$ satisfies an H_∞ performance constraint.

To establish the L_2 [0, ∞) norm bound $\gamma^2||\widetilde{\boldsymbol{w}}(t)||_2^2$, where $\tilde{\boldsymbol{w}}(t)$ is a constructed vector with norm $||\tilde{\boldsymbol{w}}|| = \sqrt{\dot{\boldsymbol{w}}^2 + \boldsymbol{v}^2}$, consider the following function:

$$J = \int_0^{+\infty}\left[\boldsymbol{y}(t)^{\mathrm{T}}\boldsymbol{y}(t) - \gamma^2\widetilde{\boldsymbol{w}}(t)^{\mathrm{T}}\widetilde{\boldsymbol{w}}(t)\right]dt$$

As the closed-loop system has quadratic stability, for arbitrary nonzero $\widetilde{\boldsymbol{w}}(t) \in L_2[0, \infty)$, under zero initial conditions, one has

$$J = \int_0^{+\infty}\left[\boldsymbol{y}(t)^{\mathrm{T}}\boldsymbol{y}(t) - \gamma^2\widetilde{\boldsymbol{w}}(t)^{\mathrm{T}}\widetilde{\boldsymbol{w}}(t) + \dot{V}(t)\right]dt - V(\infty) + V(0)$$

$$\le \int_0^{+\infty}\left\{\begin{array}{c}\boldsymbol{\eta}(t)^{\mathrm{T}}\boldsymbol{C}^{\mathrm{T}}\boldsymbol{C}\boldsymbol{\eta}(t) + 2\boldsymbol{\eta}(t)^{\mathrm{T}}\boldsymbol{C}^{\mathrm{T}}\boldsymbol{v}(t) + \boldsymbol{v}(t)^{\mathrm{T}}\boldsymbol{v}(t)\\ -\gamma^2\left[\dot{\boldsymbol{w}}(t)^{\mathrm{T}}\dot{\boldsymbol{w}}(t) + \boldsymbol{v}(t)^{\mathrm{T}}\boldsymbol{v}(t)\right] + \chi(t)^{\mathrm{T}}\overline{\boldsymbol{\Theta}}_N\chi(t)\\ -2e_\eta(t)^{\mathrm{T}}\boldsymbol{P}_2\boldsymbol{L}\boldsymbol{v}(t) + 2e_\xi(t)^{\mathrm{T}}\boldsymbol{P}_3\dot{\boldsymbol{w}}(t)\end{array}\right\}dt$$

$$\le \int_0^{+\infty}\left[\boldsymbol{\eta}(t)^{\mathrm{T}} \quad e_\eta(t)^{\mathrm{T}} \quad e_\xi(t)^{\mathrm{T}} \quad \dot{\boldsymbol{w}}(t)^{\mathrm{T}} \quad \boldsymbol{v}(t)^{\mathrm{T}}\right]$$

$$\times\begin{bmatrix}\overline{\boldsymbol{\Pi}}_{11} + \boldsymbol{C}^{\mathrm{T}}\boldsymbol{C} & \overline{\boldsymbol{\Pi}}_{12} & \boldsymbol{P}_1\boldsymbol{B} & 0 & \boldsymbol{C}^{\mathrm{T}}\\ * & \overline{\boldsymbol{\Pi}}_{22} & \overline{\boldsymbol{\Pi}}_{23} & 0 & -\boldsymbol{M}_0\\ * & * & \overline{\boldsymbol{\Pi}}_{33} & \boldsymbol{P}_3 & 0\\ * & * & * & -\gamma^2\boldsymbol{I} & 0\\ * & * & * & * & (1-\gamma^2)\boldsymbol{I}\end{bmatrix}\begin{bmatrix}\boldsymbol{\eta}(t)\\ e_\eta(t)\\ e_\xi(t)\\ \dot{\boldsymbol{w}}(t)\\ \boldsymbol{v}(t)\end{bmatrix}dt$$

According to Lemma 1.3 and inequality (Eqs. 3.22), $J < 0$ holds, i.e., $\boldsymbol{z}$(t) satisfies an H_∞ performance constraint.

Finally, it is shown that the control input is constrained, i.e., it satisfies the condition of inequality (Eqs. 3.1).

According to Lemma 1.3, the inequality (Eqs. 3.17) is equivalent to

$$\overline{\kappa}_1^2 \mathbf{1}_{1\times 6}\boldsymbol{P}_1\mathbf{1}_{6\times 1} + \overline{\kappa}_2^2 \mathbf{1}_{1\times 6}\boldsymbol{P}_2\mathbf{1}_{6\times 1} + \overline{\kappa}_3^2 \mathbf{1}_{1\times 3}\boldsymbol{P}_3\mathbf{1}_{3\times 1} < \alpha \tag{3.24}$$

Due to the fact pointed out in inequality(Eqs. 3.16), it is easily obtained that

$$\boldsymbol{\eta}(0)^{\mathrm{T}}\boldsymbol{P}_1\boldsymbol{\eta}(0) + \boldsymbol{e}_\eta(0)^{\mathrm{T}}\boldsymbol{P}_2\boldsymbol{e}_\eta(0) + \boldsymbol{e}_\xi(0)^{\mathrm{T}}\boldsymbol{P}_3\boldsymbol{e}_\xi(0) < \alpha \tag{3.25}$$

Considering inequality (Eqs. 3.16), multiplied by diag($[\boldsymbol{P}_1 \quad \boldsymbol{P}_2 \quad \boldsymbol{P}_3 \quad \boldsymbol{I} \quad \boldsymbol{I}]$) at both sides simultaneously, one has

$$\begin{bmatrix} -k_0\boldsymbol{P}_1 & 0 & 0 & 0 & \boldsymbol{K}^{\mathrm{T}} + \varpi\boldsymbol{B} \\ * & -k_0\boldsymbol{P}_2 & 0 & 0 & -\boldsymbol{K}^{\mathrm{T}} - \varpi\boldsymbol{B} \\ * & * & -k_0\boldsymbol{P}_3 & 0 & -\boldsymbol{I} \\ * & * & * & -k_0\kappa_3^{-2}\boldsymbol{I} & \boldsymbol{I} \\ * & * & * & * & -\boldsymbol{I} \end{bmatrix} < 0 \tag{3.26}$$

Using Lemma 1.3, one has

$$\begin{bmatrix} \left(\boldsymbol{K} + \varpi\boldsymbol{B}^{\mathrm{T}}\right)^{\mathrm{T}} \\ -\left(\boldsymbol{K} + \varpi\boldsymbol{B}^{\mathrm{T}}\right)^{\mathrm{T}} \\ -\boldsymbol{I} \\ \boldsymbol{I} \end{bmatrix} \begin{bmatrix} \boldsymbol{K} + \varpi\boldsymbol{B}^{\mathrm{T}} & -\left(\boldsymbol{K} + \varpi\boldsymbol{B}^{\mathrm{T}}\right) & -\boldsymbol{I} & \boldsymbol{I} \end{bmatrix}$$

$$< \begin{bmatrix} k_0\boldsymbol{P}_1 & 0 & 0 & 0 \\ * & k_0\boldsymbol{P}_2 & 0 & 0 \\ * & * & k_0\boldsymbol{P}_3 & 0 \\ * & * & * & k_0\kappa_3^{-2}\boldsymbol{I} \end{bmatrix} \tag{3.27}$$

According to Eq. (3.10), one has

$$\begin{aligned} \boldsymbol{u}(t)_2^2 &= \boldsymbol{u}(t)^{\mathrm{T}}\boldsymbol{u}(t) \\ &= [-\boldsymbol{K}\hat{\boldsymbol{\eta}}(t) - \hat{\boldsymbol{w}}(t)]^{\mathrm{T}}[-\boldsymbol{K}\hat{\boldsymbol{\eta}}(t) - \hat{\boldsymbol{w}}(t)] \\ &= \left[\boldsymbol{K}(\boldsymbol{\eta}(t) - \boldsymbol{e}_\eta(t)) + \boldsymbol{\xi}(t) - \boldsymbol{e}_\xi(t) + \varpi\boldsymbol{B}^{\mathrm{T}}(\boldsymbol{\eta}(t) - \boldsymbol{e}_\eta(t))\right]^{\mathrm{T}} \\ &\quad \left[\boldsymbol{K}(\boldsymbol{\eta}(t) - \boldsymbol{e}_\eta(t)) + \boldsymbol{\xi}(t) - \boldsymbol{e}_\xi(t) + \varpi\boldsymbol{B}^{\mathrm{T}}(\boldsymbol{\eta}(t) - \boldsymbol{e}_\eta(t))\right] \\ &= \begin{bmatrix} \boldsymbol{\eta}(t)^{\mathrm{T}} & \boldsymbol{e}_\eta(t)^{\mathrm{T}} & \boldsymbol{e}_\xi(t)^{\mathrm{T}} & \boldsymbol{\xi}(t)^{\mathrm{T}} \end{bmatrix} \begin{bmatrix} \left(\boldsymbol{K} + \varpi\boldsymbol{B}^{\mathrm{T}}\right)^{\mathrm{T}} \\ -\left(\boldsymbol{K} + \varpi\boldsymbol{B}^{\mathrm{T}}\right)^{\mathrm{T}} \\ -\boldsymbol{I} \\ \boldsymbol{I} \end{bmatrix} \end{aligned}$$

$$
\begin{aligned}
&\begin{bmatrix} \boldsymbol{K} + \varpi \boldsymbol{B}^{\mathrm{T}} & -\left(\boldsymbol{K} + \varpi \boldsymbol{B}^{\mathrm{T}}\right) & -\boldsymbol{I} & \boldsymbol{I} \end{bmatrix} \begin{bmatrix} \boldsymbol{\eta}(t) \\ e_{\eta}(t) \\ e_{\xi}(t) \\ \xi(t) \end{bmatrix} \\
&< \begin{bmatrix} \boldsymbol{\eta}(t)^{\mathrm{T}} & e_{\eta}(t)^{\mathrm{T}} & e_{\xi}(t)^{\mathrm{T}} & \xi(t)^{\mathrm{T}} \end{bmatrix} \begin{bmatrix} k_0 \boldsymbol{P}_1 & 0 & 0 & 0 \\ * & k_0 \boldsymbol{P}_2 & 0 & 0 \\ * & * & k_0 \boldsymbol{P}_3 & 0 \\ * & * & * & k_0 \kappa_3^{-2} \boldsymbol{I} \end{bmatrix} \begin{bmatrix} \boldsymbol{\eta}(t) \\ e_{\eta}(t) \\ e_{\xi}(t) \\ \xi(t) \end{bmatrix} \\
&= k_0 \left(\boldsymbol{\eta}(t)^{\mathrm{T}} \boldsymbol{P}_1 \boldsymbol{\eta}(t) + e_{\eta}(t)^{\mathrm{T}} \boldsymbol{P}_2 e_{\eta}(t) + e_{\xi}(t)^{\mathrm{T}} \boldsymbol{P}_3 e_{\xi}(t) + \kappa_3^{-2} \xi(t)^{\mathrm{T}} \xi(t) \right) \\
&\leq k_0 (V(t) + 1)
\end{aligned}
$$

According to inequalities (inequalities (Eqs. 3.23) and (Eqs. 3.25), one knows

$$
\|\boldsymbol{u}(t)\|_2^2 < k_0(\alpha + 1) = \lambda^2 \tag{3.28}
$$

This completes the proof.

3.4 Simulation test

In this section, the effectiveness of the intermediate observer-based controller (IOBC) is illustrated with simulations of a spacecraft electromagnetic docking system. In this example, for the "CM" orbit, the eccentricity $e = 0.15$, the semimajor axis $\mathrm{a} = 10{,}000\mathrm{km}$, the initial true anomaly $\theta(0) = 0$, and m_0 is only representative of a valid mass with a value of 20kg. Whatever the initial guess of m_0, the difference compared with the true value of m_{CT} can be attributed to the matrix uncertainty $\Delta \boldsymbol{B}$ or even the lumped disturbance, and the lumped disturbance can be estimated for the antidisturbance controller design. The unknown masses of the chaser and target spacecraft are assumed to be $m_{\mathrm{C}} = 40\mathrm{kg}$ and $m_{\mathrm{T}} = 100\mathrm{kg}$, respectively. Choose the initial states as $\boldsymbol{\eta}(0) = [2\mathrm{m}, 3\mathrm{m}, 4\mathrm{m}, 0\mathrm{m/s}, 0\mathrm{m/s}, 0\mathrm{m/s}]^{\mathrm{T}}$, $\boldsymbol{e}(0) = [0\mathrm{N}, 0\mathrm{N}, 0\mathrm{N}]^{\mathrm{T}}$, the gravitational parameter $\mu = 3.986 \times 10^5 \mathrm{km}^3/\mathrm{s}^2$, and the permeability of free space $\mu_0 = 4\pi \times 10^{-7} \mathrm{N{\bullet}A}^{-2}$; the magnetic moment of the target spacecraft is assumed to be $\boldsymbol{\mu}_{\mathrm{T}} = [-2.29 \quad -8.17 \quad -2.33]^{\mathrm{T}} \times 10^5 \mathrm{Am}^2$, and the control force is constrained by $\lambda = 30\mathrm{N}$. Here, an actuator additive fault is assumed to occur, i.e., $\boldsymbol{E} = \boldsymbol{B}$, with the fault signals $\boldsymbol{f}(t) = [8 \ 9 \ 4]^{\mathrm{T}} \times 10^{-2} \bar{f}(t)$, and the form of the fault signal factor $\bar{f}(t)$ is shown in Table 3.1.

For comparative purposes, two cases are studied in this section. In the first case, the proposed IOBC is applied for spacecraft electromagnetic docking.

In the second case, the disturbance observer-based controller (DOBC) introduced by Shi et al. (2019a) is employed, wherein comparison results are given to illustrate the advantages of the approach proposed in this chapter. All simulation parameters are identical in these two cases. In both cases, the maximum control force on each channel is set to be 17.321N, so that the maximum norm of control force is 30N.

In addition, to compare the control performance of IOBC and DOBC precisely, the measurement errors and external disturbances are also chosen according to Shi et al. (2019a). Then the measurement errors are assumed to be $\boldsymbol{v}(t) = 10^{-4} \times [4 \quad 5 \quad 6 \quad 0.2 \quad 0.2 \quad 0.2]^{\mathrm{T}} \sin(0.01\pi t)$. The environmental disturbance has been analyzed by Yang and Sun (2002) in detail, which includes the gravity-gradient force, aerodynamic force, and the force of the Earth's magnetic field. As pointed out in this work, the sinusoid is the prototype of periodic disturbances, and the environmental disturbance forces, such as magnetic force and gravity-gradient force, are cyclic essentially, which can be represented by sinusoids. Combined with aerodynamic force, the external disturbance force can contain a constant term. Thus, the external disturbances for the chaser and target spacecraft used in the simulations are taken to be of the following form:

$$\boldsymbol{w}_{\mathrm{C}}(t) = \begin{bmatrix} 0.1 + 0.2\sin(0.11\pi t) \\ 0.15 + 0.2\cos(0.11\pi t) \\ 0.2 + 0.3\cos\left(0.11\pi t + \pi/3\right) \end{bmatrix} \mathrm{N},$$

$$\boldsymbol{w}_{\mathrm{T}}(t) = \begin{bmatrix} 0.1 + 0.1\sin(0.11\pi t) \\ 0.2 + 0.12\cos\left(0.11\pi t + \pi/4\right) \\ 0.3 + 0.15\cos\left(0.11\pi t + \pi/6\right) \end{bmatrix} \mathrm{N}$$

Meanwhile, choose $\varpi = 200$, $\kappa_3 = 10$, $\overline{\kappa_1} = 4.0012$, $\overline{\kappa_2} = 4.0012$, $\overline{\kappa_3} = 0.032$, $\lambda = 30$, $\alpha = 7999$, then one has $k_0 = 0.1125$.

In addition, $\boldsymbol{M}_0$ is chosen as

$$\boldsymbol{M}_0 = \begin{bmatrix} 3.5321 & 0 & -0.0003 & 1.8781 & 0 & -0.0003 \\ 0 & 3.5321 & 0 & 0 & 1.8781 & 0 \\ 0.0003 & 0 & 3.5321 & 0.0003 & 0 & 1.8781 \\ 1.8895 & 0 & 0.0003 & 4.1465 & 0 & 0.0002 \\ 0 & 1.8895 & 0 & 0 & 4.1465 & 0 \\ -0.0003 & 0 & 1.8895 & -0.0002 & 0 & 4.1465 \end{bmatrix}$$

Table 3.1 The form of fault signal factor.

t/s	[0,2]	[2, 5]	[5,10]	[10,13]	[13,15]
$f(t)/\mathrm{N}$	0	$\sin(0.1\pi t)$	1	$\cos(0.1\pi t)$	0

3.4.1 Intermediate observer-based controller

Solving LMIs inequalities (Eqs. 3.15)– (Eqs. 3.17) gives

$$\boldsymbol{X}_1 = \begin{bmatrix} 0.0084 & 0 & -0.0000 & -0.0033 & 0 & 0.0000 \\ 0 & 0.0084 & 0 & 0 & -0.0033 & 0 \\ -0.0000 & 0 & 0.0084 & -0.0000 & 0 & -0.0033 \\ -0.0033 & 0 & -0.0000 & 0.0100 & 0 & -0.0000 \\ 0 & -0.0033 & 0 & 0 & 0.0100 & 0 \\ 0.0000 & 0 & -0.0033 & -0.0000 & 0 & 0.0100 \end{bmatrix}$$

$$\boldsymbol{X}_3 = \begin{bmatrix} 1.5547 & 0 & -0.0000 \\ 0 & 1.5547 & 0 \\ -0.0000 & 0 & 1.5547 \end{bmatrix}$$

$$\boldsymbol{W} = \begin{bmatrix} 0.1293 & 0 & -0.0001 & 0.3067 & 0 & 0.0003 \\ 0 & 0.1293 & 0 & 0 & 0.3067 & 0 \\ 0.0001 & 0 & 0.1293 & -0.0003 & 0 & 0.3067 \end{bmatrix}$$

$$\boldsymbol{H} = \begin{bmatrix} 0.0236 & 0 & -0.0000 & 0.0023 & 0 & 0.0000 \\ 0 & 0.0236 & 0 & 0 & 0.0023 & 0 \\ 0.0000 & 0 & 0.0236 & -0.0000 & 0 & 0.0023 \\ 0.0074 & 0 & 0.0000 & 0.0354 & 0 & 0.0000 \\ 0 & 0.0074 & 0 & 0 & 0.0354 & 0 \\ -0.0000 & 0 & 0.0074 & -0.0000 & 0 & 0.0354 \end{bmatrix}$$

Then, the controller gain matrix and the intermediate observer gain matrix are obtained as follows:

$$\begin{aligned}\boldsymbol{K} &= \boldsymbol{W}\boldsymbol{X}_1^{-1} \\ &= \begin{bmatrix} 31.1662 & 0 & 0.0028 & 40.7582 & 0 & 0.0252 \\ 0 & 31.1661 & 0 & 0 & 40.7581 & 0 \\ -0.0028 & 0 & 31.1662 & -0.0252 & 0 & 40.7582 \end{bmatrix}\end{aligned}$$

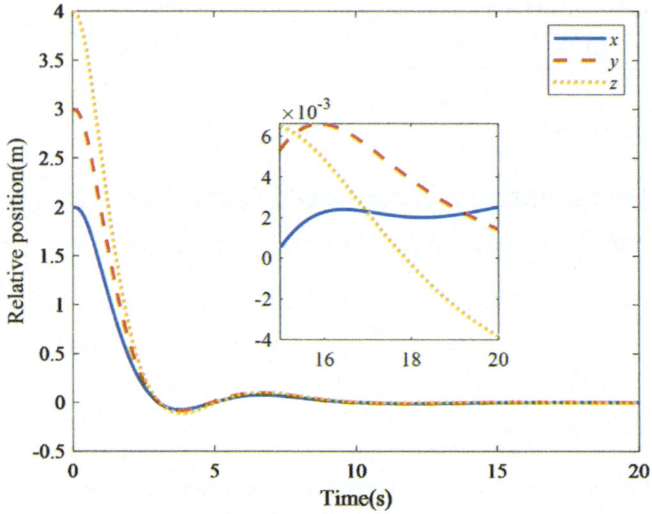

Figure 3.1 Time response of relative position.

$$
L = H(CX_1)^{-1}
= \begin{bmatrix}
3.3049 & 0 & -0.0001 & 1.3034 & 0 & -0.0003 \\
0 & 3.3049 & 0 & 0 & 1.3034 & 0 \\
0.0001 & 0 & 3.3049 & 0.0003 & 0 & 1.3034 \\
2.5759 & 0 & 0.0005 & 4.3737 & 0 & 0.0000 \\
0 & 2.5759 & 0 & 0 & 4.3737 & 0 \\
-0.0005 & 0 & 2.5759 & -0.0000 & 0 & 4.3737
\end{bmatrix}
$$

Applying the controller gain $\boldsymbol{K}$ and observer gain $\boldsymbol{L}$ to the closed-loop electromagnetic docking system expressed by inequality (Eqs. 3.15) produces the simulation results of Figs. 3.1 –3.8. As shown in Figs. 3.1 and 3.2, the relative position and velocity asymptotically converge to a small bound containing zero with steady-state errors less than 7×10^{-3}m and 4×10^{-3}m/s, respectively, and thus can be safely neglected. In addition, the relative velocity satisfies the requirement of soft docking. Figs. 3.3 and 3.4 plot the estimation errors of relative position and velocity, which suggest that they are successfully estimated within 2s and 6s, respectively. It can be concluded that the intermediate observer developed in this chapter can effectively estimate the relative motion information, with a fast convergence speed. The sphere with a radius of 0.12N in Fig. 3.5 means that the estimation

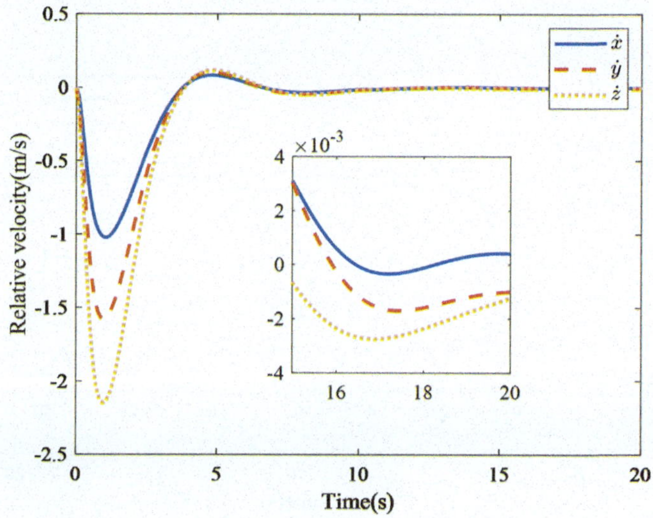

Figure 3.2 Time response of relative velocity.

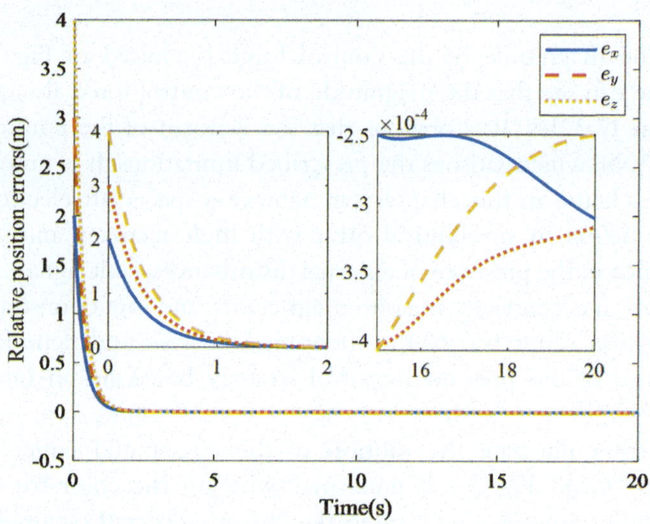

Figure 3.3 Estimation errors of relative position.

error of the lumped disturbance will converge to a small neighborhood containing the equilibrium. The blue curve denotes estimation error out of the neighborhood, while the red curve represents when the estimation error arrives in the convergence set, it will be attracted and stay in the

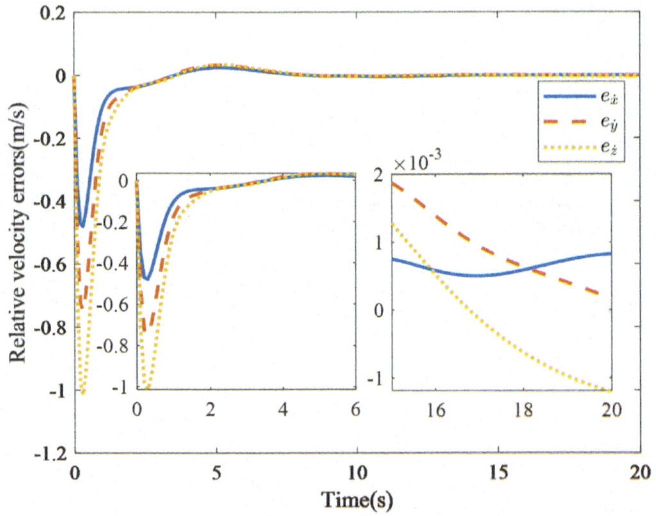

Figure 3.4 Estimation errors of relative velocity.

sphere. The magnitude of the control force is plotted in Fig. 3.6, from which one can see that the magnitude of the control force along each axis is less than 17.321N. It is obvious that the 2-norm of the control force is less than 30N, which satisfies the prescribed limitations. It is noted that the proposed scheme in this chapter can achieve a spacecraft electromagnetic docking mission in an elliptical orbit with high accuracy and very good performance in the presence of external disturbances, fault signals, unknown mass, elliptical eccentricity, measurement errors, and input constraints. Thus the simulation results verify the theoretical analysis and demonstrate the effectiveness of the presented control strategy based on an intermediate observer.

To further illustrate the motion of the chaser and target spacecraft relative to "CM," Fig. 3.7 is presented, wherein the color bar represents the simulation time. As one can see, the chaser spacecraft occupies a longer relative motion distance than the target spacecraft. This is because the mass of the target spacecraft is larger than the chaser spacecraft, which leads to the fact that "CM" is closer to the center of mass of the target spacecraft. Fig. 3.8 depicts the solved magnetic moment of the chaser spacecraft based on Eq. (1.79), from which one can see that the maximum of the magnetic moment is less than $7\times 10^4 Am^2$, which is practical and can be provided by a common electromagnetic mechanism.

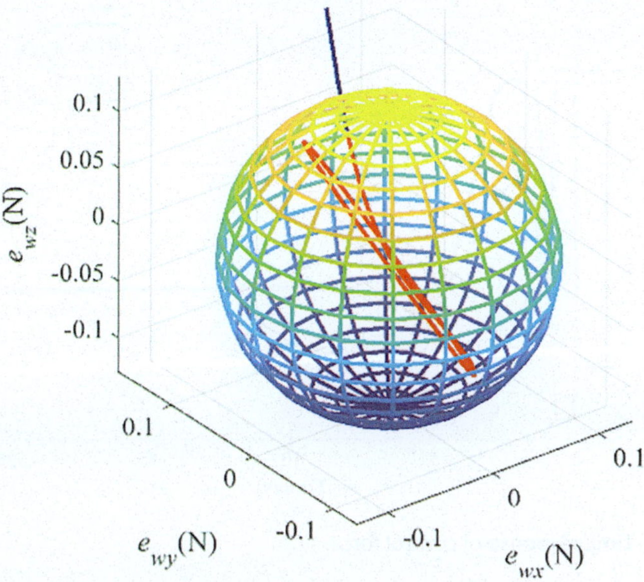

Figure 3.5 The convergence set of disturbance estimation errors.

3.4.2 Disturbance observer-based controller

In this case, the controller gain matrix $\boldsymbol{K}$ and the disturbance observer gain matrix are as follows:

$$\boldsymbol{K} = \begin{bmatrix} 0.4385 & -0.0024 & -0.0060 & 4.5770 & -0.0113 & -0.0151 \\ -0.0024 & 0.4366 & -0.0047 & -0.0114 & 4.5670 & -0.0233 \\ -0.0003 & -0.0047 & 0.4338 & -0.0153 & -0.0234 & 4.5532 \end{bmatrix}$$

$$\boldsymbol{L} = \begin{bmatrix} 0 & 0 & 0 & 2519.2 & 0.0753 & 0.1004 \\ 0 & 0 & 0 & 0.0753 & 2519.3 & 0.1508 \\ 0 & 0 & 0 & 0.1004 & 0.1508 & 2519.4 \end{bmatrix}$$

Applying this controller to the corresponding electromagnetic docking system represented by Eq. (3.14) produces the simulation results of Figs. 3.9 –3.12. Figs. 3.9 and 3.10 plot the response curves of relative position and relative velocity between the two spacecraft, from which it can be seen that the DOBC can achieve electromagnetic docking within 30s, while the IOBC stabilizes the system with better performance, within only 15 s. Fig. 3.11 depicts the comparison results of control force norm between the IOBC and DOBC schemes, which suggests that the maximum control

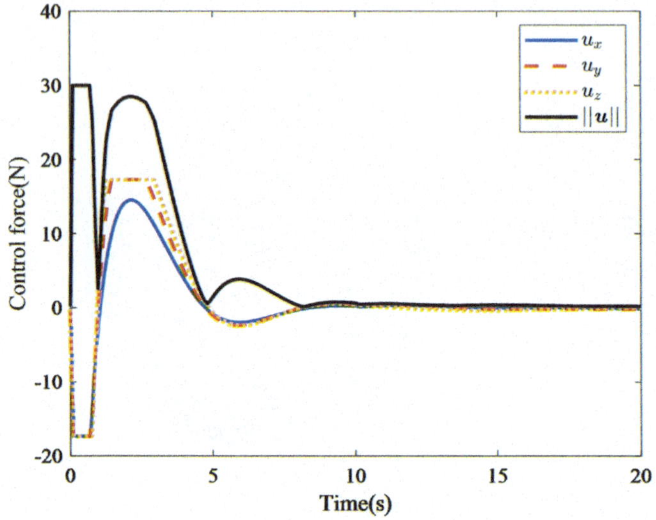

Figure 3.6 Time response of control force.

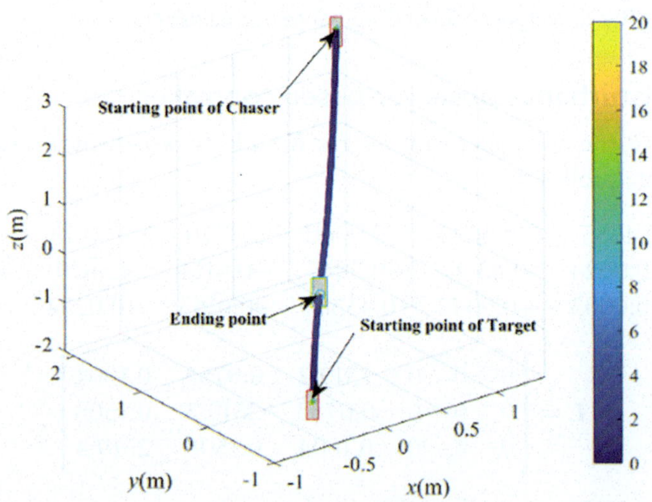

Figure 3.7 Motion of chaser and target spacecraft relative to "CM" in three-dimensional visualization.

force norm for DOBC is approximately 3.5N, much smaller than the electromagnetic force limit of 30N that can be provided. To further illustrate the advantage of the IOBC scheme proposed in this chapter, comparison results are given from the perspective of control force utilization efficiency,

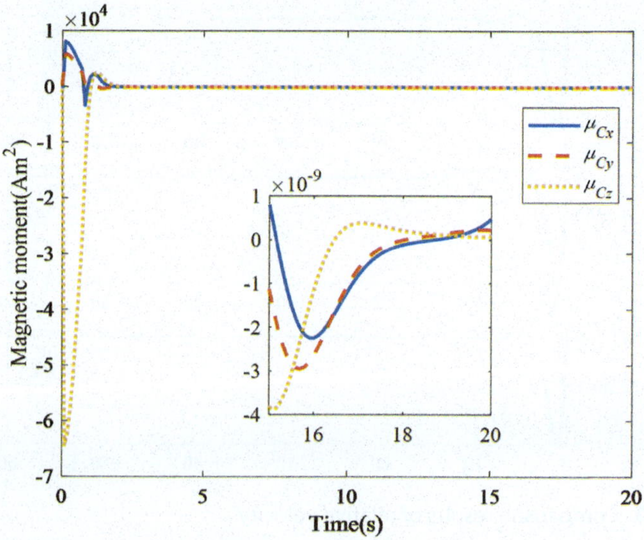

Figure 3.8 Magnetic moment of chaser spacecraft.

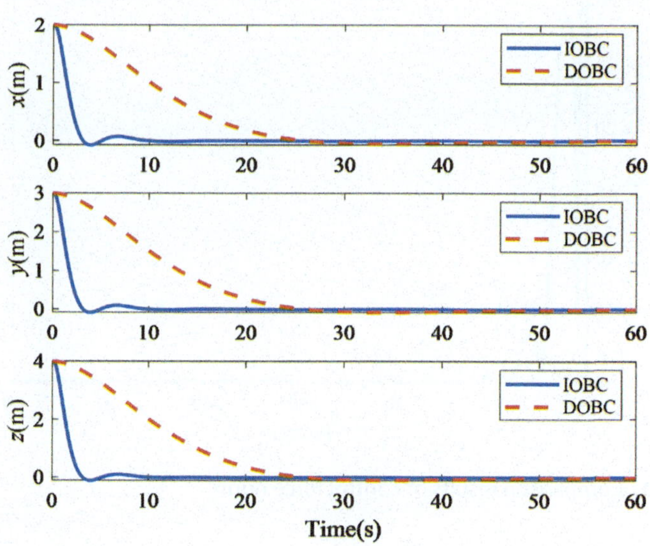

Figure 3.9 Comparison results of relative position.

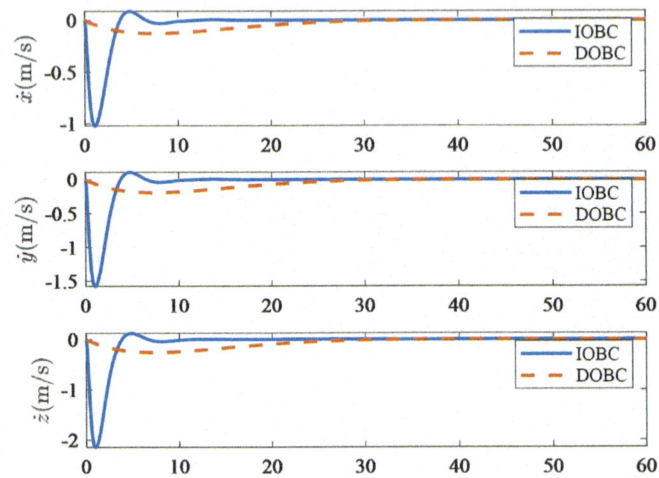

Figure 3.10 Comparison results of relative velocity.

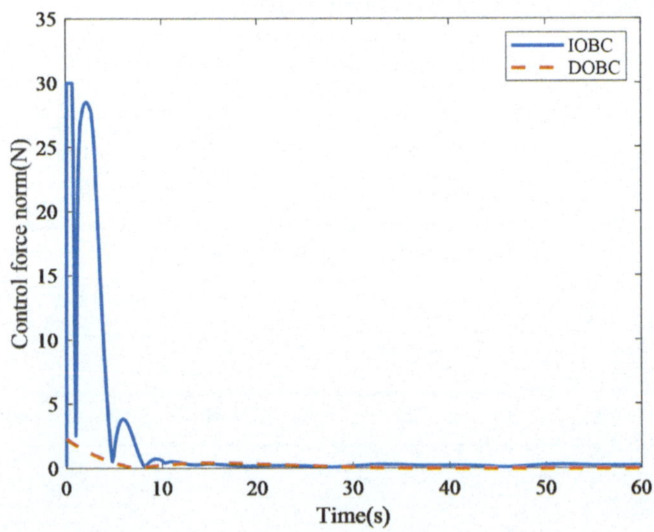

Figure 3.11 Comparison results of control force norm.

see Fig. 3.12. It is noted that the control force utilization efficiency is defined by $Y = \int_0^T (\|\boldsymbol{u}\|/\lambda)dt$, where T denotes the simulation time, and $T = 60$s is chosen in the simulation. It is observed that the proposed IOBC has a higher force utilization efficiency—approximately 6.5% than the DOBC—less than

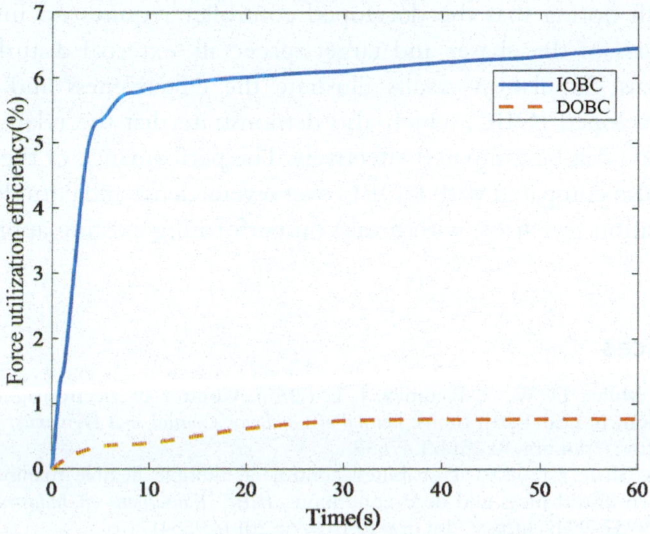

Figure 3.12 Comparison results of control force utilization efficiency.

1%, i.e., the DOBC does not make effective use of the control force that the actuator can provide, leading to a much slower speed of convergence. More importantly, the IOBC doesn't require the prior information on relative motion information, while the DOBC does. It can be concluded that, according to the measures used here, the proposed IOBC in this chapter is superior to the DOBC from the perspective of convergence time and control force utilization efficiency.

3.5 Conclusions

This chapter presents a novel intermediate observer-based control strategy for spacecraft electromagnetic docking in the presence of external disturbances, fault signals, unknown mass, elliptical eccentricity, measurement errors, and input constraints. By analyzing the effects of multiple uncertain factors, a lumped disturbance is constructed to facilitate controller design. Then, a new auxiliary variable constituted of relative motion information and the lumped disturbance is introduced for the design of the intermediate observer. Based on this, a feedback control law employing the estimates of relative motion information and the lumped disturbance is developed. Furthermore, the sufficient condition for input constraints can also be incorporated into the controller design to satisfy the actuator constraints.

It is worth noting that the developed controller requires no information on the mass of the chaser and target spacecraft, external disturbances, or fault signals. Simulation results illustrate the effectiveness and feasibility of the developed IOBC, which also demonstrate that the relative motion information can be estimated effectively. The performance of the proposed IOBC is also compared with DOBC over several dense and complex 2D and 3D simulation scenarios, with results outperforming existing approaches.

References

Ahsun, U., Miller, D. W., & Ramirez, J. L. (2010). Control of electromagnetic satellite formations in near-Earth orbits. *Journal of Guidance, Control, and Dynamics, 33*(6), 1883–1891. https://doi.org/10.2514/1.47637.

Chen, T., & Shan, J. (2020a). Distributed control of multiple flexible manipulators with unknown disturbances and dead-zone input. *IEEE Transactions on Industrial Electronics, 67*(11), 9937–9947. https://doi.org/10.1109/tie.2019.2955417.

Chen, T., & Shan, J. (2020b). Distributed tracking of a class of underactuated lagrangian systems with uncertain parameters and actuator faults. *IEEE Transactions on Industrial Electronics, 67*(5), 4244–4253. https://doi.org/10.1109/tie.2019.2922943.

Chen, T., & Wen, H. (2018). Autonomous assembly with collision avoidance of a fleet of flexible spacecraft based on disturbance observer. *Acta Astronautica, 147*, 86–96. https://doi.org/10.1016/j.actaastro.2018.03.027.

Chen, W. H., Yang, J., Guo, L., & Li, S. (2016). Disturbance-observer-based control and related methods–An overview. *IEEE Transactions on Industrial Electronics, 63*(2), 1083–1095. https://doi.org/10.1109/TIE.2015.2478397.

Elias, L. M., Kwon, D. W., Sedwick, R. J., & Miller, D. W. (2007). Electromagnetic formation flight dynamics including reaction wheel gyroscopic stiffening effects. *Journal of Guidance, Control, and Dynamics, 30*(2), 499–511. https://doi.org/10.2514/1.18679.

Huang, S. J., & Yang, G. H. (2014). Fault tolerant controller design for T-S fuzzy systems with time-varying delay and actuator faults: A k-step fault-estimation approach. *IEEE Transactions on Fuzzy Systems, 22*(6), 1526–1540. https://doi.org/10.1109/tfuzz.2014.2298053.

Kong, E.M.C. (2002). Spacecraft formation flight exploiting potential fields. Ph.D. thesis, Massachusetts Institute of Technology, 145–146.

Li, Y., & Ye, D. (2019). Near time-optimal controller based on analytical trajectory planning algorithm for satellite attitude maneuver. *Aerospace Science and Technology, 84*, 497–509. https://doi.org/10.1016/j.ast.2018.10.030.

Liu, C., Sun, Z., Shi, K., & Wang, F. (2017). Robust dynamic output feedback control for attitude stabilization of spacecraft with nonlinear perturbations. *Aerospace Science and Technology, 64*, 102–121. https://doi.org/10.1016/j.ast.2017.01.016.

Liu, C., Vukovich, G., Shi, K., & Sun, Z. (2018b). Robust fault tolerant nonfragile H_∞ attitude control for spacecraft via stochastically intermediate observer. *Advances in Space Research, 62*(9), 2631–2648. https://doi.org/10.1016/j.asr.2018.07.026.

Liu, C., Vukovich, G., Sun, Z., & Shi, K. (2018a). Observer-based fault-tolerant attitude control for spacecraft with input delay. *Journal of Guidance, Control, and Dynamics, 41*(9), 2039–2051. https://doi.org/10.2514/1.g003555.

Liu, C., Yue, X., Shi, K., & Sun, Z. (2019). Inertia-free attitude stabilization for flexible spacecraft with active vibration suppression. *International Journal of Robust and Nonlinear Control, 29*(18), 6311–6336. https://doi.org/10.1002/rnc.4742.

Narula, A., & Biggs, J. D. (2018). Fault-tolerant station-keeping on libration point orbits. *Journal of Guidance, Control, and Dynamics, 41*(4), 879–887. https://doi.org/10.2514/1.g003115.

Norman, M. C., & Peck, M. A. (2010). Stationkeeping of a flux-pinned satellite network. *Journal of Guidance, Control, and Dynamics, 33*(5), 1683–1686. https://doi.org/10.2514/1.49550.

Sanyal, A., Fosbury, A., Chaturvedi, N., & Bernstein, D. S. (2009). Inertia-free spacecraft attitude tracking with disturbance rejection and almost global stabilization. *Journal of Guidance, Control, and Dynamics, 32*(4), 1167–1178. https://doi.org/10.2514/1.41565.

Shi, K., Liu, C., & Sun, Z. (2019a). Constrained fuel-free control for spacecraft electromagnetic docking in elliptical orbits. *Acta Astronautica, 162*, 14–24. https://doi.org/10.1016/j.actaastro.2019.05.016.

Shi, K., Liu, C., Sun, Z., & Vukovich, G. (2019b). A novel robust non-fragile control approach for a class of uncertain linear systems with input constraints. *Transactions of the Institute of Measurement and Control, 41*(8), 2365–2373. https://doi.org/10.1177/0142331218799835.

Shi, P., Liu, M., & Zhang, L. (2015). Fault-tolerant sliding-mode-observer synthesis of Markovian jump systems using quantized measurements. *IEEE Transactions on Industrial Electronics, 62*(9), 5910–5918. https://doi.org/10.1109/tie.2015.2442221.

Singla, P., Subbarao, K., & Junkins, J. L. (2006). Adaptive output feedback control for spacecraft rendezvous and docking under measurement uncertainty. *Journal of Guidance, Control, and Dynamics, 29*(4), 892–902. https://doi.org/10.2514/1.17498.

Sun, L., & Zheng, Z. (2019). Saturated adaptive hierarchical fuzzy attitude-tracking control of rigid spacecraft with modeling and measurement uncertainties. *IEEE Transactions on Industrial Electronics, 66*(5), 3742–3751. https://doi.org/10.1109/TIE.2018.2856204.

Sun, L., Huo, W., & Jiao, Z. (2018). Disturbance-observer-based robust relative pose control for spacecraft rendezvous and proximity operations under input saturation. *IEEE Transactions on Aerospace and Electronic Systems, 54*(4), 1605–1617. https://doi.org/10.1109/taes.2018.2798239.

Tabatabaeipour, S. M., & Bak, T. (2014). Robust observer-based fault estimation and accommodation of discrete-time piecewise linear systems. *Journal of the Franklin Institute, 351*(1), 277–295. https://doi.org/10.1016/j.jfranklin.2013.08.021.

Weiss, A., Kolmanovsky, I., Bernstein, D. S., & Sanyal, A. (2013). Inertia-free spacecraft attitude control using reaction wheels. *Journal of Guidance, Control, and Dynamics, 36*(5), 1425–1439. https://doi.org/10.2514/1.58363.

Yang, C. D., & Sun, Y. P. (2002). Mixed H_2/H_∞ state-feedback design for microsatellite attitude control. *Control Engineering Practice, 10*(9), 951–970. https://doi.org/10.1016/s0967-0661(02)00049-7.

Zhang, J., Ye, D., Sun, Z., & Liu, C. (2018). Extended state observer based robust adaptive control on SE(3) for coupled spacecraft tracking maneuver with actuator saturation and misalignment. *Acta Astronautica, 143*, 221–233. https://doi.org/10.1016/j.actaastro.2017.11.034.

Zhang, Y. W., Yang, L. P., Zhu, Y. W., Huang, H., & Cai, W. W. (2012). Nonlinear 6-DOF control of spacecraft docking with inter-satellite electromagnetic force. *Acta Astronautica*, 77, 97–108. https://doi.org/10.1016/j.actaastro.2012.03.020.

Zhang, Y. W., Yang, L. P., Zhu, Y. W., Huang, H., & Cai, W. W. (2013). Angular-momentum management of spacecraft electromagnetic docking considering the earth's magnetic field. *Journal of Guidance, Control, and Dynamics, 36*(3), 860–869. https://doi.org/10.2514/1.57723.

Zhu, Y. W., Cai, W. W., Yang, L. P., & Huang, H. (2018). Flatness-based trajectory planning for electromagnetic spacecraft proximity operations in elliptical orbits. *Acta Astronautica, 152*, 342–351. https://doi.org/10.1016/j.actaastro.2018.08.031.

CHAPTER 4

Active disturbance rejection control for spacecraft electromagnetic docking

4.1 Introduction

Aerospace technology has grown rapidly in recent decades, which has greatly promoted the deployment of space missions. Nevertheless, it is worth mentioning that most of the current space missions usually rely on a traditional energy supply system, e.g., a gas jet, a momentum/reaction wheel, or a control moment gyroscope, etc. (Shi & Hu, 2022; Sun et al., 2017). This will inevitably cause some restrictions on the lifetime and flexibility of missions. Spacecraft electromagnetic docking is a novel technology that applies inter-craft electromagnetic force, generated by magnetic coils equipped on the neighboring target and chaser spacecraft, to manage the relative motion between them. Electromagnetic docking has distinct advantages over other docking mechanisms, e.g., it requires no propellant, nor plume contamination, with continuous, reversible, and noncontact actuation (Zhang, Yang, Zhu, Huang, & Cai, 2013). However, along with these advantages are its complex and coupled dynamics, leading to multisource complex unknown uncertainties that can influence the robust performance of the docking system.

Spacecraft autonomous docking technologies are necessary to enable on-orbit assembly strategies. In light of electromagnetic docking, the relative positions between spacecraft are then determined by electromagnetic forces produced by orthogonal coils equipped on spacecraft, while the relative orientation of the dipoles on spacecraft is controlled using reaction wheels or control moment gyros to generate the counteracting torques. As a non-contact docking scheme using the interaction of inter-craft electromagnetic forces, spacecraft electromagnetic docking technologies have been widely reported by Zhang, Yang, Zhu, Huang, & Cai, 2012 and Shi et al. (2019) and references therein. In Zhang, Yang, Zhu, Huang, & Cai, 2012, the dynamic model of relative motion is derived based on Hill's model, i.e., the docking

Spacecraft Electromagnetic Docking and Separation.
DOI: https://doi.org/10.1016/B978-0-443-49334-8.00005-7

technology was proposed for circular orbit and is not appliable to elliptical orbit to be studied in this chapter. In addition, it was required that the mass information of the docking of the two spacecraft should be known in prior. To this end, a disturbance observer-based controller was developed by Sun et al. (2020), where no information on the mass of the two docking spacecraft, external disturbances, or even orbital elements is required, but the information on relative motion should be measured accurately. The translational dynamics of spacecraft using electromagnetic forces are highly coupled and nonlinear, which further increases the challenge for high-precision docking maneuvers. Most of the published works in the field of electromagnetic docking only studied circular orbits of the spacecraft, which would fail if the target spacecraft is in a large-eccentric orbit (Biggs & Henninger, 2019; Shi et al., 2020; Sun, 2020). Since the space electromagnetic docking technology is not yet mature, controllers designed for fault-free cases may not be directly applicable to real cases. To this end, the fault-tolerant controller can guarantee desirable stability and performance properties in case of component fault/failure (Narula & Biggs, 2018; Shen et al., 2019; Van et al., 2017). Thus possible fault signals should be taken into account in the controller design. In addition to the problems mentioned above, it is often expensive or even impossible to determine the mass properties of the target spacecraft once it is noncooperative, which can lead to a loss of control precision. To alleviate this requirement, the controllers developed by Shen et al. (2015), Liu et al. (2019), and Liu et al. (2020) were inertia-free in the sense that no information on the inertia matrix is required prior. Analogous to this, for the spacecraft electromagnetic docking plant, the controllers developed by Sun (2020) and Shi et al. (2020) required no prior information on the mass of the chaser or target spacecraft. However, to the best of the authors' knowledge, there is no study in the literature accounting for the hysteresis phenomenon, which is a significant source of time delay. In reality, time delay must be taken into account as the hysteresis phenomenon is common and especially occurs in the actuator of uncertain nonlinear systems (Pepe, 2020; Su et al., 2021). The electromagnetic docking system is no exception, and the study on delayed electromagnetic docking technologies can make it closer from theory to application. Otherwise, the continuous accumulation of time delays will cause performance degradation or even lead to instability of the system, which, in turn, gives rise to the loss of some critical information.

A challenging problem arising in the domain of spacecraft electromagnetic docking is that the external disturbances, fault signals, input delay, unknown mass, and elliptical eccentricity appear simultaneously, which further increases the complexity and difficulty for spacecraft designers. Future space missions demand that the new generation of spacecraft should be able to perform on-orbit servicing missions with high precision and better robustness to the total disturbance. One way to overcome this problem is by using the active disturbance rejection control (ADRC) scheme to cope with external disturbances and other uncertainties of the controlled systems. The key idea of ADRC lies in the fact that the total disturbance can be online estimated by an extended state observer (ESO) and compensated for in the control signal (Ren et al., 2022). Thus the ADRC scheme has been widely studied and employed in many areas, e.g., vehicle or quadrotor control (Xu, Ma, Guo, Xie, & Song, 2020; Yuan et al., 2017; Yang et al., 2018), spacecraft or pneumatic motion platform control (Biggs & Brisotto, 2021; Yuan et al., 2019), and networked system control (Sun et al., 2019; Sun, 2020). In the typical ADRC framework, the system control is reduced to a simple problem of double integral plants, with the total disturbance estimated via ESO and compensated for in the control signal. In the sense, very partial model information is used, and the other known/unknown model information is attributed to the total disturbance (Sun et al., 2019; Yao & Deng, 2017). However, these control schemes actually originate from general control system design strategies or from mathematical considerations; that is, the physical structure of dynamic systems is neglected. In light of this, the conventional ADRC technique increases the risk of system instability and breeds the emergence of advanced control approaches.

In the case that the relative motion information on electromagnetic docking spacecraft is unavailable, the ESO is insufficient for effective ADRC, and an IO design is necessary. To our knowledge, no study has explicitly focused on the aforementioned issues, i.e., external disturbances, fault signals, input delay, unknown mass, and elliptical eccentricity. These gaps make it very difficult to achieve spacecraft electromagnetic docking, but through the use of well-designed controllers, these difficulties can be overcome. Motivated by the aforementioned discussions, the aim of this chapter is to develop a novel intermediate observer (IO)-based ADRC scheme without knowing exactly the external disturbances, fault signals, input delay, and mass properties to achieve high-precision electromagnetic docking for spacecraft in elliptical orbits.

The key contributions of this chapter are summarized as follows:

- Considering that the relative motion information is unavailable for spacecraft electromagnetic docking with input delay, an intermediate observer is designed to estimate the relative motion information as well as the total disturbance. Compared with the disturbance observer-based controller in Sun (2020), the relative motion information is not required to be measurable, and the input delay is taken into account for controller design.

- The framework of ADRC for delayed spacecraft electromagnetic docking systems is developed, in which a novel IO acts as a predictor to realize the simultaneous estimation of state and the total disturbance. Compared with Narula and Biggs (2018) and Weiss et al. (2013), no parameter identifications and neural or fuzzy approximations are required.

The remainder of this chapter is organized as follows. Section 4.2 presents a nonlinear model for spacecraft electromagnetic docking and some preliminaries, and the corresponding control problem is formulated. Then, an ADRC scheme is proposed in Section 4.3, and an IO as well as the stability analysis are also presented. It is followed by numerical simulations to demonstrate the performance of the proposed control strategy in Section 4.4. Finally, conclusions are drawn in Section 4.5.

4.2 Problem formulation

4.2.1 Dynamic modeling

The dynamic model is illustrated in Section 1.2, Eq. (1.78). In addition, the following important points should be noted.

In an electromagnetic docking control system, due to many factors, e.g., the reaction time of actuators and sensors, faults of actuators and sensors, hysteresis phenomenon, signal sampling and transmission, time delay always exists. The effects of these time-delay phenomena will all be reflected in the control input. Time delay will present a barrier to carrying out the required performance, and even destroy the stability of the closed-loop system. Here, $\tau(t)$ denotes time-varying delay, and one assumes that it satisfies $l_1 \leq \tau(t) \leq l_2$, $\dot{\tau}(t) \leq \vartheta$, where l_1, l_2, and ϑ are constants. As is known to all, an input delay will usually cause performance degradation and even instability. In addition, the actuator fault has been considered one of the most serious failures and usually occurs in spacecraft systems. In a spacecraft system, damage to the actuators can be caused by damage to the energized coil, the power supply systems, wiring, etc. (Chen & Shan, 2020b). Taking this issue into account, let $\boldsymbol{E}$ be a full-column-rank matrix, which has a similar structure to $\boldsymbol{B}$, denoting the distribution matrix of fault signals $f(t)$ appearing in the input.

Considering the above factors, the new form of the spacecraft electromagnetic docking system can be expressed as:

$$\begin{cases} \dot{\boldsymbol{\eta}}(t) = \boldsymbol{A}\boldsymbol{\eta}(t) + \boldsymbol{B}\boldsymbol{u}(t, t - \tau(t)) + \boldsymbol{B}\boldsymbol{w}(t) \\ \boldsymbol{y}(t) = \boldsymbol{C}\boldsymbol{\eta}(t) \end{cases} \tag{4.1}$$

where $\boldsymbol{w}(t)=\boldsymbol{B}^{\mathrm{T}}(\boldsymbol{B}\boldsymbol{B}^{\mathrm{T}})^{*}(\Delta\boldsymbol{A}\boldsymbol{\eta}(t)+\Delta\boldsymbol{B}\boldsymbol{u}(t)+(\boldsymbol{B}+\Delta\boldsymbol{B})\boldsymbol{w}_0(t)+\boldsymbol{E}\boldsymbol{f}(t))$ is regarded as the total disturbance from external disturbances, fault signals, unknown mass, and elliptical eccentricity, and where the matrix $(\boldsymbol{B}\boldsymbol{B}^{\mathrm{T}})^{*}$ is the pseudoinverse of $\boldsymbol{B}\boldsymbol{B}^{\mathrm{T}}$.

Remark 4.1 The development of controllers for spacecraft electromagnetic docking is a labor-intensive, time-consuming process. In addition, it is often expensive to determine the mass properties with high precision. Hence, the nominal representative of a valid mass m_0 is used for the controller design, then the difference between m_0 and m_{CT} will appear. In other words, the influence of the eccentricity e and m_{CT} results in the appearance of $\Delta\boldsymbol{A}$ and $\Delta\boldsymbol{B}$, respectively. It is noted that the difference between m_0 and m_{CT} only affects the relative velocity, i.e., $\dot{x}$, $\dot{y}$, and $\dot{z}$. Thus the combined uncertain term, i.e., $\Delta\boldsymbol{A}\boldsymbol{\eta}(t)+\Delta\boldsymbol{B}\boldsymbol{u}(t)+\Delta\boldsymbol{B}\boldsymbol{w}_0(t)$, will contribute to the disturbance, and the same applies to $\boldsymbol{E}\boldsymbol{f}(t)$. Together, they form the total disturbance$w(t)$.

4.2.2 Preliminaries

The subsection presents the following necessary assumptions, remarks, and lemmas, which are needed to provide theoretical support before further analysis.

Assumption 4.1 The disturbance force $\boldsymbol{w}_0(t)$ and its first-time derivative are assumed to be piecewise continuous and bounded.

Remark 4.2 Assumption 4.1 is reasonable. The main reason is that the external disturbances are primarily caused by solar radiation, geopotential anomaly, magnetic effects, and residual atmosphere, which are all continuous and bounded (Chen & Shan, 2020a; Li & Ye, 2019). In addition, because the total control authority is limited, all three available control input forces should be continuous and bounded, which is reasonable in practice. It is worth noting that the desired translational motion represented by relative position and velocity is also bounded and differentiable in practice. In particular, the derivative of $\boldsymbol{f}(\mathrm{t})$ with respect to time is often assumed to be norm bounded (Huang et al., 2014). They together result in the fact that the total disturbance force is bounded and differentiable, and the first derivative of $\boldsymbol{w}(t)$ satisfies $||\dot{\boldsymbol{w}}(t)|| \leq \kappa_1$, where $\kappa_1 \geq 0$. It should be noted that the norm of the total disturbance $\boldsymbol{w}(t)$ and its components could be unknown. Thus it is more general than the fault-tolerant sliding-mode observer method by Shi et al. (2015), which requires the preliminary knowledge of the bounds of fault signals.

Assumption 4.2 : $\text{rank}(\boldsymbol{B}, \boldsymbol{E}) = \text{rank}(\boldsymbol{B})$, and there exists a matrix $\boldsymbol{B}^*$ such that $(\boldsymbol{I} - \boldsymbol{B}\boldsymbol{B}^*)\boldsymbol{E} = 0$.

Remark 4.3 Assumption 4.2 is general by Tabatabaeipour and Bak (2014), which means that the fault appears in the actuator and can be compensated for by the control input.

4.2.3 Control objectives

The objective of this chapter is twofold:

- Design an effective control scheme to achieve the uniform ultimate boundedness of the closed-loop system in the presence of external disturbances, fault signals, input delay, unknown mass, and elliptical eccentricity;
- Design an IO such that the relative motion information as well as the total disturbance are estimated simultaneously.

4.3 Intermediate observer-based active disturbance rejection controller design

In this section, the relative motion information as well as the total disturbance should be estimated simultaneously, such that the control scheme is designed based on the estimation. The design procedure of the ADRC consists of two main steps. The first step involves constructing the IO. The next step is to design a control law that forces the system states to achieve uniform ultimate boundedness in the presence of multisource complex uncertainties. For the first step, let

$$\boldsymbol{\xi}(t) = \boldsymbol{w}(t) - \varpi \boldsymbol{B}^{\mathrm{T}} \boldsymbol{\eta}(t) \tag{4.2}$$

$$\dot{\boldsymbol{\xi}}(t) = \dot{\boldsymbol{w}}(t) - \varpi \boldsymbol{B}^{\mathrm{T}} \left[\boldsymbol{A}\boldsymbol{\eta}(t) + \boldsymbol{B}\boldsymbol{u}(t, t - \tau(t)) + \boldsymbol{B}\boldsymbol{\xi}(t) + \varpi \boldsymbol{B}\boldsymbol{B}^{\mathrm{T}} \boldsymbol{\eta}(t) \right] \tag{4.3}$$

Then, the IO is designed based on

$$\dot{\hat{\boldsymbol{\eta}}}(t) = \boldsymbol{A}\hat{\boldsymbol{\eta}}(t) + \boldsymbol{B}\boldsymbol{u}(t, t-\tau(t)) + \boldsymbol{B}\hat{\boldsymbol{w}}(t) + \boldsymbol{L}\big(\boldsymbol{y}(t) - \hat{\boldsymbol{y}}(t)\big) \tag{4.4}$$

$$\dot{\hat{\boldsymbol{\xi}}}(t) = -\varpi \boldsymbol{B}^{\mathrm{T}}\Big[\boldsymbol{A}\hat{\boldsymbol{\eta}}(t) + \boldsymbol{B}\boldsymbol{u}(t, t-\tau(t)) + \boldsymbol{B}\hat{\boldsymbol{\xi}}(t) + \varpi \boldsymbol{B}\boldsymbol{B}^{\mathrm{T}}\hat{\boldsymbol{\eta}}(t)\Big] \tag{4.5}$$

$$\hat{\boldsymbol{y}}(t) = \boldsymbol{C}\hat{\boldsymbol{\eta}}(t) \tag{4.6}$$

$$\hat{\boldsymbol{w}}(t) = \hat{\boldsymbol{\xi}}(t) + \varpi \boldsymbol{B}^{\mathrm{T}}\hat{\boldsymbol{\eta}}(t) \tag{4.7}$$

where $\hat{\boldsymbol{\eta}}(t)$, $\hat{\boldsymbol{\xi}}(t)$, $\hat{\boldsymbol{y}}(t)$, and $\hat{\boldsymbol{w}}(t)$ are the estimates of $\boldsymbol{\eta}(t)$, $\boldsymbol{\xi}(t)$, $\boldsymbol{y}(t)$, and $\boldsymbol{w}(t)$ respectively.

Based on the above estimations, the ADRC is designed as

$$\boldsymbol{u}(t-\tau(t)) = -\boldsymbol{K}\hat{\boldsymbol{\eta}}(t) - \hat{\boldsymbol{w}}(t) + \boldsymbol{B}^{*}\left(\boldsymbol{B}\boldsymbol{B}^{*}\right)^{*}\boldsymbol{A}_{d}\hat{\boldsymbol{\eta}}(t-\tau(t)) \tag{4.8}$$

where $\boldsymbol{K}$ is the controller gain matrix, which is chosen such that $\boldsymbol{A} - \boldsymbol{B}\boldsymbol{K}$ is Hurwitz, which can be guaranteed by the condition of inequality Eq. (4.12) below, $\hat{\boldsymbol{w}}(t)$ is added to the input to compensate for the effect of the total disturbance and $\hat{\boldsymbol{\eta}}(t-\tau(t))$ means the time profiles of states, which are zero when $t < \tau(t)$. In addition, the matrices $\boldsymbol{B}^{*}$ and $(\boldsymbol{B}\boldsymbol{B}^{*})^{*}$ denote the Moore–Penrose pseudoinverses of $\boldsymbol{B}$ and $\boldsymbol{B}\boldsymbol{B}^{*}$, respectively.

Denote $\boldsymbol{e}_{\eta}(t) = \boldsymbol{\eta}(t) - \hat{\boldsymbol{\eta}}(t), \boldsymbol{e}_{\xi}(t) = \boldsymbol{\xi}(t) - \hat{\boldsymbol{\xi}}(t)$ and $\boldsymbol{e}_{\eta}(t-\tau(t)) = \boldsymbol{\eta}(t-\tau(t)) - \hat{\boldsymbol{\eta}}(t-\tau(t))$, substituting Eq. (4.10) into the first of Eq. (4.3) yields

$$\begin{aligned}\dot{\boldsymbol{\eta}}(t) = {} & (\boldsymbol{A} - \boldsymbol{B}\boldsymbol{K})\boldsymbol{\eta}(t) + \boldsymbol{B}\boldsymbol{K}\boldsymbol{e}_{\eta}(t) + \boldsymbol{A}_{d}\boldsymbol{\eta}(t-\tau(t)) - \boldsymbol{A}_{d}\boldsymbol{e}_{\eta}(t-\tau(t)) \\ & + \boldsymbol{B}\boldsymbol{e}_{\xi}(t) + \varpi \boldsymbol{B}\boldsymbol{B}^{\mathrm{T}}\boldsymbol{e}_{\eta}(t)\end{aligned} \tag{4.9}$$

Then, it can be concluded that

$$\dot{\boldsymbol{e}}_{\eta}(t) = (\boldsymbol{A} - \boldsymbol{L}\boldsymbol{C})\boldsymbol{e}_{\eta}(t) + \boldsymbol{B}\boldsymbol{e}_{\xi}(t) + \varpi \boldsymbol{B}\boldsymbol{B}^{\mathrm{T}}\boldsymbol{e}_{\eta}(t) \tag{4.10}$$

$$\dot{\boldsymbol{e}}_{\xi}(t) = \dot{\boldsymbol{w}}(t) - \varpi \boldsymbol{B}^{\mathrm{T}}\Big[\boldsymbol{A}\boldsymbol{e}_{\eta}(t) + \boldsymbol{B}\boldsymbol{e}_{\xi}(t) + \varpi \boldsymbol{B}\boldsymbol{B}^{\mathrm{T}}\boldsymbol{e}_{\eta}(t)\Big] \tag{4.11}$$

Thus the objective of this chapter is to seek the controller gain matrix $\boldsymbol{K}$ and observer gain matrix $\boldsymbol{L}$ such that the states of the resulting closed-loop system represented by Eqs. (4.9)–(4.11) are uniformly ultimately bounded.

Theorem 4.1 now gives the existence conditions of the controller gain matrix $\boldsymbol{K}$ and observer gain matrix $\boldsymbol{L}$.

Theorem 4.1 The states of the closed-loop system represented by Eqs. (4.9)–(4.11) are uniformly ultimately bounded, and converge to a small residual set, if for a given symmetric positive definite matrix $\boldsymbol{P}$, scalars $\delta > 0, \vartheta > 0, \varpi > 0, \xi_1 > 0$, there exist symmetric positive definite matrices $\boldsymbol{P}_i$, $i = 1, 2, 3, 4, 5$, matrices $\boldsymbol{M} = [\boldsymbol{M}_1^{\mathrm{T}} \quad \boldsymbol{M}_2^{\mathrm{T}} \quad \boldsymbol{M}_3^{\mathrm{T}} \quad \boldsymbol{M}_4^{\mathrm{T}} \quad \boldsymbol{M}_5^{\mathrm{T}} \quad \boldsymbol{M}_6^{\mathrm{T}}]^{\mathrm{T}}$, $\boldsymbol{N} = [\boldsymbol{N}_1^{\mathrm{T}} \quad \boldsymbol{N}_2^{\mathrm{T}} \quad \boldsymbol{N}_3^{\mathrm{T}} \quad \boldsymbol{N}_4^{\mathrm{T}} \quad \boldsymbol{N}_5^{\mathrm{T}} \quad \boldsymbol{N}_6^{\mathrm{T}}]^{\mathrm{T}}$ and matrices $\boldsymbol{H}$ and $\boldsymbol{K}$, such that the following LMIs hold:

$$(\boldsymbol{A} - \boldsymbol{B}_1\boldsymbol{K})^{\mathrm{T}}\boldsymbol{P} + \boldsymbol{P}(\boldsymbol{A} - \boldsymbol{B}_1\boldsymbol{K}) < 0 \tag{4.12}$$

$$\begin{bmatrix} \boldsymbol{\Theta}_{11} & \boldsymbol{\Theta}_{12} \\ * & \boldsymbol{\Theta}_{22} \end{bmatrix} < 0 \tag{4.13}$$

where

$$\boldsymbol{\Theta}_{11} = \begin{bmatrix} \overline{\boldsymbol{\Theta}}_{11} & \overline{\boldsymbol{\Theta}}_{12} & \boldsymbol{P}_1\boldsymbol{A}_d & \overline{\boldsymbol{\Theta}}_{14} & \boldsymbol{P}_1\boldsymbol{B} & -\boldsymbol{N}_1 & \boldsymbol{M}_1 & \boldsymbol{N}_1 \\ * & \overline{\boldsymbol{\Theta}}_{22} & \boldsymbol{M}_3^{\mathrm{T}} & \overline{\boldsymbol{\Theta}}_{24} & \overline{\boldsymbol{\Theta}}_{25} & \overline{\boldsymbol{\Theta}}_{26} & \boldsymbol{M}_2 & \boldsymbol{N}_2 \\ * & * & 0 & \overline{\boldsymbol{\Theta}}_{34} & 0 & -\boldsymbol{N}_3 & \boldsymbol{M}_3 & \boldsymbol{N}_3 \\ * & * & * & \overline{\boldsymbol{\Theta}}_{44} & \overline{\boldsymbol{\Theta}}_{45} & \overline{\boldsymbol{\Theta}}_{46} & \boldsymbol{M}_4 & \boldsymbol{N}_4 \\ * & * & * & * & \overline{\boldsymbol{\Theta}}_{55} & -\boldsymbol{N}_5 & \boldsymbol{M}_5 & \boldsymbol{N}_5 \\ * & * & * & * & * & \overline{\boldsymbol{\Theta}}_{66} & \boldsymbol{M}_6 & \boldsymbol{N}_6 \\ * & * & * & * & * & * & -\frac{\delta \boldsymbol{P}_2}{l_2} & 0 \\ * & * & * & * & * & * & * & -\frac{\delta \boldsymbol{P}_2}{l_2 - l_1} \end{bmatrix}$$

and where

$$\begin{aligned}
\overline{\boldsymbol{\Theta}}_{11} &= (\boldsymbol{A} - \boldsymbol{B}\boldsymbol{K})^{\mathrm{T}}\boldsymbol{P}_1 + \boldsymbol{P}_1(\boldsymbol{A} - \boldsymbol{B}_1\boldsymbol{K}) \\
\overline{\boldsymbol{\Theta}}_{12} &= \boldsymbol{P}_1\boldsymbol{B}\boldsymbol{K} + \varpi\boldsymbol{P}_1\boldsymbol{B}\boldsymbol{B}^{\mathrm{T}} + \boldsymbol{M}_1 \\
\overline{\boldsymbol{\Theta}}_{14} &= -\boldsymbol{P}_1\boldsymbol{A}_d - \boldsymbol{M}_1 + \boldsymbol{N}_1 \\
\overline{\boldsymbol{\Theta}}_{22} &= \boldsymbol{A}^{\mathrm{T}}\boldsymbol{P}_2 - \boldsymbol{C}^{\mathrm{T}}\boldsymbol{H}^{T} + \boldsymbol{P}_2\boldsymbol{A} - \boldsymbol{H}\boldsymbol{C} + \varpi\boldsymbol{P}_2\boldsymbol{B}\boldsymbol{B}^{\mathrm{T}} + \varpi\boldsymbol{B}\boldsymbol{B}^{\mathrm{T}}\boldsymbol{P}_2 \\
&\quad + \boldsymbol{P}_4 + \boldsymbol{P}_5 + (2l_2 - l_1)\varpi\delta\left(\boldsymbol{A}^{\mathrm{T}}\boldsymbol{P}_2 - \boldsymbol{C}^{\mathrm{T}}\boldsymbol{H}^{\mathrm{T}}\right)\boldsymbol{B}\boldsymbol{B}^{\mathrm{T}} + \boldsymbol{M}_2 + \boldsymbol{M}_2^{\mathrm{T}} \\
&\quad + (2l_2 - l_1)\varpi\delta\boldsymbol{B}\boldsymbol{B}^{\mathrm{T}}(\boldsymbol{P}_2\boldsymbol{A} - \boldsymbol{H}\boldsymbol{C}) + (2l_2 - l_1)\varpi^2\delta\boldsymbol{B}\boldsymbol{B}^{\mathrm{T}}\boldsymbol{P}_2\boldsymbol{B}\boldsymbol{B}^{\mathrm{T}} \\
\overline{\boldsymbol{\Theta}}_{24} &= \boldsymbol{M}_4^{\mathrm{T}} - \boldsymbol{M}_2 + \boldsymbol{N}_2 \\
\overline{\boldsymbol{\Theta}}_{25} &= (2l_2 - l_1)\delta\left(\boldsymbol{A}^{\mathrm{T}}\boldsymbol{P}_2 - \boldsymbol{C}^{\mathrm{T}}\boldsymbol{H}^{\mathrm{T}}\right)\boldsymbol{B} + \boldsymbol{P}_2\boldsymbol{B} - \varpi\boldsymbol{A}^{\mathrm{T}}\boldsymbol{B}\boldsymbol{P}_3 \\
&\quad + (2l_2 - l_1)\varpi\boldsymbol{B}\boldsymbol{B}^{\mathrm{T}}(\delta\boldsymbol{P}_2)\boldsymbol{B} - \varpi^2\boldsymbol{B}\boldsymbol{B}^{\mathrm{T}}\boldsymbol{B}\boldsymbol{P}_3 + \boldsymbol{M}_5^{\mathrm{T}} \\
\overline{\boldsymbol{\Theta}}_{26} &= \boldsymbol{M}_6^{\mathrm{T}} - \boldsymbol{N}_2
\end{aligned}$$

$$\overline{\boldsymbol{\Theta}}_{34} = -\boldsymbol{M}_3 + \boldsymbol{N}_3$$
$$\overline{\boldsymbol{\Theta}}_{44} = -(1-\vartheta)\boldsymbol{P}_5 - \boldsymbol{M}_4 - \boldsymbol{M}_4^{\mathrm{T}} + \boldsymbol{N}_4 + \boldsymbol{N}_4^{\mathrm{T}}$$

$$\overline{\boldsymbol{\Theta}}_{45} = -\boldsymbol{M}_5^{\mathrm{T}} + \boldsymbol{N}_5^{\mathrm{T}}$$
$$\overline{\boldsymbol{\Theta}}_{46} = -\boldsymbol{M}_6^{\mathrm{T}} + \boldsymbol{N}_6^{\mathrm{T}} - \boldsymbol{N}_4$$
$$\overline{\boldsymbol{\Theta}}_{55} = (2l_2 - l_1)\boldsymbol{B}^{\mathrm{T}}(\delta\boldsymbol{P}_2)\boldsymbol{B} - \varpi\left(\boldsymbol{P}_3\boldsymbol{B}^{\mathrm{T}}\boldsymbol{B} + \boldsymbol{B}^{\mathrm{T}}\boldsymbol{B}\boldsymbol{P}_3\right)$$
$$\overline{\boldsymbol{\Theta}}_{66} = -\boldsymbol{P}_4 - \boldsymbol{N}_6 - \boldsymbol{N}_6^{\mathrm{T}}$$

and

$$\boldsymbol{\Theta}_{12} = \begin{bmatrix} 0 & \boldsymbol{P}_2\boldsymbol{A} - \boldsymbol{H}\boldsymbol{C} & 0 & 0 & 0 & 0 & 0 & 0 \\ 0 & 0 & 0 & 0 & \boldsymbol{P}_3 & 0 & 0 & 0 \end{bmatrix}^{\mathrm{T}}$$

$$\boldsymbol{\Theta}_{22} = \operatorname{diag}\left(\left[-\frac{\boldsymbol{P}_2}{\delta(2l_2-l_1)} \quad -\xi_1\boldsymbol{I}\right]\right)$$

When LMIs Eqs. (4.12) and (4.13) are feasible, one can obtain the controller gain matrix K and the observer gains the matrix $L = P_2^{-1}H$.

Proof:

As these states are distributed on different channels of the closed-loop system, one should establish a common Lyapunov function including all these states and eliminating the effect of time-varying input delay and other constraints, such that the observer-based antidisturbance controller design amounts to solving a system of LMIs.

Choose a Lyapunov function candidate as

$$\begin{aligned} V(t) &= \boldsymbol{\eta}(t)^{\mathrm{T}}\boldsymbol{P}_1\boldsymbol{\eta}(t) + \boldsymbol{e}_\eta(t)^{\mathrm{T}}\boldsymbol{P}_2\boldsymbol{e}_\eta(t) + \boldsymbol{e}_\xi(t)^{\mathrm{T}}\boldsymbol{P}_3\boldsymbol{e}_\xi(t) \\ &\quad + \int_{t-l_2}^{t} \boldsymbol{e}_\eta^{\mathrm{T}}(s)\boldsymbol{P}_4\boldsymbol{e}_\eta(s)ds + \int_{t-\tau(t)}^{t} \boldsymbol{e}_\eta^{\mathrm{T}}(s)\boldsymbol{P}_5\boldsymbol{e}_\eta(s)ds \\ &\quad + \int_{-l_2}^{0}\int_{t+\theta}^{t} \dot{\boldsymbol{e}}_\eta^{\mathrm{T}}(s)(\delta\boldsymbol{P}_2)\dot{\boldsymbol{e}}_\eta(s)dsd\theta \\ &\quad + \int_{-l_2}^{-l_1}\int_{t+\theta}^{t} \dot{\boldsymbol{e}}_\eta^{\mathrm{T}}(s)(\delta\boldsymbol{P}_2)\dot{\boldsymbol{e}}_\eta(s)dsd\theta \end{aligned}$$

Taking the derivative of $V(t)$ with respect to time yields

$$\begin{aligned} \dot{V}(t) &= \dot{\boldsymbol{\eta}}(t)^{\mathrm{T}}\boldsymbol{P}_1\boldsymbol{\eta}(t) + \boldsymbol{\eta}(t)^{\mathrm{T}}\boldsymbol{P}_1\dot{\boldsymbol{\eta}}(t) + \dot{\boldsymbol{e}}_\eta(t)^{\mathrm{T}}\boldsymbol{P}_2\boldsymbol{e}_\eta(t) + \boldsymbol{e}_\eta(t)^{\mathrm{T}}\boldsymbol{P}_2\dot{\boldsymbol{e}}_\eta(t) \\ &\quad + \dot{\boldsymbol{e}}_\xi(t)^{\mathrm{T}}\boldsymbol{P}_3\boldsymbol{e}_\xi(t) + \boldsymbol{e}_\xi(t)^{\mathrm{T}}\boldsymbol{P}_3\dot{\boldsymbol{e}}_\xi(t) + \boldsymbol{e}_\eta^{\mathrm{T}}(s)\boldsymbol{P}_4\boldsymbol{e}_\eta(s)\Big|_{t-l_2}^{t} + \boldsymbol{e}_\eta^{\mathrm{T}}(t)\boldsymbol{P}_5\boldsymbol{e}_\eta(t) \end{aligned}$$

$$
\begin{aligned}
&- (1 - \dot{\tau}(t))e_\eta^{\mathrm{T}}(t - \tau(t))\boldsymbol{P}_5 e_\eta(t - \tau(t)) + l_2 \dot{e}_\eta^{\mathrm{T}}(t)(\delta \boldsymbol{P}_2)\dot{e}_\eta(t) \\
&- \int_{t-l_2}^{t} \dot{e}_\eta^{\mathrm{T}}(s)(\delta \boldsymbol{P}_2)\dot{e}_\eta(s)ds + (l_2 - l_1)\dot{e}_\eta^{\mathrm{T}}(t)(\delta \boldsymbol{P}_2)\dot{e}_\eta(t) \\
&- \int_{t-l_2}^{t-l_1} \dot{e}_\eta^{\mathrm{T}}(s)(\delta \boldsymbol{P}_2)\dot{e}_\eta(s)ds
\end{aligned}
$$

In view of Eqs. (4.9)–(4.11), it yields

$$
\begin{aligned}
\dot{V}(t) = &\ \boldsymbol{\eta}(t)^{\mathrm{T}}\left[(\boldsymbol{A} - \boldsymbol{BK})^{\mathrm{T}}\boldsymbol{P}_1 + \boldsymbol{P}_1(\boldsymbol{A} - \boldsymbol{BK})\right]\boldsymbol{\eta}(t) \\
&+ 2\boldsymbol{\eta}(t)^{\mathrm{T}}\boldsymbol{P}_1\boldsymbol{BK}e_\eta(t) + 2\boldsymbol{\eta}(t)^{\mathrm{T}}\boldsymbol{P}_1\boldsymbol{A}_d\boldsymbol{\eta}(t - \tau(t)) \\
&- 2\boldsymbol{\eta}(t)^{\mathrm{T}}\boldsymbol{P}_1\boldsymbol{A}_d e_\eta(t - \tau(t)) + 2\boldsymbol{\eta}(t)^{\mathrm{T}}\boldsymbol{P}_1\boldsymbol{B}e_\xi(t) \\
&+ 2\varpi\boldsymbol{\eta}(t)^{\mathrm{T}}\boldsymbol{P}_1\boldsymbol{BB}^{\mathrm{T}}e_x(t) + e_\eta(t)^{\mathrm{T}}\left[(\boldsymbol{A} - \boldsymbol{LC})^{\mathrm{T}}\boldsymbol{P}_2 + \boldsymbol{P}_2(\boldsymbol{A} - \boldsymbol{LC})\right]e_\eta(t) \\
&+ 2e_\eta(t)^{\mathrm{T}}\boldsymbol{P}_2\boldsymbol{B}e_\xi(t) + e_\eta(t)^{\mathrm{T}}\left(\varpi\boldsymbol{P}_2\boldsymbol{BB}^{\mathrm{T}} + \varpi\boldsymbol{BB}^{\mathrm{T}}\boldsymbol{P}_2\right)e_\eta(t) \\
&+ 2e_\xi(t)^{\mathrm{T}}\boldsymbol{P}_3\dot{\boldsymbol{w}}(t) - \varpi e_\xi(t)^{\mathrm{T}}\left(\boldsymbol{P}_3\boldsymbol{B}^{\mathrm{T}}\boldsymbol{B} + \boldsymbol{B}^{\mathrm{T}}\boldsymbol{B}\boldsymbol{P}_3\right)e_\xi(t) \\
&- 2\varpi e_\xi(t)^{\mathrm{T}}\boldsymbol{P}_3\boldsymbol{B}^{\mathrm{T}}\left[\boldsymbol{A}e_\eta(t) + \varpi\boldsymbol{BB}^{\mathrm{T}}e_\eta(t)\right] + e_\eta^{\mathrm{T}}(t)\boldsymbol{P}_4 e_\eta(t) \\
&- e_\eta^{\mathrm{T}}(t - l_2)\boldsymbol{P}_4 e_\eta(t - l_2) + e_\eta^{\mathrm{T}}(t)\boldsymbol{P}_5 e_\eta(t) \\
&- (1 - \dot{\tau}(t))e_\eta^{\mathrm{T}}(t - \tau(t))\boldsymbol{P}_5 e_\eta(t - \tau(t)) + l_2\dot{e}_\eta^{\mathrm{T}}(t)(\delta\boldsymbol{P}_2)\dot{e}_\eta(t) \\
&- \int_{t-l_2}^{t} \dot{e}_\eta^{\mathrm{T}}(s)(\delta\boldsymbol{P}_2)\dot{e}_\eta(s)ds + (l_2 - l_1)\dot{e}_\eta^{\mathrm{T}}(t)(\delta\boldsymbol{P}_2)\dot{e}_\eta(t) \\
&- \int_{t-l_2}^{t-l_1} \dot{e}_\eta^{\mathrm{T}}(s)(\delta\boldsymbol{P}_2)\dot{e}_\eta(s)ds
\end{aligned}
$$

Define $\boldsymbol{\chi}(t) = [\boldsymbol{\eta}(t)^{\mathrm{T}}\ e_\eta(t)^{\mathrm{T}}\ \boldsymbol{\eta}(t - \tau(t))^{\mathrm{T}}\ e_\eta(t - \tau(t))^{\mathrm{T}}\ e_\xi(t)^{\mathrm{T}}\ e_\eta(t - l_2)^{\mathrm{T}}]^{\mathrm{T}}$, $\boldsymbol{M} = [\boldsymbol{M}_1^{\mathrm{T}}\ \boldsymbol{M}_2^{\mathrm{T}}\ \boldsymbol{M}_3^{\mathrm{T}}\ \boldsymbol{M}_4^{\mathrm{T}}\ \boldsymbol{M}_5^{\mathrm{T}}\ \boldsymbol{M}_6^{\mathrm{T}}]^{\mathrm{T}}$, and $\boldsymbol{N} = [\boldsymbol{N}_1^{\mathrm{T}}\ \boldsymbol{N}_2^{\mathrm{T}}\ \boldsymbol{N}_3^{\mathrm{T}}\ \boldsymbol{N}_4^{\mathrm{T}}\ \boldsymbol{N}_5^{\mathrm{T}}\ \boldsymbol{N}_6^{\mathrm{T}}]^{\mathrm{T}}$, where $\boldsymbol{M}_i$, $\boldsymbol{N}_i$, $i = 1, 2, 3, 4, 5, 6$ denote the components of $\boldsymbol{M}$ and $\boldsymbol{N}$ respectively.

According to the Schur complement lemma and the given constraint conditions, one has

$$
\begin{aligned}
\dot{V}(t) \le &\ \boldsymbol{\eta}(t)^{\mathrm{T}}\left[(\boldsymbol{A} - \boldsymbol{BK})^{\mathrm{T}}\boldsymbol{P}_1 + \boldsymbol{P}_1(\boldsymbol{A} - \boldsymbol{BK})\right]\boldsymbol{\eta}(t) + 2\boldsymbol{\eta}(t)^{\mathrm{T}}\boldsymbol{P}_1\boldsymbol{BK}e_\eta(t) \\
&+ 2\boldsymbol{\eta}(t)^{\mathrm{T}}\boldsymbol{P}_1\boldsymbol{A}_d\boldsymbol{\eta}(t - \tau(t)) \\
&- 2\boldsymbol{\eta}(t)^{\mathrm{T}}\boldsymbol{P}_1\boldsymbol{A}_d e_\eta(t - \tau(t)) + 2\boldsymbol{\eta}(t)^{\mathrm{T}}\boldsymbol{P}_1\boldsymbol{B}e_\xi(t) + 2\varpi\boldsymbol{\eta}(t)^{\mathrm{T}}\boldsymbol{P}_1\boldsymbol{BB}^{\mathrm{T}}e_\eta(t) \\
&+ e_\eta(t)^{\mathrm{T}}\left[(\boldsymbol{A} - \boldsymbol{LC})^{\mathrm{T}}\boldsymbol{P}_2 + \boldsymbol{P}_2(\boldsymbol{A} - \boldsymbol{LC})\right]e_\eta(t) + 2e_\eta(t)^{\mathrm{T}}\boldsymbol{P}_2\boldsymbol{B}e_\xi(t)
\end{aligned}
$$

$$
\begin{aligned}
&+e_\eta(t)^{\mathrm{T}}\left(\varpi \boldsymbol{P}_2\boldsymbol{B}\boldsymbol{B}^{\mathrm{T}}+\varpi \boldsymbol{B}\boldsymbol{B}^{\mathrm{T}}\boldsymbol{P}_2\right)e_\eta(t)+\xi_1^{-1}e_\xi(t)^{\mathrm{T}}\boldsymbol{P}_3\boldsymbol{P}_3e_\xi(t)+\xi_1\kappa_1^2\\
&-\varpi e_\xi(t)^{\mathrm{T}}\left(\boldsymbol{P}_3\boldsymbol{B}^{\mathrm{T}}\boldsymbol{B}+\boldsymbol{B}^{\mathrm{T}}\boldsymbol{B}\boldsymbol{P}_3\right)e_\xi(t)-2\varpi e_\xi(t)^{\mathrm{T}}\boldsymbol{P}_3\boldsymbol{B}^{\mathrm{T}}\boldsymbol{A}e_\eta(t)\\
&-2\varpi^2e_\xi(t)^{\mathrm{T}}\boldsymbol{P}_3\boldsymbol{B}^{\mathrm{T}}\boldsymbol{B}\boldsymbol{B}^{\mathrm{T}}e_\eta(t)+e_\eta^{\mathrm{T}}(t)\boldsymbol{P}_4e_\eta(t)-e_\eta^{\mathrm{T}}(t-l_2)\boldsymbol{P}_4e_\eta(t-l_2)\\
&+e_\eta^{\mathrm{T}}(t)\boldsymbol{P}_5e_\eta(t)-(1-\vartheta)e_\eta^{\mathrm{T}}(t-\tau(t))\boldsymbol{P}_5e_\eta(t-\tau(t))+l_2\dot{e}_\eta^{\mathrm{T}}(t)(\delta\boldsymbol{P}_2)\dot{e}_\eta(t)\\
&-\int_{t-\tau(t)}^{t}\dot{e}_\eta^{T}(s)(\delta\boldsymbol{P}_2)\dot{e}_\eta(s)ds+(l_2-l_1)\dot{e}_\eta^{T}(t)(\delta\boldsymbol{P}_2)\dot{e}_\eta(t)\\
&-\int_{t-l_2}^{t-\tau(t)}\dot{e}_\eta^{T}(s)(\delta\boldsymbol{P}_2)\dot{e}_\eta(s)ds+2\chi(t)^{\mathrm{T}}\\
&\quad\boldsymbol{M}\left[e_\eta(t)-e_\eta(t-\tau(t))-\int_{t-\tau(t)}^{t}\dot{e}_\eta(s)ds\right]\\
&+2\chi(t)^{\mathrm{T}}\boldsymbol{N}\left[e_\eta(t-\tau(t))-e_\eta(t-l_2)-\int_{t-l_2}^{t-\tau(t)}\dot{e}_\eta(s)ds\right]
\end{aligned}
$$

Since $l_1 \le \tau(t) \le l_2$, it can be derived that

$$
\begin{cases}
l_2\chi(t)^{\mathrm{T}}\boldsymbol{M}(\delta\boldsymbol{P}_2)^{-1}\boldsymbol{M}^{\mathrm{T}}\chi(t)\ge\tau(t)\chi(t)^{\mathrm{T}}\boldsymbol{M}(\delta\boldsymbol{P}_2)^{-1}\boldsymbol{M}^{\mathrm{T}}\chi(t)\\
(l_2-l_1)\chi(t)^{\mathrm{T}}\boldsymbol{N}(\delta\boldsymbol{P}_2)^{-1}\boldsymbol{N}^{\mathrm{T}}\chi(t)\ge(l_2-\tau(t))\chi(t)^{\mathrm{T}}\boldsymbol{N}(\delta\boldsymbol{P}_2)^{-1}\boldsymbol{N}^{\mathrm{T}}\chi(t)
\end{cases}
$$

Then,

$$
\begin{aligned}
&-\int_{t-\tau(t)}^{t}\left[\dot{e}_\eta^{\mathrm{T}}(s)(\delta\boldsymbol{P}_2)\dot{e}_\eta(s)+2\chi(t)^{\mathrm{T}}\boldsymbol{M}\dot{e}_\eta(s)\right]ds\\
&=-\int_{t-\tau(t)}^{t}\left[\chi(t)^{\mathrm{T}}\boldsymbol{M}+\dot{e}_\eta(s)^{\mathrm{T}}(\delta\boldsymbol{P}_2)\right](\delta\boldsymbol{P}_2)^{-1}\left[\boldsymbol{M}^{\mathrm{T}}\chi(t)+(\delta\boldsymbol{P}_2)\dot{e}_\eta(s)\right]ds\\
&\quad+\tau(t)\chi(t)^{\mathrm{T}}\boldsymbol{M}(\delta\boldsymbol{P}_2)^{-1}\boldsymbol{M}^{\mathrm{T}}\chi(t)\\
&\le-\int_{t-\tau(t)}^{t}\left[\chi(t)^{\mathrm{T}}\boldsymbol{M}+\dot{e}_\eta(s)^{\mathrm{T}}(\delta\boldsymbol{P}_2)\right](\delta\boldsymbol{P}_2)^{-1}\left[\boldsymbol{M}^{\mathrm{T}}\chi(t)+(\delta\boldsymbol{P}_2)\dot{e}_\eta(s)\right]ds\\
&\quad+l_2\chi(t)^{\mathrm{T}}\boldsymbol{M}(\delta\boldsymbol{P}_2)^{-1}\boldsymbol{M}^{\mathrm{T}}\chi(t)
\end{aligned}
$$

and

$$
\begin{aligned}
&-\int_{t-l_2}^{t-\tau(t)}\left[\dot{e}_\eta^{\mathrm{T}}(s)(\delta\boldsymbol{P}_2)\dot{e}_\eta(s)+2\chi(t)^{T}\boldsymbol{N}\dot{e}_\eta(s)\right]ds\\
&=-\int_{t-l_2}^{t-\tau(t)}\left[\chi(t)^{\mathrm{T}}\boldsymbol{N}+\dot{e}_\eta(s)^{\mathrm{T}}(\delta\boldsymbol{P}_2)\right](\delta\boldsymbol{P}_2)^{-1}\left[\boldsymbol{N}^{\mathrm{T}}\chi(t)+(\delta\boldsymbol{P}_2)\dot{e}_\eta(s)\right]ds
\end{aligned}
$$

$$+(l_2-\tau(t))\chi(t)^{\mathrm{T}}\boldsymbol{N}(\delta\boldsymbol{P}_2)^{-1}\boldsymbol{N}^{\mathrm{T}}\chi(t)$$

$$\leq -\int_{t-l_2}^{t-\tau(t)}\left[\chi(t)^{\mathrm{T}}\boldsymbol{N}+\dot{e}_\eta(s)^{\mathrm{T}}(\delta\boldsymbol{P}_2)\right](\delta\boldsymbol{P}_2)^{-1}\left[\boldsymbol{N}^{\mathrm{T}}\chi(t)+(\delta\boldsymbol{P}_2)\dot{e}_\eta(s)\right]ds$$

$$+(l_2-l_1)\chi(t)^{\mathrm{T}}\boldsymbol{N}(\delta\boldsymbol{P}_2)^{-1}\boldsymbol{N}^{\mathrm{T}}\chi(t)$$

According to Eq. (4.10), one knows

$$\dot{e}_\eta^{\mathrm{T}}(t)(\delta\boldsymbol{P}_2)\dot{e}_\eta(t)$$

$$=\boldsymbol{e}_\eta(t)^{\mathrm{T}}(\boldsymbol{A}-\boldsymbol{LC})^{\mathrm{T}}(\delta\boldsymbol{P}_2)(\boldsymbol{A}-\boldsymbol{LC})\boldsymbol{e}_\eta(t)+2\boldsymbol{e}_\eta(t)^{\mathrm{T}}(\boldsymbol{A}-\boldsymbol{LC})^{\mathrm{T}}(\delta\boldsymbol{P}_2)\boldsymbol{B}\boldsymbol{e}_\xi(t)$$

$$+\varpi\boldsymbol{e}_\eta(t)^{\mathrm{T}}\left[(\boldsymbol{A}-\boldsymbol{LC})^{\mathrm{T}}(\delta\boldsymbol{P}_2)\boldsymbol{BB}^{\mathrm{T}}+\boldsymbol{BB}^{\mathrm{T}}(\delta\boldsymbol{P}_2)(\boldsymbol{A}-\boldsymbol{LC})\right]\boldsymbol{e}_\eta(t)$$

$$+\boldsymbol{e}_\xi(t)^{\mathrm{T}}\boldsymbol{B}^{\mathrm{T}}(\delta\boldsymbol{P}_2)\boldsymbol{B}\boldsymbol{e}_\xi(t)2\varpi\boldsymbol{e}_\xi(t)^{\mathrm{T}}\boldsymbol{B}^{\mathrm{T}}(\delta\boldsymbol{P}_2)\boldsymbol{BB}^{\mathrm{T}}\boldsymbol{e}_\eta(t)$$

$$+\varpi^2\boldsymbol{e}_\eta(t)\boldsymbol{BB}^{\mathrm{T}}(\delta\boldsymbol{P}_2)\boldsymbol{BB}^{\mathrm{T}}\boldsymbol{e}_\eta(t)$$

Therefore the first derivative of the Lyapunov function becomes

$$\dot{V}(t)\leq\chi(t)^{\mathrm{T}}\begin{bmatrix}\boldsymbol{\Pi}_{11} & \boldsymbol{\Pi}_{12} & \boldsymbol{P}_1\boldsymbol{A}_d & -\boldsymbol{P}_1\boldsymbol{A}_d & \boldsymbol{P}_1\boldsymbol{B} & 0\\ * & \boldsymbol{\Pi}_{22} & 0 & 0 & \boldsymbol{\Pi}_{25} & 0\\ * & * & 0 & 0 & 0 & 0\\ * & * & * & -(1-\vartheta)\boldsymbol{P}_5 & 0 & 0\\ * & * & * & * & \boldsymbol{\Pi}_{55} & 0\\ * & * & * & * & * & -\boldsymbol{P}_4\end{bmatrix}\chi(t)$$

$$+\xi_1\kappa_1^2+l_2\dot{e}_\eta^{\mathrm{T}}(t)(\delta\boldsymbol{P}_2)\dot{e}_\eta(t)+(l_2-l_1)\dot{e}_\eta^{\mathrm{T}}(t)(\delta\boldsymbol{P}_2)\dot{e}_\eta(t)$$

$$+2\chi(t)^{\mathrm{T}}\boldsymbol{M}\left[\boldsymbol{e}_\eta(t)-\boldsymbol{e}_\eta(t-\tau(t))\right]+2\chi(t)^{\mathrm{T}}\boldsymbol{N}\left[\boldsymbol{e}_\eta(t-\tau(t))-\boldsymbol{e}_\eta(t-l_2)\right]$$

$$-\int_{t-\tau(t)}^{t}\left[\chi(t)^{\mathrm{T}}\boldsymbol{M}+\dot{e}_\eta(s)(\delta\boldsymbol{P}_2)\right](\delta\boldsymbol{P}_2)^{-1}\left[\boldsymbol{M}^{\mathrm{T}}\chi(t)+(\delta\boldsymbol{P}_2)^{\mathrm{T}}\dot{e}_\eta(s)^{\mathrm{T}}\right]ds$$

$$-\int_{t-l_2}^{t-\tau(t)}\left[\chi(t)^{\mathrm{T}}\boldsymbol{N}+\dot{e}_\eta(s)(\delta\boldsymbol{P}_2)\right](\delta\boldsymbol{P}_2)^{-1}\left[\boldsymbol{N}^{\mathrm{T}}\chi(t)+(\delta\boldsymbol{P}_2)^{\mathrm{T}}\dot{e}_\eta(s)^{\mathrm{T}}\right]ds$$

$$+l_2\chi(t)^{\mathrm{T}}\boldsymbol{M}(\delta\boldsymbol{P}_2)^{-1}\boldsymbol{M}^{\mathrm{T}}\chi(t)+(l_2-l_1)\chi(t)^{\mathrm{T}}\boldsymbol{N}(\delta\boldsymbol{P}_2)^{-1}\boldsymbol{N}^{\mathrm{T}}\chi(t)$$

where

$$\boldsymbol{\Pi}_{11}=(\boldsymbol{A}-\boldsymbol{BK})^{\mathrm{T}}\boldsymbol{P}_1+\boldsymbol{P}_1(\boldsymbol{A}-\boldsymbol{BK})$$

$$\boldsymbol{\Pi}_{12}=\boldsymbol{P}_1\boldsymbol{BK}+\varpi\boldsymbol{P}_1\boldsymbol{BB}^{\mathrm{T}}$$

$$\boldsymbol{\Pi}_{22}=(\boldsymbol{A}-\boldsymbol{LC})^{\mathrm{T}}\boldsymbol{P}_2+\boldsymbol{P}_2(\boldsymbol{A}-\boldsymbol{LC})$$

$$+\varpi\boldsymbol{P}_2\boldsymbol{BB}^{\mathrm{T}}+\varpi\boldsymbol{BB}^{\mathrm{T}}\boldsymbol{P}_2+\boldsymbol{P}_4+\boldsymbol{P}_5$$

$$\boldsymbol{\Pi}_{25}=\boldsymbol{P}_2\boldsymbol{B}-\varpi\boldsymbol{A}^{\mathrm{T}}\boldsymbol{BP}_3-\varpi^2\boldsymbol{BB}^{\mathrm{T}}\boldsymbol{BP}_3$$

$$\boldsymbol{\Pi}_{55} = -\varpi\left(\boldsymbol{P}_3\boldsymbol{B}^{\mathrm{T}}\boldsymbol{B} + \boldsymbol{B}^{\mathrm{T}}\boldsymbol{B}\boldsymbol{P}_3\right) + \xi_1^{-1}\boldsymbol{P}_3\boldsymbol{P}_3$$

Namely,

$$\dot{V}(t) \le \boldsymbol{\chi}(t)^{\mathrm{T}} \begin{bmatrix} \boldsymbol{\Pi}_{11} & \boldsymbol{\Pi}_{12} & \boldsymbol{P}_1\boldsymbol{A}_d & -\boldsymbol{P}_1\boldsymbol{A}_d & \boldsymbol{P}_1\boldsymbol{B} & 0 \\ * & \widetilde{\boldsymbol{\Pi}}_{22} & 0 & 0 & \widetilde{\boldsymbol{\Pi}}_{25} & 0 \\ * & * & 0 & 0 & 0 & 0 \\ * & * & * & -(1-\vartheta)\boldsymbol{P}_5 & 0 & 0 \\ * & * & * & * & \sim\boldsymbol{\Pi}_{55} & 0 \\ * & * & * & * & * & -\boldsymbol{P}_4 \end{bmatrix} \boldsymbol{\chi}(t)$$

$$+ \xi_1\kappa_1^2 + 2\boldsymbol{\chi}(t)^{\mathrm{T}}\boldsymbol{M}\left[\boldsymbol{e}_\eta(t) - \boldsymbol{e}_\eta(t-\tau(t))\right] + l_2\boldsymbol{\chi}(t)^{\mathrm{T}}\boldsymbol{M}(\delta\boldsymbol{P}_2)^{-1}\boldsymbol{M}^{\mathrm{T}}\boldsymbol{\chi}(t)$$
$$+ 2\boldsymbol{\chi}(t)^{\mathrm{T}}\boldsymbol{N}\left[\boldsymbol{e}_\eta(t-\tau(t)) - \boldsymbol{e}_\eta(t-l_2)\right] + (l_2 - l_1)\boldsymbol{\chi}(t)^{\mathrm{T}}\boldsymbol{N}(\delta\boldsymbol{P}_2)^{-1}\boldsymbol{N}^{\mathrm{T}}\boldsymbol{\chi}(t)$$

where

$$\widetilde{\boldsymbol{\Pi}}_{22} = \boldsymbol{\Pi}_{22} + (2l_2 - l_1)(\boldsymbol{A}-\boldsymbol{LC})^{\mathrm{T}}(\delta\boldsymbol{P}_2)(\boldsymbol{A}-\boldsymbol{LC}) + (2l_2-l_1)\varpi^2\boldsymbol{B}\boldsymbol{B}^{\mathrm{T}}(\delta\boldsymbol{P}_2)\boldsymbol{B}\boldsymbol{B}^{\mathrm{T}}$$
$$+ (2l_2 - l_1)\varpi(\boldsymbol{A} - \boldsymbol{LC})^{\mathrm{T}}(\delta\boldsymbol{P}_2)\boldsymbol{B}\boldsymbol{B}^{\mathrm{T}} + (2l_2 - l_1)\varpi\boldsymbol{B}\boldsymbol{B}^{\mathrm{T}}(\delta\boldsymbol{P}_2)(\boldsymbol{A} - \boldsymbol{LC})$$
$$\widetilde{\boldsymbol{\Pi}}_{25} = \boldsymbol{\Pi}_{25} + (2l_2 - l_1)(\boldsymbol{A} - \boldsymbol{LC})^{\mathrm{T}}(\delta\boldsymbol{P}_2)\boldsymbol{B} + (2l_2 - l_1)\varpi\boldsymbol{B}\boldsymbol{B}^{\mathrm{T}}(\delta\boldsymbol{P}_2)\boldsymbol{B}$$
$$\widetilde{\boldsymbol{\Pi}}_{55} = \boldsymbol{\Pi}_{55} + (2l_2 - l_1)\boldsymbol{B}^{\mathrm{T}}(\delta\boldsymbol{P}_2)\boldsymbol{B}$$

Then,

$$\dot{V}(t) \le \boldsymbol{\chi}(t)^{\mathrm{T}}\left\{\boldsymbol{\Theta}_N + l_2\boldsymbol{M}(\delta\boldsymbol{P}_2)^{-1}\boldsymbol{M}^{\mathrm{T}} + (l_2 - l_1)\boldsymbol{N}(\delta\boldsymbol{P}_2)^{-1}\boldsymbol{N}^{\mathrm{T}}\right\}\boldsymbol{\chi}(t) + \xi_1\kappa_1^2$$

where

$$\boldsymbol{\Theta}_N = \begin{bmatrix} \boldsymbol{\Theta}_{N11} & \boldsymbol{\Theta}_{N12} \\ * & \boldsymbol{\Theta}_{N22} \end{bmatrix}$$

and where

$$\boldsymbol{\Theta}_{N11} = \begin{bmatrix} \boldsymbol{\Pi}_{11} & \boldsymbol{\Pi}_{12} + \boldsymbol{M}_1 & \boldsymbol{P}_1\boldsymbol{A}_d \\ * & \widetilde{\boldsymbol{\Pi}}_{22} + \boldsymbol{M}_2 + \boldsymbol{M}_2^{\mathrm{T}} & \boldsymbol{M}_3^{\mathrm{T}} \\ * & * & 0 \end{bmatrix}$$

$$\boldsymbol{\Theta}_{N12} = \begin{bmatrix} -\boldsymbol{P}_1\boldsymbol{A}_d - \boldsymbol{M}_1 + \boldsymbol{N}_1 & \boldsymbol{P}_1\boldsymbol{B} & -\boldsymbol{N}_1 \\ \boldsymbol{M}_4^{\mathrm{T}} - \boldsymbol{M}_2 + \boldsymbol{N}_2 & \widetilde{\boldsymbol{\Pi}}_{25} + \boldsymbol{M}_5^{\mathrm{T}} & \boldsymbol{M}_6^{\mathrm{T}} - \boldsymbol{N}_2 \\ -\boldsymbol{M}_3 + \boldsymbol{N}_3 & 0 & -\boldsymbol{N}_3 \end{bmatrix}$$

$$\boldsymbol{\Theta}_{N22} = \begin{bmatrix} -(1-\vartheta)\boldsymbol{P}_5 - \boldsymbol{M}_4 - \boldsymbol{M}_4^{\mathrm{T}} + \boldsymbol{N}_4 + \boldsymbol{N}_4^{\mathrm{T}} & -\boldsymbol{M}_5^{\mathrm{T}} + \boldsymbol{N}_5^{\mathrm{T}} & -\boldsymbol{M}_6^{\mathrm{T}} + \boldsymbol{N}_6^{\mathrm{T}} - \boldsymbol{N}_4 \\ * & \widetilde{\boldsymbol{\Pi}}_{55} & -\boldsymbol{N}_5 \\ * & * & -\boldsymbol{P}_4 - \boldsymbol{N}_6 - \boldsymbol{N}_6^{\mathrm{T}} \end{bmatrix}$$

Denote $\boldsymbol{\Theta}_P = -\{\boldsymbol{\Theta}_N + l_2\boldsymbol{M}(\delta\boldsymbol{P}_2)^{-1}\boldsymbol{M}^{\mathrm{T}} + (l_2 - l_1)\boldsymbol{N}(\delta\boldsymbol{P}_2)^{-1}\boldsymbol{N}^{\mathrm{T}}\}$, one knows $\dot{V}(t) \leq -\chi(t)^{\mathrm{T}}\boldsymbol{\Theta}_P\chi(t) + \xi_1\kappa_1^2$. Let $\gamma = \xi_1\kappa_1^2$, it can be concluded that if $\boldsymbol{\Theta}_P > 0$, then

$$\dot{V}(t) \leq -\lambda_{\min}(\boldsymbol{\Theta}_P)||\chi||^2 + \gamma$$

Furthermore, define a set S described by

$$S = \left\{\chi(t) | \lambda_{\min}(\boldsymbol{\Theta}_P)||\chi||^2 > \gamma\right\}$$

Let $\bar{S}$ be the complement of S. It follows that $\dot{V}(t) < 0$, $\chi(t) \in S$, and the state $\chi(t)$ will converge asymptotically to $\bar{S}$. With Lemma 1.3, $\boldsymbol{\Theta}_P > 0$ equals the following inequality:

$$\begin{bmatrix} \boldsymbol{\Theta}_N & \boldsymbol{M} & \boldsymbol{N} \\ * & -\frac{\delta\boldsymbol{P}_2}{l_2} & 0 \\ * & * & -\frac{\delta\boldsymbol{P}_2}{l_2 - l_1} \end{bmatrix} < 0$$

According to Lemma 1.3, and defining $\boldsymbol{H} = \boldsymbol{P}_2\boldsymbol{L}$, one can obtain

$$\begin{bmatrix} \boldsymbol{\Theta}_{11} & \boldsymbol{\Theta}_{12} \\ * & \boldsymbol{\Theta}_{22} \end{bmatrix} < 0$$

where

$$\boldsymbol{\Theta}_{11} = \begin{bmatrix} \overline{\boldsymbol{\Theta}}_{11} & \overline{\boldsymbol{\Theta}}_{12} & \boldsymbol{P}_1\boldsymbol{A}_d & \overline{\boldsymbol{\Theta}}_{14} & \boldsymbol{P}_1\boldsymbol{B} & -\boldsymbol{N}_1 & \boldsymbol{M}_1 & \boldsymbol{N}_1 \\ * & \overline{\boldsymbol{\Theta}}_{22} & \boldsymbol{M}_3^{\mathrm{T}} & \overline{\boldsymbol{\Theta}}_{24} & \overline{\boldsymbol{\Theta}}_{25} & \overline{\boldsymbol{\Theta}}_{26} & \boldsymbol{M}_2 & \boldsymbol{N}_2 \\ * & * & 0 & \overline{\boldsymbol{\Theta}}_{34} & 0 & -\boldsymbol{N}_3 & \boldsymbol{M}_3 & \boldsymbol{N}_3 \\ * & * & * & \overline{\boldsymbol{\Theta}}_{44} & \overline{\boldsymbol{\Theta}}_{45} & \overline{\boldsymbol{\Theta}}_{46} & \boldsymbol{M}_4 & \boldsymbol{N}_4 \\ * & * & * & * & \overline{\boldsymbol{\Theta}}_{55} & -\boldsymbol{N}_5 & \boldsymbol{M}_5 & \boldsymbol{N}_5 \\ * & * & * & * & * & \overline{\boldsymbol{\Theta}}_{66} & \boldsymbol{M}_6 & \boldsymbol{N}_6 \\ * & * & * & * & * & * & -\frac{\delta\boldsymbol{P}_2}{l_2} & 0 \\ * & * & * & * & * & * & * & -\frac{\delta\boldsymbol{P}_2}{l_2 - l_1} \end{bmatrix}$$

and where

$$\overline{\boldsymbol{\Theta}}_{11} = (\boldsymbol{A} - \boldsymbol{B}\boldsymbol{K})^{\mathrm{T}}\boldsymbol{P}_1 + \boldsymbol{P}_1(\boldsymbol{A} - \boldsymbol{B}_1\boldsymbol{K})$$
$$\overline{\boldsymbol{\Theta}}_{12} = \boldsymbol{P}_1\boldsymbol{B}\boldsymbol{K} + \varpi\boldsymbol{P}_1\boldsymbol{B}\boldsymbol{B}^{\mathrm{T}} + \boldsymbol{M}_1$$
$$\overline{\boldsymbol{\Theta}}_{14} = -\boldsymbol{P}_1\boldsymbol{A}_d - \boldsymbol{M}_1 + \boldsymbol{N}_1$$
$$\begin{aligned} \overline{\boldsymbol{\Theta}}_{22} = {} & \boldsymbol{A}^{\mathrm{T}}\boldsymbol{P}_2 - \boldsymbol{C}^{\mathrm{T}}\boldsymbol{H}^{\mathrm{T}} + \boldsymbol{P}_2\boldsymbol{A} - \boldsymbol{H}\boldsymbol{C} + \varpi\boldsymbol{P}_2\boldsymbol{B}\boldsymbol{B}^{\mathrm{T}} + \varpi\boldsymbol{B}\boldsymbol{B}^{\mathrm{T}}\boldsymbol{P}_2 \\ & + \boldsymbol{P}_4 + \boldsymbol{P}_5 + (2l_2 - l_1)\varpi\delta\left(\boldsymbol{A}^{\mathrm{T}}\boldsymbol{P}_2 - \boldsymbol{C}^{\mathrm{T}}\boldsymbol{H}^{\mathrm{T}}\right)\boldsymbol{B}\boldsymbol{B}^{\mathrm{T}} + \boldsymbol{M}_2 + \boldsymbol{M}_2^{\mathrm{T}} \end{aligned}$$

$$+\,(2l_2-l_1)\varpi\delta\boldsymbol{B}\boldsymbol{B}^{\mathrm{T}}(\boldsymbol{P}_2\boldsymbol{A}-\boldsymbol{H}\boldsymbol{C})+(2l_2-l_1)\varpi^2\delta\boldsymbol{B}\boldsymbol{B}^{\mathrm{T}}\boldsymbol{P}_2\boldsymbol{B}\boldsymbol{B}^{\mathrm{T}}$$

$$\overline{\boldsymbol{\Theta}}_{24}=\boldsymbol{M}_4^{\mathrm{T}}-\boldsymbol{M}_2+\boldsymbol{N}_2$$

$$\overline{\boldsymbol{\Theta}}_{25}=(2l_2-l_1)\delta\left(\boldsymbol{A}^{\mathrm{T}}\boldsymbol{P}_2-\boldsymbol{C}^{\mathrm{T}}\boldsymbol{H}^{\mathrm{T}}\right)\boldsymbol{B}+\boldsymbol{P}_2\boldsymbol{B}-\varpi\boldsymbol{A}^{\mathrm{T}}\boldsymbol{B}\boldsymbol{P}_3$$

$$+\,(2l_2-l_1)\varpi\delta\boldsymbol{B}\boldsymbol{B}^{\mathrm{T}}\boldsymbol{P}_2\boldsymbol{B}-\varpi^2\boldsymbol{B}\boldsymbol{B}^{\mathrm{T}}\boldsymbol{B}\boldsymbol{P}_3+\boldsymbol{M}_5^{\mathrm{T}}$$

$$\overline{\boldsymbol{\Theta}}_{26}=\boldsymbol{M}_6^{\mathrm{T}}-\boldsymbol{N}_2$$

$$\overline{\boldsymbol{\Theta}}_{34}=-\boldsymbol{M}_3+\boldsymbol{N}_3$$

$$\overline{\boldsymbol{\Theta}}_{44}=-(1-\vartheta)\boldsymbol{P}_5-\boldsymbol{M}_4-\boldsymbol{M}_4^{\mathrm{T}}+\boldsymbol{N}_4+\boldsymbol{N}_4^{\mathrm{T}}$$

$$\overline{\boldsymbol{\Theta}}_{45}=-\boldsymbol{M}_5^{\mathrm{T}}+\boldsymbol{N}_5^{\mathrm{T}}$$

$$\overline{\boldsymbol{\Theta}}_{46}=-\boldsymbol{M}_6^{\mathrm{T}}+\boldsymbol{N}_6^{\mathrm{T}}-\boldsymbol{N}_4$$

$$\overline{\boldsymbol{\Theta}}_{55}=(2l_2-l_1)\delta\boldsymbol{B}^{\mathrm{T}}\boldsymbol{P}_2\boldsymbol{B}-\varpi\left(\boldsymbol{P}_3\boldsymbol{B}^{\mathrm{T}}\boldsymbol{B}+\boldsymbol{B}^{\mathrm{T}}\boldsymbol{B}\boldsymbol{P}_3\right)$$

$$\overline{\boldsymbol{\Theta}}_{66}=-\boldsymbol{P}_4-\boldsymbol{N}_6-\boldsymbol{N}_6^{\mathrm{T}}$$

and

$$\boldsymbol{\Theta}_{12}=\begin{bmatrix}0 & \boldsymbol{P}_2\boldsymbol{A}-\boldsymbol{H}\boldsymbol{C} & 0 & 0 & 0 & 0 & 0 & 0\\ 0 & 0 & 0 & 0 & P_3 & 0 & 0 & 0\end{bmatrix}^{\mathrm{T}}$$

$$\boldsymbol{\Theta}_{22}=\mathrm{diag}\left(-\frac{\boldsymbol{P}_2}{\delta(2l_2-l_1)}\quad -\xi_1\boldsymbol{I}\right)$$

This completes the proof of Theorem 4.1.

4.4 Simulation test

In this section, the effectiveness of the IO-based controller is illustrated with simulations of a spacecraft electromagnetic docking system. In this example, for the "CM" orbit, the eccentricity $e = 0.15$, the semimajor axis a = 10,000km, the initial true anomaly $\theta(0) = 0$, and m_0 is only representative of a valid mass with a value of 20kg. Whatever the initial guess of m_0, the difference compared with the true value of m_{CT} can be attributed to the matrix uncertainty ΔB or even the total disturbance, and the total disturbance can be estimated for the antidisturbance controller design. The unknown masses of the chaser and target spacecraft are assumed to be $m_{\mathrm{C}} = 40$kg and $m_{\mathrm{T}} = 100$kg, respectively. Choose the initial states as $\eta(0) = [2\mathrm{m}, 3\mathrm{m}, 4\mathrm{m}, 0\mathrm{m/s}, 0\mathrm{m/s}, 0\mathrm{m/s}]^{\mathrm{T}}$, $e_\eta(0) = \eta(0)$, the gravitational parameter $\mu = 3.986 \times 10^5\mathrm{km}^3/\mathrm{s}^2$, and the permeability of free space $\mu_0 = 4\pi \times 10^{-7}\mathrm{N{\cdot}A}^{-2}$; the magnetic moment of the target spacecraft is assumed to be $\boldsymbol{\mu}_{\mathrm{T}} = [-2.29 \quad -8.17 \quad -2.33]^{\mathrm{T}} \times 10^5\mathrm{Am}^2$. Here, an

Table 4.1 The form of fault signal factor.

t/s	[0,2]	[2, 5]	[5,10]	[10,13]	[13,50]
$\overline{f}(t)$/N	0	$\sin(0.1\pi t)$	1	$\cos(0.1\pi t)$	0

actuator additive fault is assumed to occur, i.e., $\boldsymbol{E} = \boldsymbol{B}$, with the fault signals $f(t) = [8\ 9\ 4]^{\mathrm{T}} \times 10^{-2}\overline{f}(t)$, and the form of the fault signal factor $\overline{f}(t)$ is shown in Table 4.1.

The external disturbances for the chaser and target spacecraft used in the simulation are taken to be of the following form:

$$\boldsymbol{w}_{\mathrm{C}}(t) = \begin{bmatrix} 0.1 + 0.2\sin(0.11\pi t) \\ 0.15 + 0.2\cos(0.11\pi t) \\ 0.2 + 0.3\cos\left(0.11\pi t + \pi/3\right) \end{bmatrix} \mathrm{N}$$

$$\boldsymbol{w}_{\mathrm{T}}(t) = \begin{bmatrix} 0.1 + 0.1\sin(0.11\pi t) \\ 0.2 + 0.12\cos\left(0.11\pi t + \pi/4\right) \\ 0.3 + 0.15\cos\left(0.11\pi t + \pi/6\right) \end{bmatrix} \mathrm{N}$$

Meanwhile, choose

$$\boldsymbol{A}_d = 0.25\boldsymbol{A},\ \xi_1 = 0.001,\ l_1 = 0.2,\ l_2 = 0.6,\ \vartheta = 0.3,\ \varpi = 1,\ \delta = 0.1,$$

$$\widehat{\boldsymbol{P}} = 0.1 \times \mathrm{diag}\left([2\ 1\ 3\ 4\ 2\ 5]\right)$$

Using the Yet Another LMI Parser (YALMIP) toolbox with solver MOSEK Optimization Tool to solve inequality Eq. (4.12), one gets

$$\boldsymbol{K} = \begin{bmatrix} 10 & 0 & 0 & 25 & 0 & -0.0003 \\ 0 & 10 & 0 & 0 & 50 & 0 \\ 0 & 0 & 12 & -0.0048 & 0 & 20 \end{bmatrix}$$

Solving inequality Eq. (4.13) and using $\boldsymbol{L} = \boldsymbol{P}_2^{-1}\boldsymbol{H}$ yields

$$\boldsymbol{L} = \begin{bmatrix} 3.2718 & 0.0000 & 0.0000 & 0.9974 & -0.0000 & -0.0000 \\ 0.0000 & 3.2804 & -0.0000 & 0.0000 & 1.0001 & -0.0000 \\ -0.0000 & 0.0000 & 3.2550 & -0.0000 & 0.0000 & 0.9958 \\ 0.6167 & 0.0000 & 0.0002 & 10.0436 & -0.0000 & -0.0008 \\ -0.0000 & 0.5718 & 0.0000 & -0.0000 & 8.1550 & 0.0000 \\ 0.0003 & 0.0000 & 0.8801 & -0.0009 & 0.0000 & 9.7757 \end{bmatrix}$$

Applying the controller and observer to the closed-loop electromagnetic docking system expressed byEqs.(4.9)-(4.11) produces the simulation results of Figs. 4.1 –4.8. As shown in Figs. 4.1– 4.2, the relative position and velocity

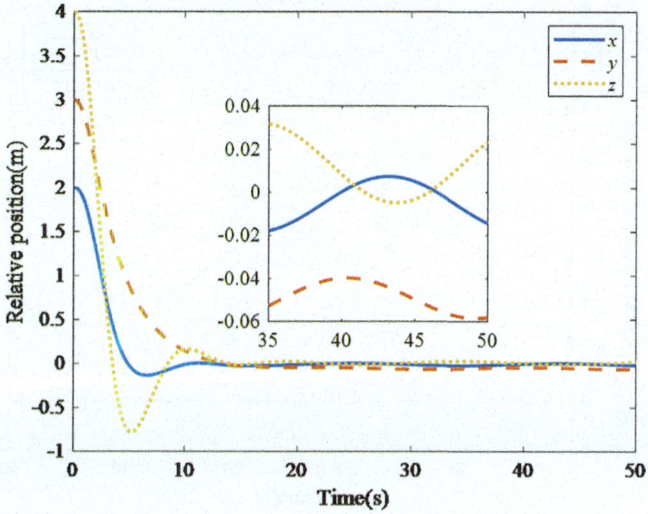

Figure 4.1 Time response of relative position.

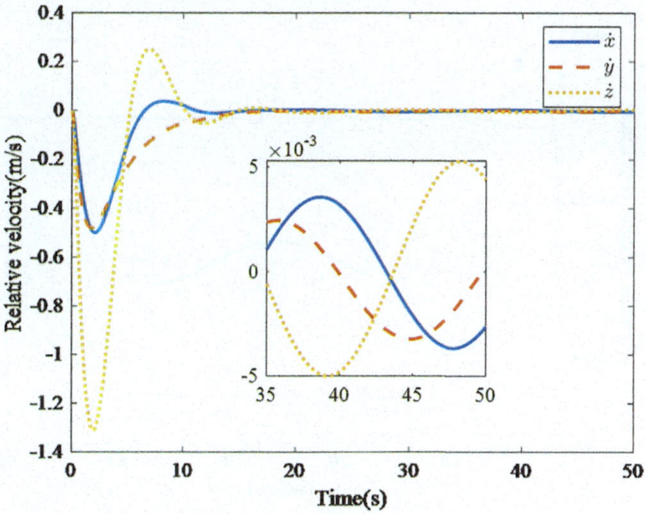

Figure 4.2 Time response of relative velocity.

asymptotically converge to a small bound containing zero within 20s, with steady-state errors less than 6×10^{-2}m and 5×10^{-3}m/s, respectively, thus can be safely neglected. In addition, the relative velocity satisfies the requirement of soft docking. Figs. 4.3–4.4 plot the estimation errors of relative position

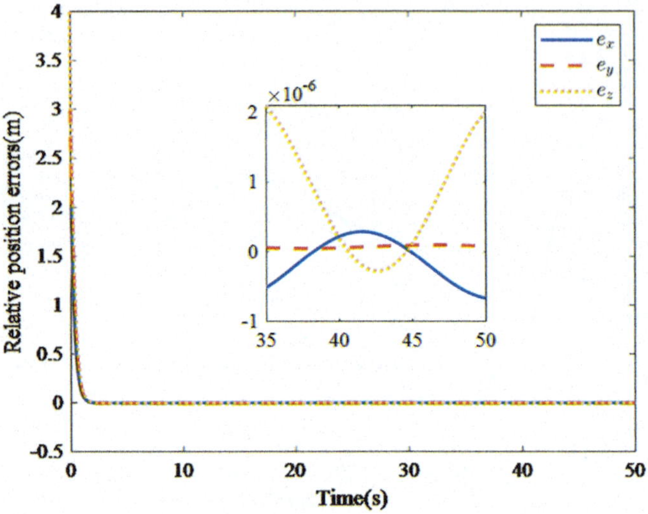

Figure 4.3 Estimation errors of relative position.

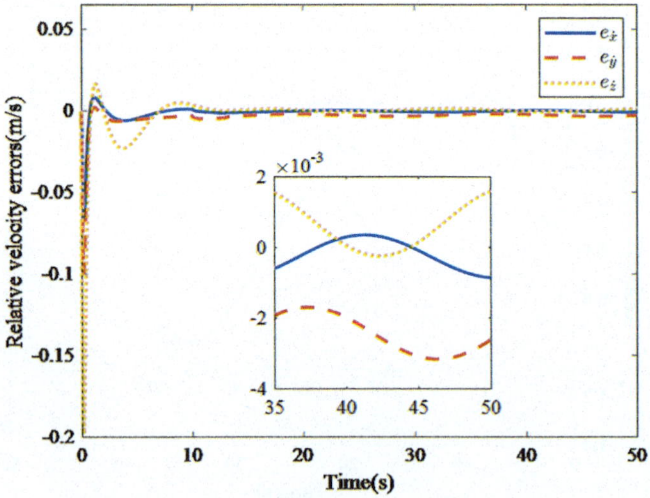

Figure 4.4 Estimation errors of relative velocity.

and velocity, which suggest that they are successfully estimated with steady-state errors less than 2×10^{-6}m and 4×10^{-3}m/s, respectively. It can be concluded that the IO developed in this chapter can effectively estimate the relative motion information, with a fast convergence speed. The sphere

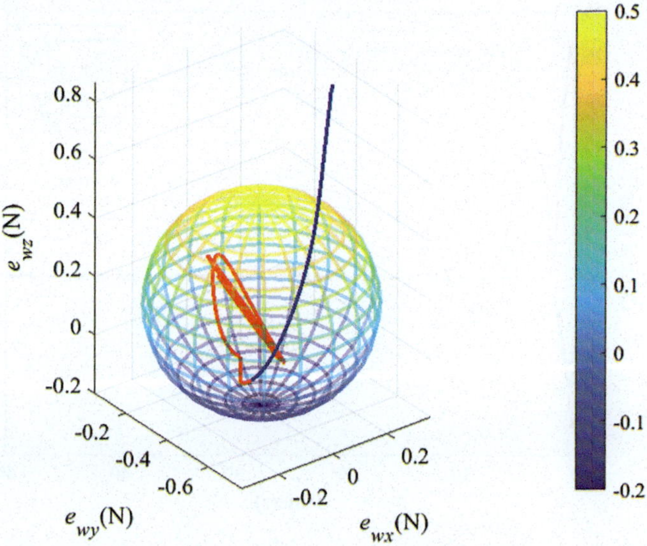

Figure 4.5 Estimation errors convergence set of total disturbance.

with a radius of 0.35N in Fig. 4.5 means that the estimation error of total disturbance will converge to a small neighborhood containing the equilibrium, wherein the color-bar represents the magnitude of disturbance estimation errors. The blue curve denotes estimation error out of the neighborhood, while the red curve represents when the estimation error arrives in the convergence set, it will be attracted and stay in the sphere. The magnitude of control force is plotted in Fig. 4.6, from which one can see that the magnitude of control force along each axis is less than 18N. It is obvious that the 2-norm of the control force is less than 28N. It is noted that the proposed scheme in this chapter can achieve a spacecraft electromagnetic docking mission in an elliptical orbit with high accuracy and very good performance in the presence of external disturbances, fault signals, input delay, unknown mass, and elliptical eccentricity. Thus the simulation results verify the theoretical analysis and demonstrate the effectiveness of the presented ADRC scheme.

To further illustrate the motion of chaser and target spacecraft relative to "CM," Fig. 4.7 is presented, wherein the color bar represents the simulation time. As one can see, the chaser spacecraft occupies a longer relative motion distance than the target spacecraft. This is because the "CM" is closer to the center of mass of the target spacecraft, as the mass of the target spacecraft is larger than that of the chaser spacecraft. Fig. 4.8 depicts the solved magnetic

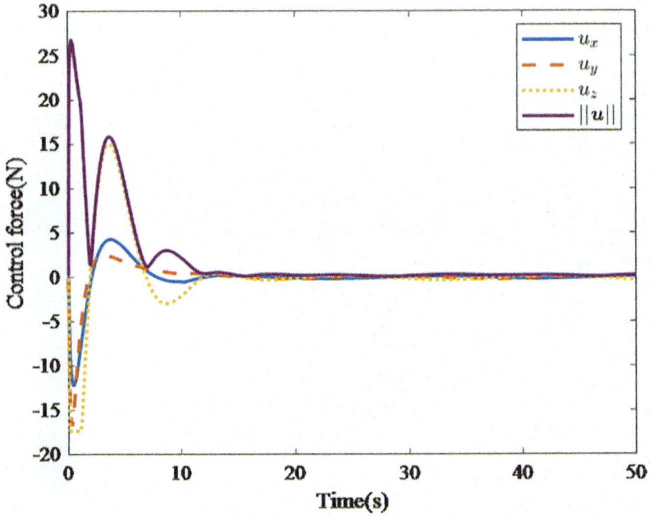

Figure 4.6 Time response of control force.

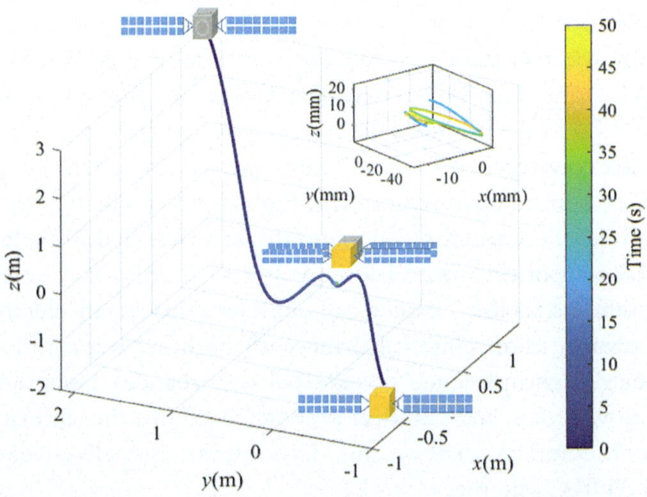

Figure 4.7 Motion of chaser and target spacecraft relative to "CM" in three-dimensional visualization.

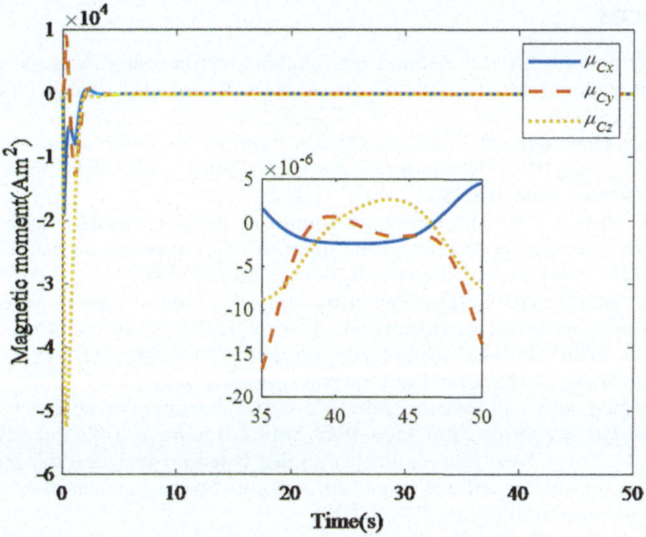

Figure 4.8 Magnetic moment of chaser spacecraft.

moment of the chaser spacecraft based on Eq. (1.79), from which one can see that the maximum of the magnetic moment is less than $5.5 \times 10^4 \mathrm{Am}^2$, which is practical and can be provided by a common electromagnetic mechanism.

4.5 Conclusions

In this work, the control problem of spacecraft electromagnetic docking was investigated, where multisource complex disturbances, e.g., external disturbances, fault signals, input delay, unknown mass, and elliptical eccentricity, were considered simultaneously. By analyzing the effects of these disturbances, a total disturbance was constructed to facilitate controller design. Then, an intermediate variable consisted of relative motion information, and the total disturbance was introduced for the design of the IO. Based on this, an ADRC law employing the estimates of relative motion information and the total disturbance was proposed. It is noted that the ADRC scheme requires no information on the mass properties of the docking spacecraft, external disturbances, and fault signals. Under the proposed control algorithm, the closed-loop system was capable of achieving uniform ultimate boundedness. Rigorous analysis and numerical simulations validated the effectiveness and feasibility of the developed IO-based ADRC scheme. The research on the coordinated control problem of multiple spacecraft electromagnetic docking is recommended for future work.

References

Biggs, J. D., & Brisotto, S. (2021). Robust spacecraft rendezvous using a variable speed control moment gyro and thruster. *Aerospace Science and Technology, 112*, 106644. https://doi.org/10.1016/j.ast.2021.106644.

Biggs, J. D., & Henninger, H. C. (2019). Motion planning on a class of 6-D Lie groups via a covering map. *IEEE Transactions on Automatic Control, 64*(9), 3544–3554. https://doi.org/10.1109/tac.2018.2885241.

Chen, T., & Shan, J. (2020a). Distributed control of multiple flexible manipulators with unknown disturbances and dead-zone input. *IEEE Transactions on Industrial Electronics, 67*(11), 9937–9947. https://doi.org/10.1109/tie.2019.2955417.

Chen, T., & Shan, J. (2020b). Distributed tracking of a class of underactuated lagrangian systems with uncertain parameters and actuator faults. *IEEE Transactions on Industrial Electronics, 67*(5), 4244–4253. https://doi.org/10.1109/tie.2019.2922943.

Huang, S. J., & Yang, G. H. (2014). Fault tolerant controller design for T-S fuzzy systems with time-varying delay and actuator faults: A k-step fault-estimation approach. *IEEE Transactions on Fuzzy Systems, 22*(6), 1526–1540. https://doi.org/10.1109/tfuzz.2014.2298053.

Li, Y., & Ye, D. (2019). Near time-optimal controller based on analytical trajectory planning algorithm for satellite attitude maneuver. *Aerospace Science and Technology, 84*, 497–509. https://doi.org/10.1016/j.ast.2018.10.030.

Liu, C., Shi, K., Yue, X., & Sun, Z. (2020). Inertia-free saturated output feedback attitude stabilization for uncertain spacecraft. *International Journal of Robust and Nonlinear Control, 30*(13), 5101–5121. https://doi.org/10.1002/rnc.5044.

Liu, C., Yue, X., Shi, K., & Sun, Z. (2019). Inertia-free attitude stabilization for flexible spacecraft with active vibration suppression. *International Journal of Robust and Nonlinear Control, 29*(18), 6311–6336. https://doi.org/10.1002/rnc.4742.

Narula, A., & Biggs, J. D. (2018). Fault-tolerant station-keeping on libration point orbits. *Journal of Guidance, Control, and Dynamics, 41*(4), 879–887. https://doi.org/10.2514/1.g003115.

Pepe, P. (2020). Discrete-time systems with constrained time delays and delay-dependent Lyapunov functions. *IEEE Transactions on Automatic Control, 65*(4), 1724–1730. https://doi.org/10.1109/tac.2019.2934391.

Ren, C., Ding, Y., Hu, L., Liu, J., Ju, Z., & Ma, S. (2022). Active disturbance rejection control of euler-lagrange systems exploiting internal damping. *IEEE Transactions on Cybernetics, 52*(6), 4334–4345. https://doi.org/10.1109/TCYB.2020.3026190.

Shen, Q., Wang, D., Zhu, S., & Poh, E. K. (2015). Inertia-free fault-tolerant spacecraft attitude tracking using control allocation. *Automatica, 62*, 114–121. https://doi.org/10.1016/j.automatica.2015.09.027.

Shen, Q., Yue, C., Goh, C. H., & Wang, D. (2019). Active fault-tolerant control system design for spacecraft attitude maneuvers with actuator saturation and faults. *IEEE Transactions on Industrial Electronics, 66*(5), 3763–3772. https://doi.org/10.1109/tie.2018.2854602.

Shi, K., Liu, C., Biggs, J. D., Sun, Z., & Yue, X. (2020). Observer-based control for spacecraft electromagnetic docking. *Aerospace Science and Technology, 99*, 105759. https://doi.org/10.1016/j.ast.2020.105759.

Shi, K., Liu, C., & Sun, Z. (2019). Constrained fuel-free control for spacecraft electromagnetic docking in elliptical orbits. *Acta Astronautica, 162*, 14–24. https://doi.org/10.1016/j.actaastro.2019.05.016.

Shi, P., Liu, M., & Zhang, L. (2015). Fault-tolerant sliding-mode-observer synthesis of markovian jump systems using quantized measurements. *IEEE Transactions on Industrial Electronics, 62*(9), 5910–5918. https://doi.org/10.1109/tie.2015.2442221.

Shi, Y., & Hu, Q. (2022). Event-driven connectivity-preserving coordinated control for multiple spacecraft systems with a distance-dependent dynamic graph. *IEEE Transactions on Cybernetics, 52*(11), 12551–12560. https://doi.org/10.1109/tcyb.2021.3072139.

Su, X., Liu, Z., Zhang, Y., & Chen, P. (2021). Event-triggered adaptive fuzzy tracking control for uncertain nonlinear systems preceded by unknown Prandtl-Ishlinskii hysteresis. *IEEE Transactions on Cybernetics, 51*(6), 2979–2992. https://doi.org/10.1109/tcyb.2019.2949022.

Sun, J., Yang, J., Li, S., & Zheng, W. (2020). Output-based dynamic event-triggered mechanisms for disturbance rejection control of networked nonlinear systems. *IEEE Transactions on Cybernetics, 50*(5), 1978–1988. https://doi.org/10.1109/tcyb.2018.2877413.

Sun, J., Yang, J., Li, S., & Zheng, W. X. (2019). Sampled-data-based event-triggered active disturbance rejection control for disturbed systems in networked environment. *IEEE Transactions on Cybernetics, 49*(2), 556–566. https://doi.org/10.1109/TCYB.2017.2780625.

Sun, L. (2020). Adaptive fault-tolerant constrained control of cooperative spacecraft rendezvous and docking. *IEEE Transactions on Industrial Electronics, 67*(4), 3107–3115. https://doi.org/10.1109/TIE.2019.2913826.

Sun, L., Huo, W., & Jiao, Z. (2017). Adaptive backstepping control of spacecraft rendezvous and proximity operations with input saturation and full-state constraint. *IEEE Transactions on Industrial Electronics, 64*(1), 480–492. https://doi.org/10.1109/TIE.2016.2609399.

Tabatabaeipour, S. M., & Bak, T. (2014). Robust observer-based fault estimation and accommodation of discrete-time piecewise linear systems. *Journal of the Franklin Institute, 351*(1), 277–295. https://doi.org/10.1016/j.jfranklin.2013.08.021.

Van, M., Ge, S. S., & Ren, H. (2017). Finite time fault tolerant control for robot manipulators using time delay estimation and continuous nonsingular fast terminal sliding mode control. *IEEE Transactions on Cybernetics, 47*(7), 1681–1693. https://doi.org/10.1109/tcyb.2016.2555307.

Weiss, A., Kolmanovsky, I., Bernstein, D. S., & Sanyal, A. (2013). Inertia-free spacecraft attitude control using reaction wheels. *Journal of Guidance, Control, and Dynamics, 36*(5), 1425–1439. https://doi.org/10.2514/1.58363.

Xu, L. X., Ma, H. J., Guo, D., Xie, A. H., & Song, D. L. (2020). Backstepping sliding-mode and cascade active disturbance rejection control for a quadrotor UAV. *IEEE/ASME Transactions on Mechatronics, 25*(6), 2743–2753. https://doi.org/10.1109/tmech.2020.2990582.

Yang, H., Cheng, L., Xia, Y., & Yuan, Y. (2018). Active disturbance rejection attitude control for a dual closed-loop quadrotor under gust wind. *IEEE Transactions on Control Systems Technology, 26*(4), 1400–1405. https://doi.org/10.1109/TCST.2017.2710951.

Yao, J., & Deng, W. (2017). Active disturbance rejection adaptive control of hydraulic servo systems. *IEEE Transactions on Industrial Electronics, 64*(10), 8023–8032. https://doi.org/10.1109/tie.2017.2694382.

Yuan, Y., Wang, Z., Yu, Y., Guo, L., & Yang, H. (2019). Active disturbance rejection control for a pneumatic motion platform subject to actuator saturation: An extended state observer approach. *Automatica, 107*, 353–361. https://doi.org/10.1016/j.automatica.2019.05.056.

Yuan, Y., Zhang, P., Wang, Z., Guo, L., & Yang, H. (2017). Active disturbance rejection control for the ranger neutral buoyancy vehicle: A delta operator approach. *IEEE Transactions on Industrial Electronics, 64*(12), 9410–9420. https://doi.org/10.1109/TIE.2017.2711538.

Zhang, Y. W., Yang, L. P., Zhu, Y. W., Huang, H., & Cai, W. W. (2013). Angular-momentum management of spacecraft electromagnetic docking considering the earth's magnetic field. *Journal of Guidance, Control, and Dynamics, 36*(3), 860–869. https://doi.org/10.2514/1.57723.

Zhang, Y. W., Yang, L. P., Zhu, Y. W., Huang, H., & Cai, W. W. (2012). Nonlinear 6-DOF control of spacecraft docking with inter-satellite electromagnetic force. *Acta Astronautica, 77*(1), 97–108. https://doi.org/10.1016/j.actaastro.2012.03.020.

CHAPTER 5

Disturbance observer-based control for spacecraft electromagnetic docking

5.1 Introduction

Traditional thruster-based docking mechanisms may require continuous expenditure of fuel that can contaminate the sensitive sensors on board, and mission lifespan also becomes dependent on the fuel available (Ahsun et al., 2010; Hallaj & Assadian, 2015). Compared with conventional methods, spacecraft docking technology with intercraft electromagnetic force has many inherent advantages, such as no propellant consumption, no plume contamination, and continuous, reversible, and synchronous controllability, which suggests that it can be applied in many on-orbit servicing missions (Zhang et al., 2012, Zhang et al., 2013). These missions include a variety of space programs, such as large spacecraft construction, fuel supply, repair service, and formation flight. One of the major topics investigated in the field of electromagnetic force application in space missions is electromagnetic formation flight (EMFF), and much literature has addressed this problem, with a wealth of well-understood methods and algorithms (Kong et al., 2004; Zhang & Huang, 2016). In EMFF, each spacecraft is equipped with three large orthogonal coils that can generate electromagnetic forces. The relative positions of the spacecraft are then determined by these electromagnetic forces, while the relative orientation of the dipoles is controlled using reaction wheels or control moment gyros to generate counteracting torques (Kong, 2002). However, electromagnetic docking differs from EMFF since it faces a more serious nonlinear problem with the relative distance between the target and chaser spacecraft decreasing gradually in the docking process. Therefore robust control capability should be a top priority in electromagnetic docking missions.

Due to electromagnetic forces in use, the translational dynamics of spacecraft are highly coupled and nonlinear (Song et al., 2015). The accurate control for electromagnetic docking or EMFF is an interesting issue, which

Spacecraft Electromagnetic Docking and Separation.
DOI: https://doi.org/10.1016/B978-0-443-49334-8.00011-2

has attracted more attention, (e.g., Huang et al., 2016; Zhang et al., 2011) and the references therein. Despite the vast range of available techniques, however, the development of controllers for spacecraft electromagnetic docking is a labor-intensive and time-consuming process. In addition, the uncertain parameters often appear due to various actuator and measurement limitations, which will bring great difficulties to controller design (Chen & Chen, 2018). For applications in rapid response spacecraft, it is of great interest to employ control algorithms that are robust to inexact knowledge of the spacecraft mass and other uncertainties. In addition, it is often expensive to determine mass properties with high precision. To alleviate this requirement, the attitude controllers developed in Sanyal et al. (2009) and Weiss et al. (2013) are inertia-free in the sense that no prior information on the inertia matrix is required. Analogous to this, for electromagnetic docking missions, it is of interest to develop such a controller that requires no mass information on the chaser or target spacecraft.

Most of the existing works in the field of electromagnetic force applications only discussed circular orbits (e.g., Ahsun et al., 2010; Hallaj & Assadian, 2015; Zhang et al., 2015), which cannot be applied directly to elliptical orbits, especially for orbits with large eccentricity. For autonomous rendezvous and docking missions, the chaser and target spacecraft can fly in elliptical orbits, however, challenges arise for high-precision tasks if controllers for circular orbits are applied. Similar to traditional docking systems, electromagnetic docking systems are also subject to external disturbances and input magnitude and rate constraints (MRCs) resulting from physical limits of the onboard actuators. Due to the fact that the spacecraft's electromagnetic docking technology is not mature until now, the controllers designed for the fault-free case, which do not take the possibility of fault occurrence into consideration, may not be applicable in practice. In contrast to this, the fault-tolerant controller can guarantee desirable stability and performance properties even in the event of a component fault/failure. Thus the possible fault signals should be taken into account in the process of controller design (Chen & Shan, 2019; Liu et al., 2018). A challenging problem that arises in this domain is that the external disturbances, fault signals, and mass properties may be uncertain or unknown, which further increases the complexity and difficulty in controller design. Future space missions demand that the new generation of spacecraft should be able to perform on-orbit servicing missions with high precision and better robustness to external disturbances and model uncertainties. One way to overcome this problem is to estimate the disturbance for a control action to be taken, to compensate for the influence of the disturbance (Chen, Yang, Guo, & Li, 2016). Thus disturbance observer-based control has been widely studied and explored in practical applications in recent decades, due to its particular characteristics to deal with the external disturbances and model uncertainties (Hu et al., 2014; Wu et al., 2010). In Chen and Wen (2018), a disturbance observer-based distributed controller is designed by treating the flexible dynamics as disturbances acting on the rigid motion of the flexible spacecraft. It is noted that the effect of the vibration of flexible appendages is regarded as a source of the lumped disturbance. In Sun and Zheng (2018) and Li et al. (2018), two types of nonlinear disturbance observer were employed to estimate the lumped disturbance, and the corresponding feedback control strategy is applied for driving a chaser spacecraft to rendezvous and dock with a target spacecraft in space. The above two disturbance observer-based controllers (DOBCs) are interesting, but the observer and controller are separately designed, resulting in computational complexity, and the control performance is insufficient for high-precision docking missions. A disturbance-observer-based controller considering input MRCs was presented in Liu et al. (2018), where only one-dimensional motion was considered. It is also worth noting that this method is conservative, for it sets the Lyapunov variable involved in estimation errors to be an identity matrix. Subsequently, an attitude tracking controller is designed based on the backstepping and finite-time disturbance observer techniques, and the input magnitude and rate saturation are also considered (Zou et al., 2019).

As noted by Elias et al. (2007), the limited maximum current density inevitably imposes an input magnitude constraint, and an adaptive reaching law-based sliding mode control method was proposed in Huang et al. (2016) to solve this problem. However, from a practical viewpoint, that is, the rate of change of the maximum current is also limited, control constraints on input rate should also be taken into account for safe operation of actuators (Akella et al., 2005; Leonessa et al., 2009). Thus it is necessary to develop effective control schemes for the spacecraft electromagnetic docking system in the presence of external disturbances, fault signals, unknown mass information, elliptical eccentricity, and input MRCs.

In particular, no study, to our knowledge, has simultaneously considered the aforementioned issues, that is, external disturbances, fault signals, unknown mass, elliptical eccentricity, and input MRCs. This chapter aims to develop a novel constrained fuel-free control scheme without knowing exactly the external disturbances, fault signals, and mass properties to achieve high-precision control of electromagnetic docking for spacecraft in elliptical orbits. The controller is designed based on linear matrix inequalities (LMIs), which have become a very effective technique in the field of stability and control, with global optimality and numerical reliability, multiobjective design capability, and mature software packages. This chapter documents several key contributions made to the field of spacecraft electromagnetic docking. First, the relative motion information and the estimation errors of lumped disturbances from external disturbances, fault signals, unknown mass, and elliptical eccentricity can converge to the vicinity of the expected states with high precision. Second, explicit bounds of relative motion information and estimation errors of the lumped disturbance are obtained in the process of Lyapunov analysis. Third, the observer and controller gains can be obtained simultaneously by solving LMIs, with the input MRCs satisfying given limits. Fourth, no parameter identification and no fuzzy or neural nonlinear approximations for the mass of the chaser and target spacecraft are required, which decreases the complexity of controller design dramatically. Fifth, the developed control scheme is not only applicable to spacecraft electromagnetic docking systems subject to input MRCs, but also easily extendable to a more general class of second-order systems with input MRCs.

The rest of this chapter is organized as follows. Section 5.2 presents the mathematical models for electromagnetic docking systems of the chaser and target spacecraft, where external disturbances, fault signals, unknown mass information, and orbital eccentricity. A corresponding lumped disturbance is defined, and the corresponding control problem is formulated. Then, the main results of this work are presented, where a novel disturbance observer-based control scheme with input MRCs is proposed, and a corresponding stability analysis is performed via a Lyapunov approach in Section 5.3. In Section 5.4, it is followed by the numerical simulations to demonstrate the validity and performance of the proposed strategy. Finally, conclusions are drawn.

5.2 Problem formulation

By approximating the coils on each spacecraft to a steerable dipole, the electromagnetic force between the two spacecraft can be calculated by Eq. (1.79).

The far-field model, that is, Eq. (1.79), of electromagnetic force is derived based on the dipole hypothesis. It is only accurate when the relative distance between the two spacecraft is not less than a certain threshold, which is related to the mechanism configuration, such as six to eight times the coil radius, as described in Zhang, Yang, Zhu, Huang, & Cai, 2013). When the two spacecraft are very close to each other, the resulting error from the far-field model can be regarded as a source of external disturbances. Due to inherent nonlinearities and couplings, the dynamics and control problems associated with electromagnetic docking are difficult, especially for elliptical orbits. Thus the electromagnetic force can be first regarded as the control input, which will result in the solution of magnetic moments with details below. According to Zhang et al., 2011) and Fabacher et al. (2017), the magnetic moment vector of the target spacecraft can be assumed to be a known vector, as the information on relative motion, that is, $\widetilde{\boldsymbol{\eta}}$, can be obtained at each time instant, so that the matrix $\Xi(\boldsymbol{\mu}_{\mathrm{T}}, \widetilde{\boldsymbol{\eta}})$can be calculated. According to Eq. (1.79), the solution of the magnetic moment vector of the chaser spacecraft can be obtained via the electromagnetic force $\boldsymbol{F}_{\mathrm{eC}}$ and the matrix $\Xi(\boldsymbol{\mu}_{\mathrm{T}}, \widetilde{\boldsymbol{\eta}})$.

Since the performance of a spacecraft docking system can be severely impaired by improper actuator actions, actuator fault has been considered to be one of the most critical challenges to be solved. Taking this issue into account, let $\boldsymbol{E}$ with full-column rank, which has a similar structure to $\boldsymbol{B}$, denote the distribution matrix of fault signals $f(t)$ appearing in the input. Specifically, it represents the process fault if $\boldsymbol{E} \neq \boldsymbol{B}$, and it also represents the actuator fault if $\boldsymbol{E} = \boldsymbol{B}$.

According to the dynamic model introduced in Section 1.2 and considering the actuator fault, Eq. (1.78) can be converted into the following state-space form:

$$\dot{\boldsymbol{\eta}}(t) = (\boldsymbol{A} + \Delta \boldsymbol{A})\boldsymbol{\eta}(t) + (\boldsymbol{B} + \Delta \boldsymbol{B})(\boldsymbol{u}(t) + \boldsymbol{w}_0(t)) + \boldsymbol{E}\boldsymbol{f}(t) \tag{5.1}$$

Eq. (5.1) can be converted into the following form:

$$\dot{\boldsymbol{\eta}}(t) = \boldsymbol{A}\boldsymbol{\eta}(t) + \boldsymbol{B}\boldsymbol{u}(t) + \boldsymbol{B}\boldsymbol{w}(t) \tag{5.2}$$

where $\boldsymbol{w}(t) = \boldsymbol{B}^{\mathrm{T}}(\boldsymbol{B}\boldsymbol{B}^{\mathrm{T}})^{*}(\Delta \boldsymbol{A}\boldsymbol{\eta}(t) + \Delta \boldsymbol{B}\boldsymbol{u}(t) + (\boldsymbol{B} + \Delta \boldsymbol{B})\boldsymbol{w}_0(t) + \boldsymbol{E}\boldsymbol{f}(t))$ is regarded as the lumped disturbance from external disturbances, fault signals, unknown mass, and elliptical eccentricity, and where the matrix $(\boldsymbol{B}\boldsymbol{B}^{\mathrm{T}})^{*}$ is the pseudoinverse of $\boldsymbol{B}\boldsymbol{B}^{\mathrm{T}}$.

Assumption 5.1 The lumped disturbance $\boldsymbol{w}(t)$ and its first time derivative is assumed to be piecewise continuous and bounded, that is, $||\boldsymbol{w}(t)|| \leq \kappa_1$, $||\dot{\boldsymbol{w}}(t)|| \leq \kappa_2$, where $\kappa_{1,2} > 0$.

Remark 5.1 Assumption 5.1 is reasonable. The main reason is that the external disturbances primarily include magnetic force, solar radiation pressure, and aerodynamic force, which are all continuous and bounded. In addition, because the total control authority is limited, all three available control input forces should be continuous and bounded, which is reasonable in practice. It is worth noting that the desired translational motion represented by relative position and velocity is also bounded and differentiable in practice. In particular, the derivative of $\boldsymbol{f}(t)$ with respect to time is often assumed to be norm-bounded (Huang et al., 2014). They together result in the fact that the lumped disturbance force is bounded and differentiable, and there exist two positive constants $\tilde{\kappa}_1$ and $\tilde{\kappa}_2$, which can be regarded as the maximums of the lumped disturbance force $||\boldsymbol{w}(t)||$ and its first derivative $||\dot{\boldsymbol{w}}(t)||$. Although the two positive constants $\tilde{\kappa}_1$ and $\tilde{\kappa}_2$ are difficult to be exactly known, there must be two other positive constants satisfying $\kappa_1 \geq \tilde{\kappa}_1$ and $\kappa_2 \geq \tilde{\kappa}_2$. Thus $||\boldsymbol{w}(t)|| \leq \kappa_1$and $||\dot{\boldsymbol{w}}(t)|| \leq \kappa_2$ are satisfied.

The objective of this chapter is to construct a disturbance observer-based controller $\boldsymbol{u}(t)$ such that the states of the resulting closed-loop system are uniformly ultimately bounded in the face of external disturbances, fault signals, unknown mass, elliptical eccentricity, and input MRCs.

5.3 Disturbance observer-based controller design

5.3.1 Controller design

To achieve the objective, an auxiliary variable is first introduced as

$$\boldsymbol{z}(t) = \hat{\boldsymbol{w}}(t) - \boldsymbol{L}\boldsymbol{\eta}(t) \tag{5.3}$$

where $\hat{\boldsymbol{w}}(t)$ is the estimate of $\boldsymbol{w}(t)$, $\boldsymbol{L}$ is the disturbance observer gain matrix.

Then, to estimate the lumped disturbance, a disturbance observer is designed based on

$$\begin{cases} \dot{\boldsymbol{z}}(t) = -\boldsymbol{L}(\boldsymbol{A}\boldsymbol{\eta}(t) + \boldsymbol{B}\boldsymbol{u}(t) + \boldsymbol{B}\hat{\boldsymbol{w}}(t)) \\ \hat{\boldsymbol{w}}(t) = \boldsymbol{z}(t) + \boldsymbol{L}\boldsymbol{\eta}(t) \end{cases} \tag{5.4}$$

Based on the disturbance observer designed, the controller is designed as

$$\boldsymbol{u}(t) = -\boldsymbol{K}\boldsymbol{\eta}(t) - \hat{\boldsymbol{w}}(t) \tag{5.5}$$

where $\boldsymbol{K}$ is the controller gain matrix.

Denoting $\boldsymbol{e}(t) = \boldsymbol{w}(t) - \hat{\boldsymbol{w}}(t)$and $\boldsymbol{\xi}(t) = \left[\boldsymbol{\eta}(t)^{\mathrm{T}} \quad \boldsymbol{e}(t)^{\mathrm{T}}\right]^{\mathrm{T}}$, one has

$$\dot{\boldsymbol{\xi}}(t) = \boldsymbol{A}_{\xi}\boldsymbol{\xi}(t) + \boldsymbol{B}_{\xi}\dot{\boldsymbol{w}}(t) \tag{5.6}$$

where

$$A_\xi = \begin{bmatrix} A - BK & B \\ 0 & -LB \end{bmatrix}, B_\xi = \begin{bmatrix} 0 \\ I \end{bmatrix}$$

Theorem 5.1 The states of the closed-loop system represented by Eq. (5.6) are uniformly ultimately bounded, and the input MRCs, that is, $||u|| \leq \lambda_1$ and$||\dot{u}|| \leq \lambda_2$, are satisfied, if for positive scalars $\lambda_1, \lambda_2, \xi_1, \xi_2, \xi_3, \xi_4$ and the known initial conditions $\eta(0)$ and $e(0)$, there exist symmetric positive definite matrices X_1, X_2, and matrices W and H, such that the following LMIs hold:

$$\begin{bmatrix} AX_1 - BW + X_1A^{\mathrm{T}} - W^{\mathrm{T}}B^{\mathrm{T}} + \xi_2X_1 & BX_2 & 0 \\ * & -H^{\mathrm{T}} - H + \xi_2X_2 & X_2 \\ * & * & -\xi_1^{-1}I \end{bmatrix} \leq 0 \tag{5.7}$$

$$\begin{bmatrix} \xi_2^{-1}\xi_1^{-1}\kappa_2^2 - \varpi & \eta(0)^{\mathrm{T}} & e(0)^{\mathrm{T}} \\ * & -X_1 & 0 \\ * & * & -X_2 \end{bmatrix} \leq 0 \tag{5.8}$$

$$\begin{bmatrix} -k_0X_1 & 0 & 0 & W^{\mathrm{T}} \\ * & -k_0X_2 & 0 & -X_2 \\ * & * & -k_0\kappa_1^{-2}I & I \\ * & * & * & -I \end{bmatrix} \leq 0 \tag{5.9}$$

$$\begin{bmatrix} -\frac{(1+\xi_3^{-1})}{(1+\xi_3)}k_0k_1X_2 & X_2B^{\mathrm{T}} & H^{\mathrm{T}} \\ * & -(1+\xi_4)^{-1}k_0^{-1}X_1 & 0 \\ * & * & -(1+\xi_4^{-1})^{-1}I \end{bmatrix} \leq 0 \tag{5.10}$$

$$\begin{bmatrix} -k_1X_1 & X_1A^{\mathrm{T}} - W^{\mathrm{T}}B^{\mathrm{T}} \\ * & -X_1 \end{bmatrix} \leq 0 \tag{5.11}$$

where $k_0 = (\varpi+1)^{-1}\lambda_1^2$ and $k_1 = (\varpi + 1)\lambda_2^2/((1 + \xi_3^{-1})\lambda_1^2\varpi)$.

When LMIs inequalities (5.7)–(5.11) are feasible, and one can obtain the controller gain matrix $K = WX_1^{-1}$ and the observer gain matrix $L = H(BX_2)^{\mathrm{T}}[(BX_2)(BX_2)^{\mathrm{T}}]^*$, where the matrix $[(BX_2)(BX_2)^{\mathrm{T}}]^*$ is the pseudoinverse of $(BX_2)(BX_2)^{\mathrm{T}}$.

Proof The proof of Theorem 5.1 can be divided into three steps: First, the closed-loop uniformly ultimately bounded stabilization is shown, and then the input magnitude constraint is proved to be satisfied; finally, the input rate constraint is further proved.

First, it is shown that the states of the closed-loop system are uniformly ultimately bounded.

Choose a Lyapunov function candidate as

$$V(t) = \boldsymbol{\eta}(t)^{\mathrm{T}} \boldsymbol{X}_1^{-1} \boldsymbol{\eta}(t) + \boldsymbol{e}^{\mathrm{T}}(t) \boldsymbol{X}_2^{-1} \boldsymbol{e}(t) \tag{5.12}$$

Taking the derivative of $V(t)$ with respect to time yields

$$\begin{aligned}
\dot{V}(t) &= \dot{\boldsymbol{\eta}}(t)^{\mathrm{T}} \boldsymbol{X}_1^{-1} \boldsymbol{\eta}(t) + \boldsymbol{\eta}(t)^{\mathrm{T}} \boldsymbol{X}_1^{-1} \dot{\boldsymbol{\eta}}(t) + \dot{\boldsymbol{e}}(t)^{\mathrm{T}} \boldsymbol{X}_2^{-1} \boldsymbol{e}(t) + \boldsymbol{e}(t)^{\mathrm{T}} \boldsymbol{X}_2^{-1} \dot{\boldsymbol{e}}(t) \\
&= \boldsymbol{\eta}(t)^{\mathrm{T}} \left[(\boldsymbol{A} - \boldsymbol{BK})^{\mathrm{T}} \boldsymbol{X}_1^{-1} + \boldsymbol{X}_1^{-1} (\boldsymbol{A} - \boldsymbol{BK}) \right] \boldsymbol{\eta}(t) + 2\boldsymbol{\eta}(t)^{\mathrm{T}} \boldsymbol{X}_1^{-1} \boldsymbol{Be}(t) \\
&\quad - \boldsymbol{e}(t)^{\mathrm{T}} \left(\boldsymbol{B}^{\mathrm{T}} \boldsymbol{L}^{\mathrm{T}} \boldsymbol{X}_2^{-1} + \boldsymbol{X}_2^{-1} \boldsymbol{LB} \right) \boldsymbol{e}(t) + 2\boldsymbol{e}(t)^{\mathrm{T}} \dot{\boldsymbol{w}}(t)
\end{aligned}$$

With Lemma 1.1 and the given constraint conditions, one has

$$\begin{aligned}
\dot{V}(t) &\le \boldsymbol{\eta}(t)^{\mathrm{T}} \left[(\boldsymbol{A} - \boldsymbol{BK})^{\mathrm{T}} \boldsymbol{X}_1^{-1} + \boldsymbol{X}_1^{-1} (\boldsymbol{A} - \boldsymbol{BK}) \right] \boldsymbol{\eta}(t) + 2\boldsymbol{\eta}(t)^{\mathrm{T}} \boldsymbol{X}_1^{-1} \boldsymbol{Be}(t) \\
&\quad - \boldsymbol{e}(t)^{\mathrm{T}} \left(\boldsymbol{B}^{\mathrm{T}} \boldsymbol{L}^{\mathrm{T}} \boldsymbol{X}_2^{-1} + \boldsymbol{X}_2^{-1} \boldsymbol{LB} \right) \boldsymbol{e}(t) + \xi_1 \boldsymbol{e}(t)^{\mathrm{T}} \boldsymbol{e}(t) + \xi_1^{-1} \dot{\boldsymbol{w}}(t)^{\mathrm{T}} \dot{\boldsymbol{w}}(t) \\
&\le \boldsymbol{\eta}(t)^{\mathrm{T}} \left[(\boldsymbol{A} - \boldsymbol{BK})^{\mathrm{T}} \boldsymbol{X}_1^{-1} + \boldsymbol{X}_1^{-1} (\boldsymbol{A} - \boldsymbol{BK}) \right] \boldsymbol{\eta}(t) + 2\boldsymbol{\eta}(t)^{\mathrm{T}} \boldsymbol{X}_1^{-1} \boldsymbol{Be}(t) \\
&\quad - \boldsymbol{e}(t)^{\mathrm{T}} \left[\left(\boldsymbol{B}^{\mathrm{T}} \boldsymbol{L}^{\mathrm{T}} \boldsymbol{X}_2^{-1} + \boldsymbol{X}_2^{-1} \boldsymbol{LB} \right) - \xi_1 \boldsymbol{I} \right] \boldsymbol{e}(t) + \xi_1^{-1} \kappa_2^2 \\
&= \boldsymbol{\xi}(t)^{\mathrm{T}} \boldsymbol{\Theta} \boldsymbol{\xi}(t) + \xi_1^{-1} \kappa_2^2
\end{aligned}$$

where

$$\boldsymbol{\Theta} = \begin{bmatrix} (\boldsymbol{A} - \boldsymbol{BK})^{\mathrm{T}} \boldsymbol{X}_1^{-1} + \boldsymbol{X}_1^{-1} (\boldsymbol{A} - \boldsymbol{BK}) & \boldsymbol{X}_1^{-1} \boldsymbol{B} \\ * & -\left(\boldsymbol{B}^{\mathrm{T}} \boldsymbol{L}^{\mathrm{T}} \boldsymbol{X}_2^{-1} + \boldsymbol{X}_2^{-1} \boldsymbol{LB} \right) + \xi_1 \boldsymbol{I} \end{bmatrix}$$

Defining $\boldsymbol{W} = \boldsymbol{KX}_1$ and $\boldsymbol{H} = \boldsymbol{LBX}_2$, multiplied by $\mathrm{diag}([\boldsymbol{X}_1^{-1} \quad \boldsymbol{X}_2^{-1} \quad \boldsymbol{I}])$ and its transpose on both sides of inequality (5.7) simultaneously, inequality (5.13) can be obtained.

$$\begin{bmatrix} (\boldsymbol{A} - \boldsymbol{BK})^{\mathrm{T}} \boldsymbol{X}_1^{-1} + \boldsymbol{X}_1^{-1} (\boldsymbol{A} - \boldsymbol{BK}) + \xi_2 \boldsymbol{X}_1^{-1} & \boldsymbol{X}_1^{-1} \boldsymbol{B} & 0 \\ * & -\boldsymbol{B}^{\mathrm{T}} \boldsymbol{L}^{\mathrm{T}} \boldsymbol{X}_2^{-1} - \boldsymbol{X}_2^{-1} \boldsymbol{LB} + \xi_2 \boldsymbol{X}_2^{-1} & \boldsymbol{I} \\ * & * & -\xi_1^{-1} \boldsymbol{I} \end{bmatrix} \le 0 \tag{5.13}$$

With Lemma 1.3, one has

$$\begin{bmatrix} (\boldsymbol{A} - \boldsymbol{BK})^{\mathrm{T}} \boldsymbol{X}_1^{-1} + \boldsymbol{X}_1^{-1} (\boldsymbol{A} - \boldsymbol{BK}) & \boldsymbol{X}_1^{-1} \boldsymbol{B} \\ * & -\boldsymbol{B}^{\mathrm{T}} \boldsymbol{L}^{\mathrm{T}} \boldsymbol{X}_2^{-1} - \boldsymbol{X}_2^{-1} \boldsymbol{LB} + \xi_1 \boldsymbol{I} \end{bmatrix} + \xi_2 \begin{bmatrix} \boldsymbol{X}_1^{-1} & 0 \\ * & \boldsymbol{X}_2^{-1} \end{bmatrix} \le 0 \tag{5.14}$$

Then, one has

$$\begin{aligned}\dot{V}(t) &\le \boldsymbol{\xi}(t)^{\mathrm{T}}\boldsymbol{\Theta}\boldsymbol{\xi}(t) + \xi_1^{-1}\kappa_2^2 \\ &\le -\xi_2\boldsymbol{\xi}(t)^{\mathrm{T}}\begin{bmatrix}\boldsymbol{X}_1^{-1} & 0 \\ * & \boldsymbol{X}_2^{-1}\end{bmatrix}\boldsymbol{\xi}(t) + \xi_1^{-1}\kappa_2^2\end{aligned} \tag{5.15}$$

that is,

$$\dot{V}(t) + \xi_2 V(t) \le \xi_1^{-1}\kappa_2^2 \tag{5.16}$$

With Lemma 1.2, the states of the closed-loop system are uniformly ultimately bounded under controller Eq. (5.5).

Next it is shown that the input magnitude constraint, that is, $||\boldsymbol{u}|| \le \lambda_1$, is satisfied.

Multiplying inequality (5.16)by $e^{\xi_2 t}$ yields

$$\frac{d}{dt}\left(V(t)\mathrm{e}^{\xi_2 t}\right) \le \xi_1^{-1}\kappa_2^2\mathrm{e}^{\xi_2 t} \tag{5.17}$$

Integrating inequality (5.17) over $[0, t]$ leads to

$$0 \le V(t) \le V(0)\mathrm{e}^{-\xi_2 t} - \xi_2^{-1}\xi_1^{-1}\kappa_2^2\mathrm{e}^{-\xi_2 t} + \xi_2^{-1}\xi_1^{-1}\kappa_2^2 \le V(0) + \xi_2^{-1}\xi_1^{-1}\kappa_2^2 \tag{5.18}$$

where $V(0) = \boldsymbol{\eta}(0)^{\mathrm{T}}\boldsymbol{X}_1^{-1}\boldsymbol{\eta}(0) + \boldsymbol{e}^{\mathrm{T}}(0)\boldsymbol{X}_2^{-1}\boldsymbol{e}(0)$.

According to Lemma 1.3, the LMI (5.8) is equivalent to

$$\xi_2^{-1}\xi_1^{-1}\kappa_2^2 + \boldsymbol{\eta}(0)^{\mathrm{T}}\boldsymbol{X}_1^{-1}\boldsymbol{\eta}(0) + \boldsymbol{e}(0)^{\mathrm{T}}\boldsymbol{X}_2^{-1}\boldsymbol{e}(0) \le \overline{\varpi} \tag{5.19}$$

Multiplying $\mathrm{diag}([\boldsymbol{X}_1^{-1} \quad \boldsymbol{X}_2^{-1} \quad \boldsymbol{I} \quad \boldsymbol{I}])$ and its transpose on both sides of inequality (5.9) yields

$$\begin{bmatrix} -k_0\boldsymbol{X}_1^{-1} & 0 & 0 & \boldsymbol{K}^{\mathrm{T}} \\ * & -k_0\boldsymbol{X}_2^{-1} & 0 & -\boldsymbol{I} \\ * & * & -k_0\kappa_1^{-2}\boldsymbol{I} & \boldsymbol{I} \\ * & * & * & -\boldsymbol{I} \end{bmatrix} \le 0 \tag{5.20}$$

With Lemma 1.3, one has

$$\begin{bmatrix} \boldsymbol{K}^{\mathrm{T}} \\ -\boldsymbol{I} \\ \boldsymbol{I} \end{bmatrix} \begin{bmatrix} \boldsymbol{K} & -\boldsymbol{I} & \boldsymbol{I} \end{bmatrix} \leq \begin{bmatrix} k_0 \boldsymbol{X}_1^{-1} & 0 & 0 \\ * & k_0 \boldsymbol{X}_2^{-1} & 0 \\ * & * & k_0 \kappa_1^{-2} \boldsymbol{I} \end{bmatrix} \tag{5.21}$$

Then,

$$\begin{aligned}
||\boldsymbol{u}(t)||_2^2 &= \boldsymbol{u}(t)^{\mathrm{T}} \boldsymbol{u}(t) \\
&= [-\boldsymbol{K}\boldsymbol{\eta}(t) - \hat{\boldsymbol{w}}(t)]^{\mathrm{T}} [-\boldsymbol{K}\boldsymbol{\eta}(t) - \hat{\boldsymbol{w}}(t)] \\
&= [\boldsymbol{K}\boldsymbol{\eta}(t) + \boldsymbol{w}(t) - \boldsymbol{e}(t)]^{\mathrm{T}} [\boldsymbol{K}\boldsymbol{\eta}(t) + \boldsymbol{w}(t) - \boldsymbol{e}(t)] \\
&= \begin{bmatrix} \boldsymbol{\eta}(t)^{\mathrm{T}} & \boldsymbol{e}(t)^{\mathrm{T}} & \boldsymbol{w}(t)^{\mathrm{T}} \end{bmatrix} \begin{bmatrix} \boldsymbol{K}^{\mathrm{T}} \\ -\boldsymbol{I} \\ \boldsymbol{I} \end{bmatrix} \begin{bmatrix} \boldsymbol{K} & -\boldsymbol{I} & \boldsymbol{I} \end{bmatrix} \begin{bmatrix} \boldsymbol{\eta}(t) \\ \boldsymbol{e}(t) \\ \boldsymbol{w}(t) \end{bmatrix} \\
&\leq \begin{bmatrix} \boldsymbol{\eta}(t)^{\mathrm{T}} & \boldsymbol{e}(t)^{\mathrm{T}} & \boldsymbol{w}(t)^{\mathrm{T}} \end{bmatrix} \begin{bmatrix} k_0 \boldsymbol{X}_1^{-1} & 0 & 0 \\ * & k_0 \boldsymbol{X}_2^{-1} & 0 \\ * & * & k_0 \kappa_1^{-2} \boldsymbol{I} \end{bmatrix} \begin{bmatrix} \boldsymbol{\eta}(t) \\ \boldsymbol{e}(t) \\ \boldsymbol{w}(t) \end{bmatrix} \\
&\leq k_0 \left(\boldsymbol{\eta}(t)^{\mathrm{T}} X_1^{-1} \boldsymbol{\eta}(t) + \boldsymbol{e}^{\mathrm{T}}(t) \boldsymbol{X}_2^{-1} \boldsymbol{e}^{\mathrm{T}}(t) + \kappa_1^{-2} \boldsymbol{w}(t)^{\mathrm{T}} \boldsymbol{w}(t) \right) \\
&\leq k_0 (V(t) + 1)
\end{aligned} \tag{5.22}$$

According to inequalities Eqs. (5.8) and (5.18), one knows

$$||\boldsymbol{u}(t)||_2^2 \leq k_0(\varpi + 1) = \lambda_1^2 \tag{5.23}$$

Finally, it is shown that the input rate constraint, that is, $||\dot{\boldsymbol{u}}|| \leq \lambda_2$, is satisfied.

With $\boldsymbol{H} = \boldsymbol{L}\boldsymbol{B}\boldsymbol{X}_2$, multiplying diag($[\boldsymbol{X}_2^{-1} \quad \boldsymbol{I} \quad \boldsymbol{I}]$)and its transpose on both sides of the inequality (5.10) yields

$$\begin{bmatrix} -\frac{(1+\xi_3^{-1})}{(1+\xi_3)} k_0 k_1 \boldsymbol{X}_2^{-1} & \boldsymbol{B}^{\mathrm{T}} & \boldsymbol{B}^{\mathrm{T}} \boldsymbol{L}^{\mathrm{T}} \\ * & -(1+\xi_4)^{-1} k_0^{-1} \boldsymbol{X}_1 & 0 \\ * & * & -(1+\xi_4^{-1})^{-1} \boldsymbol{I} \end{bmatrix} \leq 0 \tag{5.24}$$

According to Lemma 1.3, the inequality (5.24) is equivalent to

$$(1+\xi_4^{-1}) \boldsymbol{B}^{\mathrm{T}} \boldsymbol{L}^{\mathrm{T}} \boldsymbol{L} \boldsymbol{B} + (1+\xi_4) k_0 \boldsymbol{B}^{\mathrm{T}} \boldsymbol{X}_1^{-1} \boldsymbol{B} \leq \frac{(1+\xi_3^{-1})}{(1+\xi_3)} k_0 k_1 \boldsymbol{X}_2^{-1} \tag{5.25}$$

Multiplying diag($[\boldsymbol{X}_1^{-1} \quad \boldsymbol{I}]$)and its transpose on both sides of inequality (5.9) yields

$$\begin{bmatrix} -k_1 \boldsymbol{X}_1^{-1} & \boldsymbol{A}^{\mathrm{T}} - \boldsymbol{K}^{\mathrm{T}} \boldsymbol{B}^{\mathrm{T}} \\ * & -\boldsymbol{X}_1 \end{bmatrix} \leq 0 \tag{5.26}$$

With Lemma 1.3, one has

$$(\boldsymbol{A}-\boldsymbol{B}\boldsymbol{K})^{\mathrm{T}}\boldsymbol{X}_1^{-1}(\boldsymbol{A}-\boldsymbol{B}\boldsymbol{K}) \le k_1\boldsymbol{X}_1^{-1} \tag{5.27}$$

Inequality (5.21) leads to

$$\boldsymbol{K}^{\mathrm{T}}\boldsymbol{K} \le k_0\boldsymbol{K}_1^{-1} \tag{5.28}$$

With inequalities (5.27) and (5.28), one has

$$(\boldsymbol{A}-\boldsymbol{B}\boldsymbol{K})^{\mathrm{T}}\boldsymbol{K}^{\mathrm{T}}\boldsymbol{K}(\boldsymbol{A}-\boldsymbol{B}\boldsymbol{K}) \le k_0k_1\boldsymbol{X}_1^{-1} \tag{5.29}$$

With inequalities (5.25) and (5.28), one has

$$\left(1+\xi_4^{-1}\right)\boldsymbol{B}^{\mathrm{T}}\boldsymbol{L}^{\mathrm{T}}\boldsymbol{L}\boldsymbol{B}+(1+\xi_4)\boldsymbol{B}^{\mathrm{T}}\boldsymbol{K}^{\mathrm{T}}\boldsymbol{K}\boldsymbol{B} \le \frac{\left(1+\xi_3^{-1}\right)}{(1+\xi_3)}k_0k_1\boldsymbol{X}_2^{-1} \tag{5.30}$$

Considering Eqs. (5.5) and (5.6), one has

$$\begin{aligned}
\dot{\boldsymbol{u}}(t) &= -\boldsymbol{K}\dot{\boldsymbol{\eta}}(t)-\dot{\hat{\boldsymbol{w}}}(t)\\
&= -\boldsymbol{K}((\boldsymbol{A}-\boldsymbol{B}\boldsymbol{K})\boldsymbol{\eta}(t)+\boldsymbol{B}\boldsymbol{e}(t))-\dot{\boldsymbol{w}}(t)-\boldsymbol{L}\boldsymbol{B}\boldsymbol{e}(t)+\dot{\boldsymbol{w}}(t)\\
&= -\boldsymbol{K}(\boldsymbol{A}-\boldsymbol{B}\boldsymbol{K})\boldsymbol{\eta}(t)-(\boldsymbol{K}\boldsymbol{B}+\boldsymbol{L}\boldsymbol{B})\boldsymbol{e}(t)
\end{aligned} \tag{5.31}$$

Then,

$$\begin{aligned}
\|\dot{\boldsymbol{u}}(t)\|_2^2 &= \dot{\boldsymbol{u}}(t)^{\mathrm{T}}\dot{\boldsymbol{u}}(t)\\
&= [-\boldsymbol{K}(\boldsymbol{A}-\boldsymbol{B}\boldsymbol{K})\boldsymbol{\eta}(t)-(\boldsymbol{K}\boldsymbol{B}+\boldsymbol{L}\boldsymbol{B})\boldsymbol{e}(t)]^{\mathrm{T}}[-\boldsymbol{K}(\boldsymbol{A}-\boldsymbol{B}\boldsymbol{K})\boldsymbol{\eta}(t)-(\boldsymbol{K}\boldsymbol{B}+\boldsymbol{L}\boldsymbol{B})\boldsymbol{e}(t)]\\
&= \boldsymbol{\eta}(t)^{\mathrm{T}}(\boldsymbol{A}-\boldsymbol{B}\boldsymbol{K})^{\mathrm{T}}\boldsymbol{K}^{\mathrm{T}}\boldsymbol{K}(\boldsymbol{A}-\boldsymbol{B}\boldsymbol{K})\boldsymbol{\eta}(t)+2\boldsymbol{\eta}(t)^{\mathrm{T}}(\boldsymbol{A}-\boldsymbol{B}\boldsymbol{K})^{\mathrm{T}}\boldsymbol{K}^{\mathrm{T}}(\boldsymbol{K}\boldsymbol{B}+\boldsymbol{L}\boldsymbol{B})\boldsymbol{e}(t)\\
&\quad +\boldsymbol{e}(t)^{\mathrm{T}}(\boldsymbol{K}\boldsymbol{B}+\boldsymbol{L}\boldsymbol{B})^{\mathrm{T}}(\boldsymbol{K}\boldsymbol{B}+\boldsymbol{L}\boldsymbol{B})\boldsymbol{e}(t)\\
&\le \left(1+\xi_3^{-1}\right)\boldsymbol{\eta}(t)^{\mathrm{T}}(\boldsymbol{A}-\boldsymbol{B}\boldsymbol{K})^{\mathrm{T}}\boldsymbol{K}^{\mathrm{T}}\boldsymbol{K}(\boldsymbol{A}-\boldsymbol{B}\boldsymbol{K})\boldsymbol{\eta}(t)+(1+\xi_3)\boldsymbol{e}(t)^{\mathrm{T}}(\boldsymbol{K}\boldsymbol{B}+\boldsymbol{L}\boldsymbol{B})^{\mathrm{T}}(\boldsymbol{K}\boldsymbol{B}+\boldsymbol{L}\boldsymbol{B})\boldsymbol{e}(t)\\
&\le \left(1+\xi_3^{-1}\right)k_0k_1\boldsymbol{\eta}(t)^{\mathrm{T}}\boldsymbol{X}_1^{-1}\boldsymbol{\eta}(t)+(1+\xi_3)\boldsymbol{e}(t)^{\mathrm{T}}\left(\left(1+\xi_4^{-1}\right)\boldsymbol{B}^{\mathrm{T}}\boldsymbol{L}^{\mathrm{T}}\boldsymbol{L}\boldsymbol{B}+(1+\xi_4)\boldsymbol{B}^{\mathrm{T}}\boldsymbol{K}^{\mathrm{T}}\boldsymbol{K}\boldsymbol{B}\right)\boldsymbol{e}(t)\\
&\le \left(1+\xi_3^{-1}\right)k_0k_1\boldsymbol{\eta}(t)^{\mathrm{T}}\boldsymbol{X}_1^{-1}\boldsymbol{\eta}(t)+\left(1+\xi_3^{-1}\right)k_0k_1\boldsymbol{e}(t)^{\mathrm{T}}\boldsymbol{X}_2^{-1}\boldsymbol{e}(t)\\
&= \left(1+\xi_3^{-1}\right)k_0k_1V(t)
\end{aligned} \tag{5.32}$$

According to inequalities (5.18) and (5.19), one knows

$$\|\dot{\boldsymbol{u}}(t)\|_2^2 < \left(1+\xi_3^{-1}\right)k_0k_1\varpi = \lambda_2^2 \tag{5.33}$$

This completes the proof.

5.3.2 Linear quadratic regulator controller and optimal sliding mode controller formulation

To illustrate the advantage of DOBC, the linear quadratic regulator controller (LQRC) and the optimal sliding mode controller (OSMC) are chosen as the reference groups. The following is the formulation of LQRC and OSMC.

For Eq. (5.2) ($\boldsymbol{w}(t) = 0$), the following quadratic performance index is considered for optimization purposes,

$$J(\boldsymbol{\eta}(t), \boldsymbol{u}(t)) = \frac{1}{2}\int_0^{\infty}\left(\boldsymbol{\eta}(t)^{\mathrm{T}}\boldsymbol{Q}(t)\boldsymbol{\eta}(t) + \boldsymbol{u}(t)^{\mathrm{T}}\boldsymbol{R}(t)\boldsymbol{u}(t)\right)dt \tag{5.34}$$

where $\boldsymbol{Q}(t)$ is a positive semidefinite matrix and $\boldsymbol{R}(t)$ is a positive definite matrix. It can be seen that this performance index consists of two parts, that is, state quadratic and control quadratic items. In order to improve control performance and reduce energy consumption, the above two weighting matrices can be selected in a compromise.

For LQRC, it is expressed as

$$\boldsymbol{u}_{\mathrm{LQR}} = -\boldsymbol{R}(t)^{-1}\boldsymbol{B}^{\mathrm{T}}\boldsymbol{P}(t)\boldsymbol{\eta}(t) \tag{5.35}$$

where $\boldsymbol{P}(t)$ is determined by the following Riccati equation, that is,

$$\boldsymbol{A}^{\mathrm{T}}\boldsymbol{P}(t) + \boldsymbol{P}(t)\boldsymbol{A} - \boldsymbol{P}(t)\boldsymbol{B}\boldsymbol{R}(t)^{-1}\boldsymbol{B}^{\mathrm{T}}\boldsymbol{P}(t) + \boldsymbol{Q}(t) = 0 \tag{5.36}$$

To deal with the lumped disturbance, based on LQRC, the OSMC is proposed, with the integral sliding surface considered as

$$\boldsymbol{s}(\boldsymbol{\eta}, t) = \boldsymbol{B}^{\mathrm{T}}(\boldsymbol{\eta}(t) - \boldsymbol{\eta}(0)) - \boldsymbol{B}^{\mathrm{T}}\int_0^t\left(\boldsymbol{A} - \boldsymbol{B}\boldsymbol{R}(\tau)^{-1}\boldsymbol{B}^{\mathrm{T}}\boldsymbol{P}(\tau)\right)\boldsymbol{\eta}(\tau)d\tau \tag{5.37}$$

An equivalent control input can be obtained with $\dot{s}(\eta, t) = 0$, that is,

$$\boldsymbol{u}_{\mathrm{OSM}_{eq}} = -\boldsymbol{R}(t)^{-1}\boldsymbol{B}^{\mathrm{T}}\boldsymbol{P}(t)\boldsymbol{\eta}(t) \tag{5.38}$$

which is the same as the LQR controller (5.35). In order to make the control system robust against the lumped disturbance, the OSMC is designed as

$$\boldsymbol{u}_{\mathrm{OSM}} = -\boldsymbol{R}(t)^{-1}\boldsymbol{B}^{\mathrm{T}}\boldsymbol{P}(t)\boldsymbol{\eta}(t) - \left(\boldsymbol{B}^{\mathrm{T}}\boldsymbol{B}\right)^{-1}\tilde{\boldsymbol{k}}\mathrm{sgn}(\boldsymbol{s}) \tag{5.39}$$

where $\tilde{\boldsymbol{k}} = \mathrm{diag}([\tilde{k}_1 \quad \tilde{k}_2 \quad \tilde{k}_3])$ is the gain matrix. Choose a Lyapunov function candidate as $\tilde{V} = \frac{1}{2}\boldsymbol{s}^{\mathrm{T}}\boldsymbol{s}$, then

$$\begin{aligned}
\dot{\tilde{V}} &= \boldsymbol{s}^{\mathrm{T}}\dot{\boldsymbol{s}} \\
&= \boldsymbol{s}^{\mathrm{T}}\left(\boldsymbol{B}^{\mathrm{T}}\dot{\boldsymbol{\eta}}(t) - \boldsymbol{B}^{\mathrm{T}}\left(\boldsymbol{A} - \boldsymbol{B}\boldsymbol{R}(t)^{-1}\boldsymbol{B}^{\mathrm{T}}\boldsymbol{P}(t)\right)\boldsymbol{\eta}(t)\right) \\
&= \boldsymbol{s}^{\mathrm{T}}\left(\boldsymbol{B}^{\mathrm{T}}\left(-\boldsymbol{B}\boldsymbol{R}(t)^{-1}\boldsymbol{B}^{\mathrm{T}}\boldsymbol{P}(t)\boldsymbol{\eta}(t) - \boldsymbol{B}\left(\boldsymbol{B}^{\mathrm{T}}\boldsymbol{B}\right)^{-1}\tilde{\boldsymbol{k}}\mathrm{sgn}(\boldsymbol{s}) + \boldsymbol{B}\boldsymbol{w}(t)\right)\right)
\end{aligned}$$

$$
\begin{aligned}
&- \boldsymbol{B}^{\mathrm{T}}\left(-\boldsymbol{B}\boldsymbol{R}(t)^{-1}\boldsymbol{B}^{\mathrm{T}}\boldsymbol{P}(t)\right)\boldsymbol{\eta}(t)\Big) \\
&= \boldsymbol{s}^{\mathrm{T}}\left(-\tilde{\boldsymbol{k}}\mathrm{sgn}(\boldsymbol{s}) + \boldsymbol{B}^{\mathrm{T}}\boldsymbol{B}\boldsymbol{w}(t)\right) \\
&= \sum_{i=1}^{3}\left(|s_i|\left(-\tilde{k}_i + \frac{1}{m_0^2}\mathrm{sgn}(s_i)w_i(t)\right)\right) \\
&\leq \sum_{i=1}^{3}\left(|s_i|\left(-\tilde{k}_i + \frac{1}{m_0^2}\kappa_1\right)\right)
\end{aligned}
\tag{5.40}
$$

By choosing $\tilde{k}_i \geq \kappa_1/m_0^2$, the asymptotically stability of the closed-loop system can be guaranteed. To reduce the chattering phenomenon, the sign function $\mathrm{sgn}(\boldsymbol{s})$ can be replaced by a continuous function $\boldsymbol{\theta}(\boldsymbol{s}) = [\theta(s_1) \quad \theta(s_2) \quad \theta(s_3)]^{\mathrm{T}}$, with $\Theta(s_i) = s_i/(|s_i| + \delta)$, where δ is a very small positive constant. Then, the OSMC becomes

$$
\boldsymbol{u}_{\mathrm{OSM}} = -\boldsymbol{R}(t)^{-1}\boldsymbol{B}^{\mathrm{T}}\boldsymbol{P}(t)\boldsymbol{\eta}(t) - \left(\boldsymbol{B}^{\mathrm{T}}\boldsymbol{B}\right)^{-1}\tilde{\boldsymbol{k}}\boldsymbol{\theta}(\boldsymbol{s})
\tag{5.41}
$$

This completes the problem formulations for LQRC and OSMC.

5.4 Simulation test

In this section, the effectiveness of the disturbance observer-based controller (DOBC) is illustrated with simulations of a spacecraft electromagnetic docking system. In this example, for the "CM" orbit, the eccentricity $e = 0.15$, the semimajor axis $a = 10{,}000$km, the initial true anomaly $\Theta(0) = 0$, and m_0 is only representative of a valid mass with a value of 20kg. The unknown mass of the chaser and target spacecraft are assumed to be $m_C = 40$kg and $m_T = 100$kg, respectively. Choose the initial states as $\eta(0) = [2\mathrm{m}, 3\mathrm{m}, 4\mathrm{m}, 0\mathrm{m/s}, 0\mathrm{m/s}, 0\mathrm{m/s}]^{\mathrm{T}}$, $\boldsymbol{e}(0) = [0\mathrm{N}, 0\mathrm{N}, 0\mathrm{N}]^{\mathrm{T}}$, the gravitational parameter $\mu = 3.986 \times 10^5 \mathrm{km}^3/\mathrm{s}^2$, and the permeability of free space $\mu_0 = 4\pi \times 10^{-7}\mathrm{N{\bullet}A}^{-2}$; the magnetic moment of the target spacecraft is assumed to be a constant, that is, $\mu_T = [-2.29 \quad -8.17 \quad -2.33]^{\mathrm{T}} \times 10^5 \mathrm{Am}^2$, and the control force is constrained by $\lambda_1 = 30$N, $\lambda_2 = 300$N/s. Here, an actuator additive fault is assumed to occur, that is, $\boldsymbol{E} = \boldsymbol{B}$, with the fault signals $\boldsymbol{f}(t) = [8\ 9\ 4]^{\mathrm{T}} \times 10^{-2}\bar{f}(t)$, and the form of the fault signal factor $\bar{f}(t)$ is shown in Table 5.1.

Table 5.1 The form of fault signal factor.

t/s	[0,10]	[10,45]	[45,60]	[60,75]	[75,120]
$f(t)/\mathrm{Nm}$	0	$\sin(0.1\pi t)$	1	$\cos(0.1\pi t)$	0

The external disturbances for the chaser and target spacecraft used in the simulation are taken to be of the following form:

$$w_{\mathrm{C}}(t) = \begin{bmatrix} 0.1 + 0.2\sin(0.11\pi t) \\ 0.15 + 0.2\cos(0.11\pi t) \\ 0.2 + 0.3\cos\left(0.11\pi t + \pi/3\right) \end{bmatrix} \mathrm{N},$$

$$w_{\mathrm{T}}(t) = \begin{bmatrix} 0.1 + 0.1\sin(0.11\pi t) \\ 0.2 + 0.12\cos\left(0.11\pi t + \pi/4\right) \\ 0.3 + 0.15\cos\left(0.11\pi t + \pi/6\right) \end{bmatrix} \mathrm{N}$$

Meanwhile, choose $\xi_1 = 100$, $\xi_2 = \xi_3 = \xi_4 = 0.2$, $\kappa_1 = 10$, $\kappa_2 = 3$, and $\varpi = 0.5$, then one obtains $k_0 = 600$ and $k_1 = 50$.

With the YALMIP toolbox, solving LMIs inequalities (5.7)–(5.11) gives

$$\mathbf{X}_1 = 10^3 \times \begin{bmatrix} 2.3600 & 0.0333 & 0.0443 & -0.2639 & -0.0029 & -0.0053 \\ 0.0333 & 2.3871 & 0.0659 & -0.0029 & -0.2663 & -0.0057 \\ 0.0443 & 0.0659 & 2.4254 & -0.0024 & -0.0057 & -0.2696 \\ -0.2639 & -0.0029 & -0.0024 & 0.0499 & 0.0004 & 0.0006 \\ -0.0029 & -0.2663 & -0.0057 & 0.0004 & 0.0503 & 0.0009 \\ -0.0053 & -0.0057 & -0.2696 & 0.0006 & 0.0009 & 0.0508 \end{bmatrix}$$

$$\mathbf{X}_2 = \begin{bmatrix} 1.0065 & -0.0001 & -0.0001 \\ -0.0001 & 1.0065 & -0.0001 \\ -0.0001 & -0.0001 & 1.0064 \end{bmatrix}$$

$$\boldsymbol{W} = \begin{bmatrix} -173.4021 & -1.5066 & -2.0331 & 112.6525 & 0.8749 & 1.2809 \\ -1.3433 & -174.2471 & -2.4976 & 0.8552 & 113.3569 & 1.7161 \\ -1.6581 & -2.4123 & -175.4908 & 1.0207 & 1.7058 & 114.3443 \end{bmatrix}$$

$$\boldsymbol{H} = \begin{bmatrix} 126.7806 & -0.0033 & -0.0044 \\ -0.0033 & 126.7779 & -0.0066 \\ -0.0044 & -0.0066 & 126.7741 \end{bmatrix}$$

Thus $\varpi - \boldsymbol{x}(0)^{\mathrm{T}}\mathbf{X}_1^{-1}\boldsymbol{x}(0) + \boldsymbol{e}(0)^{\mathrm{T}}\mathbf{X}_2^{-1}\boldsymbol{e}(0) + \xi_2^{-1}\xi_1^{-1}\kappa_2^2 = 0.0218 > 0$ is satisfied. Then, the controller gain matrix and the disturbance observer gain

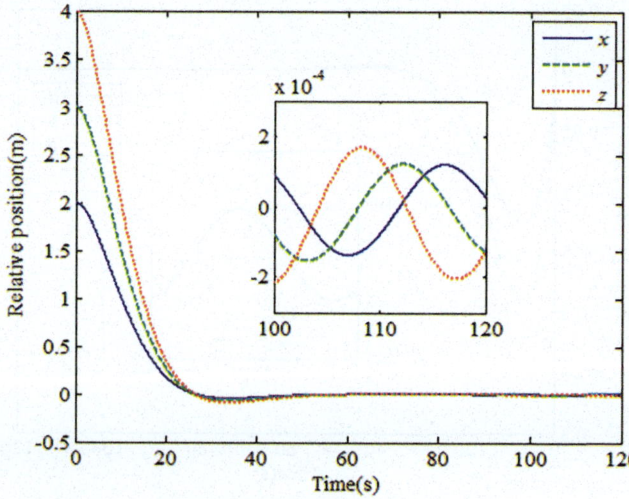

Figure 5.1 Time response of relative position.

matrix are obtained as follows

$$\boldsymbol{K} = \boldsymbol{W}\boldsymbol{X}_1^{-1} = \begin{bmatrix} 0.4385 & -0.0024 & -0.0060 & 4.5770 & -0.0113 & -0.0151 \\ -0.0024 & 0.4366 & -0.0047 & -0.0114 & 4.5670 & -0.0233 \\ -0.0003 & -0.0047 & 0.4338 & -0.0153 & -0.0234 & 4.5532 \end{bmatrix}$$

$$\boldsymbol{L} = \boldsymbol{H}(\boldsymbol{B}\boldsymbol{X}_2)^{\mathrm{T}}\left[(\boldsymbol{B}\boldsymbol{X}_2)(\boldsymbol{B}\boldsymbol{X}_2)^{\mathrm{T}}\right]^{*} = \begin{bmatrix} 0 & 0 & 0 & 2519.2 & 0.0753 & 0.1004 \\ 0 & 0 & 0 & 0.0753 & 2519.3 & 0.1508 \\ 0 & 0 & 0 & 0.1004 & 0.1508 & 2519.4 \end{bmatrix}$$

Applying the controller gain $\boldsymbol{K}$ and observer gain $\boldsymbol{L}$ to the closed-loop electromagnetic docking system expressed by Eq. (5.6) produces the simulation results of Figs. 5.1–5.6. As shown in Figs. 5.1 and 5.2, the relative position and velocity asymptotically converge to a small bound containing zero with little overshooting and steady-state errors less than 2.2×10^{-4}m and 0.75×10^{-4}m/s, respectively, thus can be safely neglected. In addition, the relative velocity satisfies the requirement of soft docking. The lumped disturbance and its estimation are plotted in Fig. 5.3, from which one can see that the observed lumped disturbance is nearly the same as the real value, which is both less than 1N. The sphere with a radius of 8×10^{-4}N in Fig. 5.4 means that the estimation error of the lumped disturbance will converge to

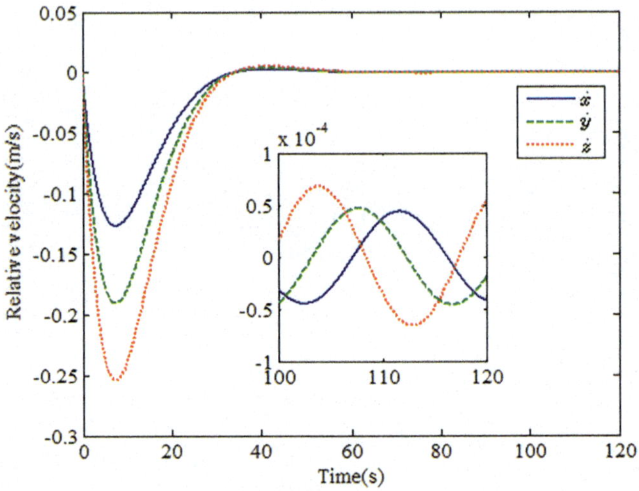

Figure 5.2 Time response of relative velocity.

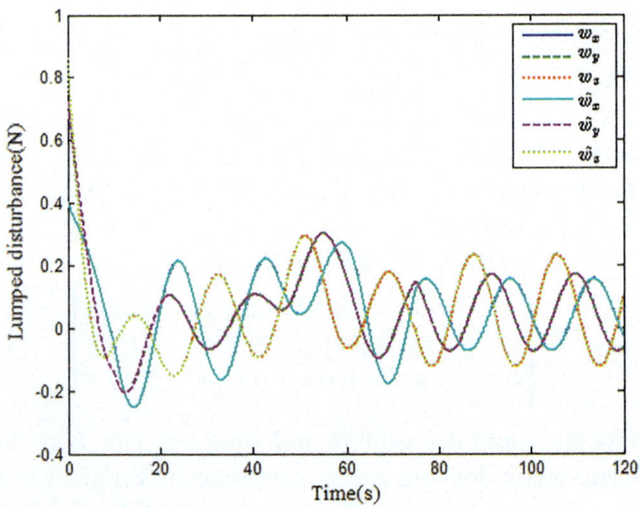

Figure 5.3 Lumped disturbance and its estimation.

a small neighborhood containing the equilibrium. The blue curve denotes estimation error out of the neighborhood, while the red curve represents that when the estimation error arrives in the convergence set, it will be attracted and stay in the sphere. The magnitude and rate of control force are plotted in Figs. 5.5 and 5.6, from which one can see that the magnitude and rate of control force along each axis are less than 3N and 0.6N/s, respectively. It is

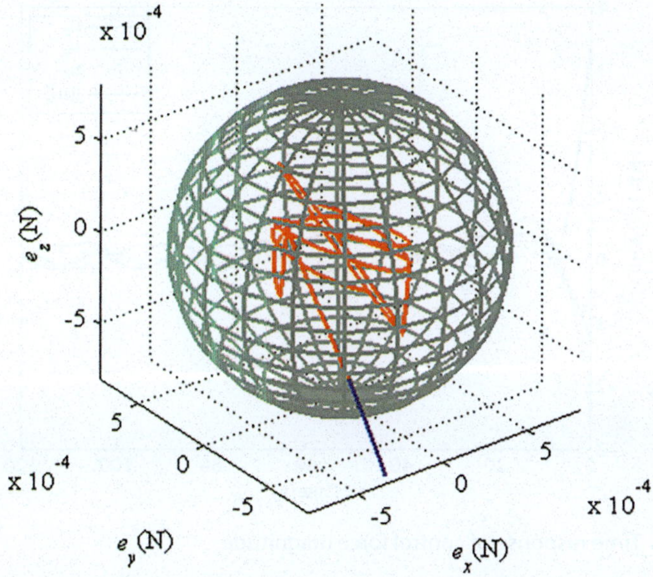

Figure 5.4 The convergence set of disturbance estimation error.

obvious that the norms of magnitude and rate of control force are less than 4N and 0.8N/s, respectively, which satisfy their limit constraints.

To further illustrate the motion of chaser and target spacecraft relative to "CM," Fig. 5.7 is presented, wherein the color bar represents the simulation time. As one can see, the relative motion distance of the chaser spacecraft is longer than that of the target spacecraft. This is because the mass of the target spacecraft is larger than the chaser spacecraft, which leads to the fact that "CM" is closer to the center of mass of the target spacecraft. Fig. 5.8 depicts the solved magnetic moment of the chaser spacecraft, from which one can see that the maximum of the magnetic moment is less than $7000Am^2$ with a small magnitude of 10^{-13} in steady state, which is practical and can be provided by a common electromagnetic mechanism.

For comparative purposes, three different controllers are employed for simulations: the benchmark controller (BC) proposed in Huang et al, (2014) without controller perturbations, the LQRC presented in Zhang, Yang, Zhu, Huang, & Cai, 2012), and the OSMC proposed in Imani and Beigzadeh (2016). The details of LQRC and OSMC can be seen in Appendix A. In these cases, the maximum control force on each channel is set to be 20N, so that the norm of the control force is approximately 30N.

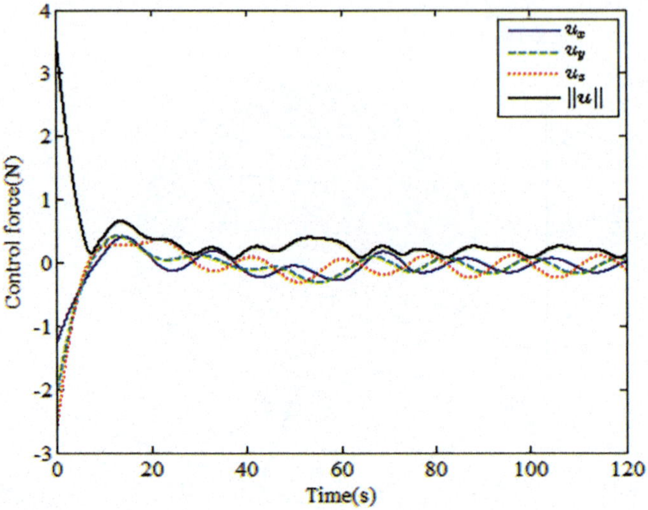

Figure 5.5 Time response of control force magnitude.

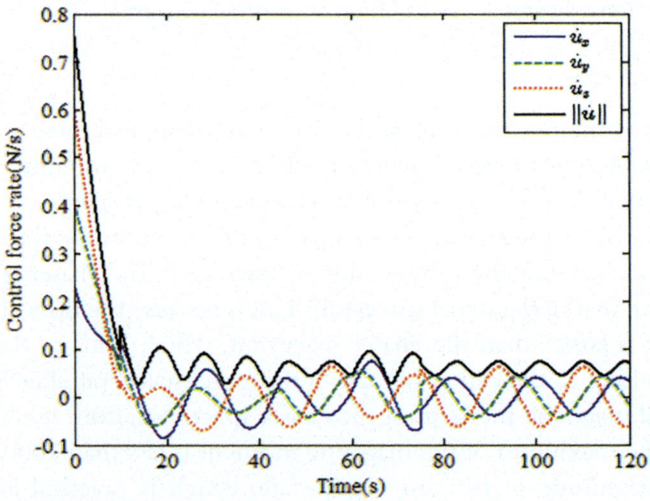

Figure 5.6 Time response of control force rate.

When the BC is adopted, the controller gain matrix $\boldsymbol{K}$ is

$$K = \begin{bmatrix} -10.4455 & 0 & 0.0113 & -23.0524 & 0 & 0 \\ 0 & -10.4455 & 0 & 0 & -23.0524 & 0 \\ -0.0113 & 0 & -10.4456 & 0 & 0 & -23.0524 \end{bmatrix}$$

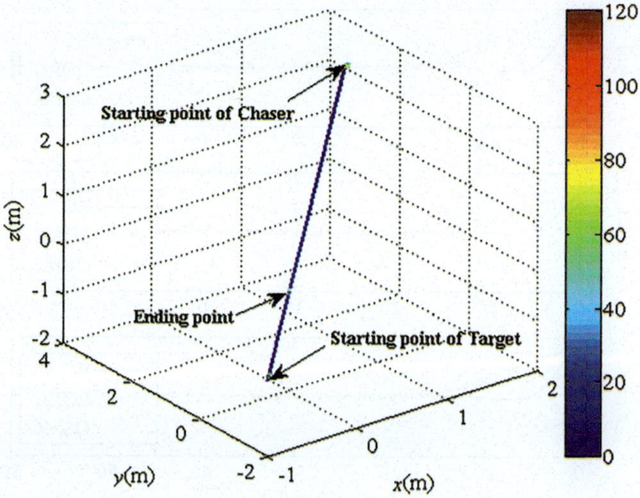

Figure 5.7 Motion of chaser and target spacecraft relative to "CM" in three-dimensional visualization.

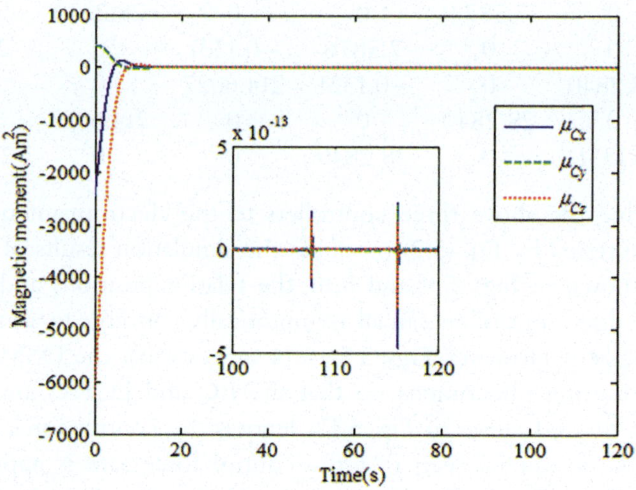

Figure 5.8 Magnetic moment of chaser spacecraft.

When the LQRC and OSMC are adopted, choose $\mathbf{Q} = \boldsymbol{I}_{6\times 6}, \mathbf{R} = 2\boldsymbol{I}_{3\times 3}, \widetilde{\boldsymbol{k}} = 0.025\boldsymbol{I}_{3\times 3}, \delta = 10^{-7}$, where $\boldsymbol{I}$ is an identity matrix, then one can obtain

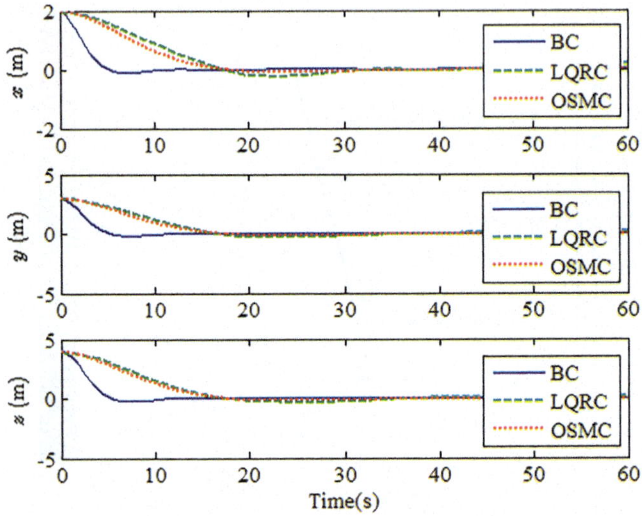

Figure 5.9 Time response of relative position using different controllers.

$$P = \begin{bmatrix} 7.5874 & 0 & 0 & 28.2840 & 0 & 0.1331 \\ 0 & 7.5874 & 0 & 0 & 28.2840 & 0 \\ 0 & 0 & 7.5876 & -0.1331 & 0 & 28.2849 \\ 28.2840 & 0 & -0.1331 & 214.6027 & 0 & 0 \\ 0 & 28.2840 & 0 & 0 & 214.6027 & 0 \\ 0.1331 & 0 & 28.2849 & 0 & 0 & 214.6063 \end{bmatrix}$$

Applying the above three controllers to the electromagnetic docking system expressed by Eq. (5.2) produces the simulation results of Figs. 5.9–5.14. As shown in Figs. 5.9 and 5.10, the relative positions and velocities under the three controllers can all asymptotically converge to a small bound containing zero. However, Figs. 5.11–5.14 suggests that the OSMC leads to severe chattering phenomena, so that the BC and LQRC are better. In particular, the red curve in Fig. 5.12 denotes the control force rate using the OSMC. It can be seen that the control force rate is approximately 2×10^5N/s, much larger than the predefined limit of 300N/s. As the OSMC is also a sliding mode control method, the chattering phenomenon in control input is very common, which will lead to a rapid variation in control force, that is to say, a very large control force rate will appear. The results shown in Figs. 5.13 and 5.14 demonstrate two things. First, the BC requires a larger control force magnitude and magnetic moment of the chaser spacecraft than the LQRC. Second, the LQRC performs better, giving better results.

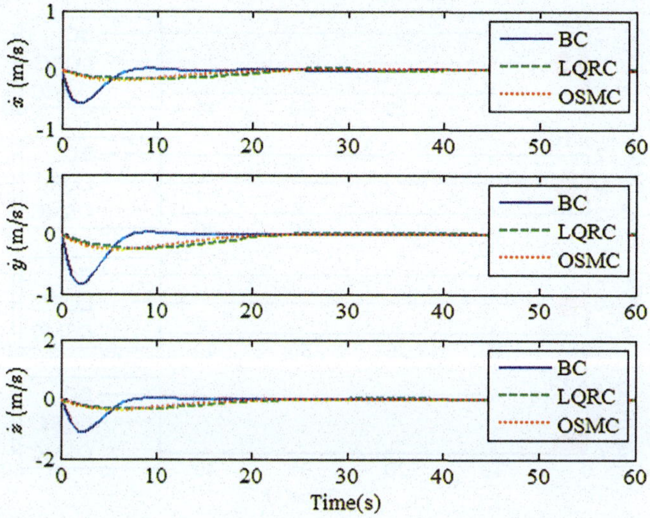

Figure 5.10 Time response of relative velocity using different controllers.

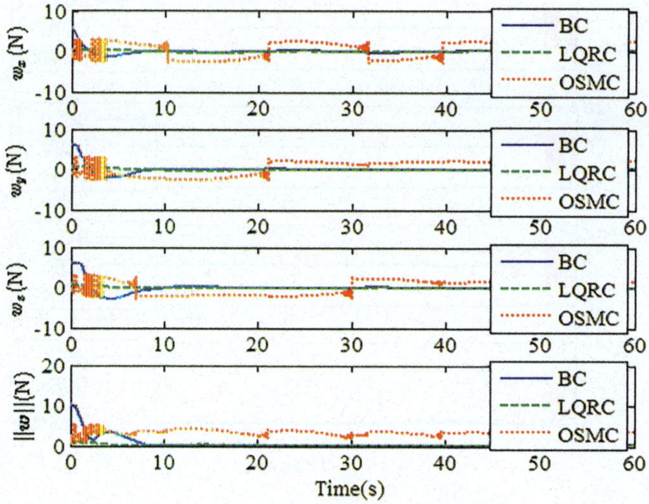

Figure 5.11 Lumped disturbance using different controllers.

In particular, Fig. 5.13 indicates that the initial magnitude of control force along each axis for BC can reach 20N with a norm more than 30N, and Fig. 5.14 implies that the magnetic moment of the chaser spacecraft can reach approximately $0.75 \times 10^5 \text{Am}^2$, both of which are much larger than

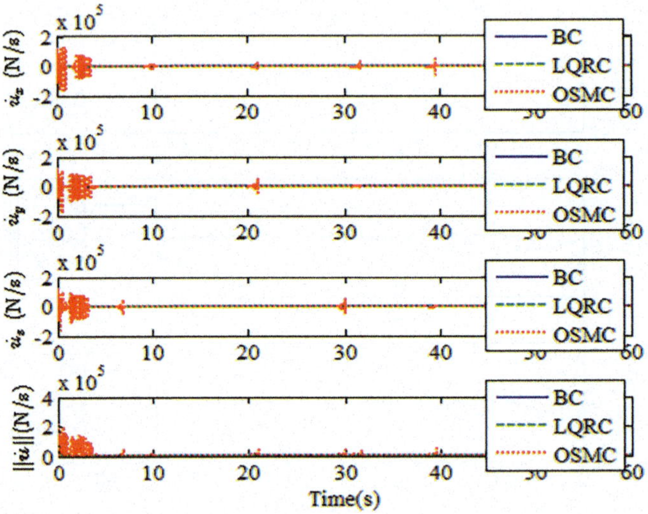

Figure 5.12 Time response of control force rate using different controllers.

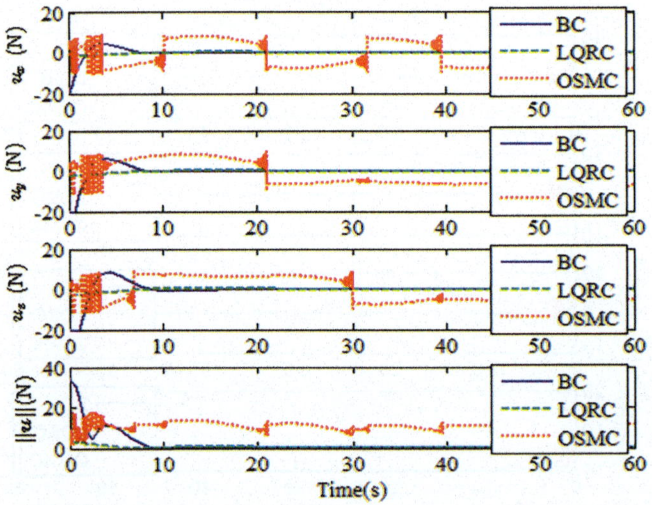

Figure 5.13 Time response of control force magnitude using different controllers.

the DOBC. In short, the LQRC is better than the other two controllers, that is, BC and OSMC, here.

Next the performance of LQRC and the proposed DOBC is compared, from the point of view of control accuracy, control force magnitude, and magnetic moment of chaser spacecraft. For control accuracy, the simulation

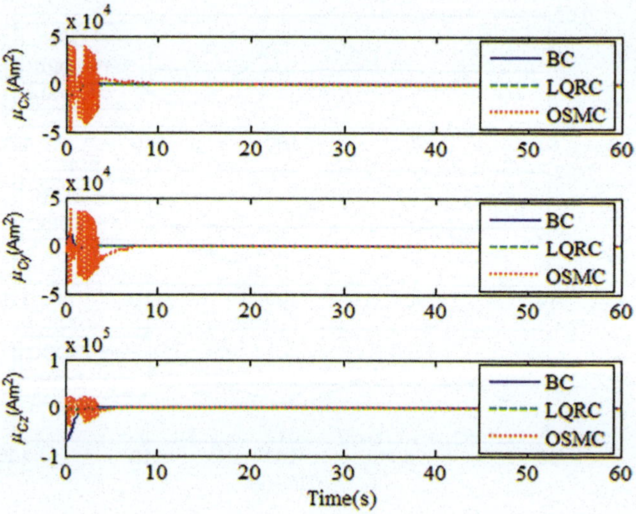

Figure 5.14 Magnetic moment of chaser spacecraft using different controllers.

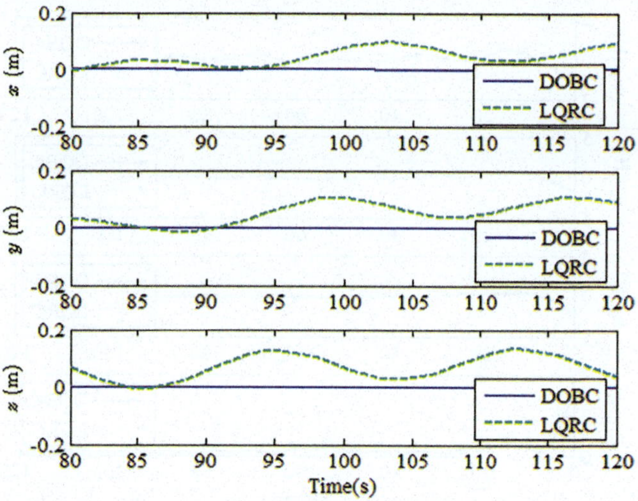

Figure 5.15 Time response of relative position using DOBC and LQRC. *DOBC*, disturbance observer-based controller; *LQRC*, linear quadratic regulator controller.

results of relative positions and velocities from 80s to 120s are taken as an example. This produces the simulation results of Figs. 5.15–5.18. As shown in Figs. 5.15 and 5.16, the steady-state errors of relative position and velocity for LQRC are approximately 0.1m and 0.02m/s respectively. Figs. 5.1 and

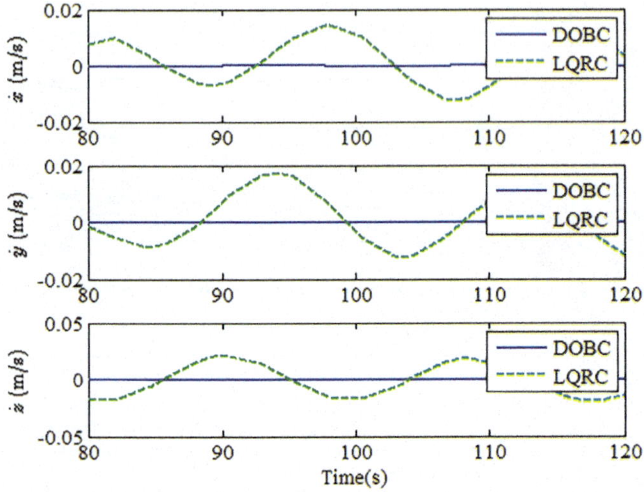

Figure 5.16 Time response of relative velocity using DOBC and LQRC. *DOBC*, disturbance observer-based controller; *LQRC*, linear quadratic regulator controller.

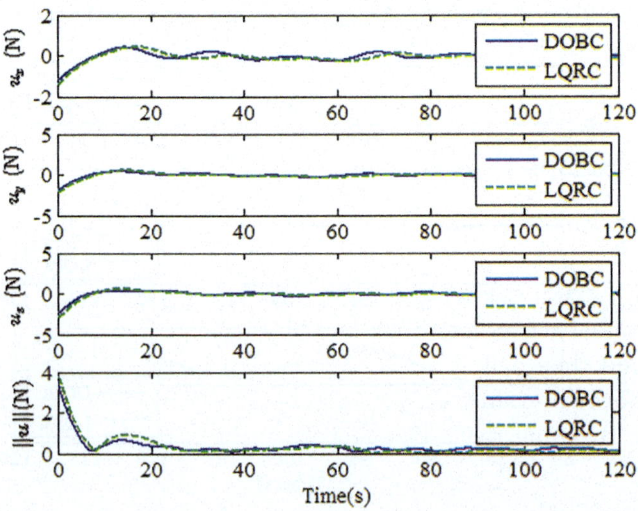

Figure 5.17 Time response of control force magnitude using DOBC and LQRC. *DOBC*, disturbance observer-based controller; *LQRC*, linear quadratic regulator controller.

5.2 have suggested that the steady-state errors for DOBC are less than 2.2×10^{-4}m and 0.75×10^{-4}m/s, respectively. The control force magnitude and magnetic moment of the chaser spacecraft are plotted in Figs. 5.17 and 5.18, from which one can see that response curves for DOBC and LQRC

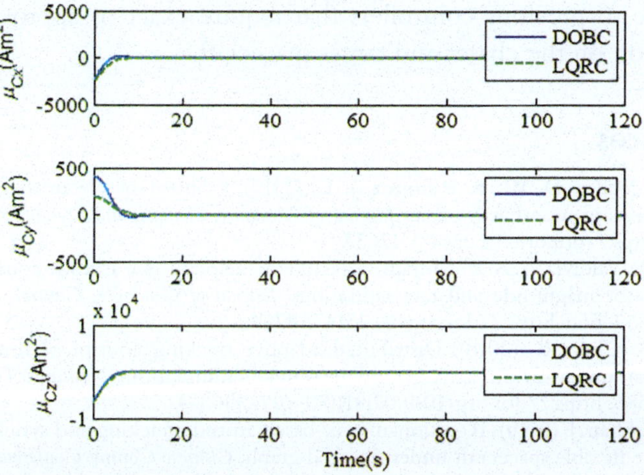

Figure 5.18 Magnetic moment of chaser spacecraft using DOBC and LQRC. *DOBC*, disturbance observer-based controller; *LQRC*, linear quadratic regulator controller.

have little difference. Thus it can be concluded that the DOBC is superior to LQRC from the perspective of control accuracy.

5.5 Conclusions

This chapter presents a novel disturbance observer-based control strategy for spacecraft electromagnetic docking in the presence of external disturbances, fault signals, unknown mass, elliptical eccentricity, and input MRCs. A new auxiliary variable is introduced first in the design of the disturbance observer, and then a new closed-loop system, which consists of relative motion information and estimation errors of the lumped disturbance, is established. By using the presented disturbance observer-based controller, the states of the closed-loop system converge asymptotically to the vicinity of the expected states with very good performance. More importantly, the effects of input MRCs and multiple unknown perturbations, such as external disturbances, fault signals, unknown mass, and elliptical eccentricity, are considered simultaneously in the design and analysis. Simulation results illustrate the effectiveness and feasibility of the proposed disturbance observer-based controller, which also demonstrates that the relative motion distances between the chaser and target spacecraft are affected by the ratio between their mass. The control scheme presented in this chapter can provide an alternative to previous controllers that require exact or approximate mass information on the chaser and target spacecraft.

References

Ahsun, U., Miller, D. W., & Ramirez, J. L. (2010). Control of electromagnetic satellite formations in near-Earth orbits. *Journal of Guidance, Control, and Dynamics, 33*(6), 1883–1891. https://doi.org/10.2514/1.47637.

Akella, M. R., Valdivia, A., & Kotamraju, G. R. (2005). Velocity-free attitude controllers subject to actuator magnitude and rate saturations. *Journal of Guidance, Control, and Dynamics, 28*(4), 659–666. https://doi.org/10.2514/1.8784.

Chen, T., & Chen, G. (2018). Distributed adaptive tracking control of multiple flexible spacecraft under various actuator and measurement limitations. *Nonlinear Dynamics, 91*(3), 1571–1586. https://doi.org/10.1007/s11071-017-3965-4.

Chen, T., & Shan, J. (2019). Rotation-matrix-based attitude tracking and synchronization of multiple flexible spacecraft under directed graph. *Chinese Control Conference, CCC*, 301–306. https://doi.org/10.23919/ChiCC.2019.8866116.

Chen, T., & Wen, H. (2018). Autonomous assembly with collision avoidance of a fleet of flexible spacecraft based on disturbance observer. *Acta Astronautica, 147*, 86–96. https://doi.org/10.1016/j.actaastro.2018.03.027.

Chen, W. H., Yang, J., Guo, L., & Li, S. (2016). Disturbance-observer-based control and related methods–An overview. *IEEE Transactions on Industrial Electronics, 63*(2), 1083–1095. https://doi.org/10.1109/TIE.2015.2478397.

Elias, L. M., Kwon, D. W., Sedwick, R. J., & Miller, D. W. (2007). Electromagnetic formation flight dynamics including reaction wheel gyroscopic stiffening effects. *Journal of Guidance, Control, and Dynamics, 30*(2), 499–511. https://doi.org/10.2514/1.18679.

Fabacher, E., Lizy-Destrez, S., Alazard, D., Ankersen, F., & Profizi, A. (2017). Guidance of magnetic space tug. *Advances in Space Research, 60*(1), 14–27. https://doi.org/10.1016/j.asr.2017.03.042.

Hallaj, M. A. A., & Assadian, N. (2015). Tethered satellite system control using electromagnetic forces and reaction wheels. *Acta Astronautica, 117*, 390–401. https://doi.org/10.1016/j.actaastro.2015.08.015.

Hu, Q., Li, B., & Qi, J. (2014). Disturbance observer based finite-time attitude control for rigid spacecraft under input saturation. *Aerospace Science and Technology, 39*, 13–21. https://doi.org/10.1016/j.ast.2014.08.009.

Huang, S. J., & Yang, G. H. (2014). Fault tolerant controller design for T-S fuzzy systems with time-varying delay and actuator faults: A k-step fault-estimation approach. *IEEE Transactions on Fuzzy Systems, 22*(6), 1526–1540. https://doi.org/10.1109/tfuzz.2014.2298053.

Huang, X., Zhang, C., Lu, H., & Li, M. (2016). Adaptive reaching law based sliding mode control for electromagnetic formation flight with input saturation. *Journal of the Franklin Institute, 353*(11), 2398–2417. https://doi.org/10.1016/j.jfranklin.2016.04.004.

Imani, A., & Beigzadeh, B. (2016). Robust control of spacecraft rendezvous on elliptical orbits: Optimal sliding mode and backstepping sliding mode approaches. *Proceedings of the Institution of Mechanical Engineers, Part G: Journal of Aerospace Engineering, 230*(10), 1975–1989. https://doi.org/10.1177/0954410015620710.

Kong, E.M.C. (2002). Spacecraft formation flight exploiting potential fields. Ph.D. thesis, Massachusetts Institute of Technology, 145–146.https://dspace.mit.edu/handle/1721.1/16835

Kong, E. M. C., Kwon, D. W., Schweighart, S. A., Elias, L. M., Sedwick, R. J., & Miller, D. W. (2004). Electromagnetic formation flight for multisatellite arrays. *Journal of Spacecraft and Rockets, 41*(4), 659–666. https://doi.org/10.2514/1.2172.

Leonessa, A., Haddad, W. M., Hayakawa, T., & Morel, Y. (2009). Adaptive control for nonlinear uncertain systems with actuator amplitude and rate saturation constraints. *International Journal of Adaptive Control and Signal Processing, 23*(1), 73–96. https://doi.org/10.1002/acs.1065.

Li, Q, Yuan, J., Zhang, B, & Wang, H. (2018). Disturbance observer based control for spacecraft proximity operations with path constraint. *Aerospace Science and Technology, 79*, 154–163. https://doi.org/10.1016/j.ast.2018.05.042.

Liu, C., Vukovich, G., Sun, Z., & Shi, K. (2018). Observer-based fault-tolerant attitude control for spacecraft with input delay. *Journal of Guidance, Control, and Dynamics, 41*(9), 2039–2051. https://doi.org/10.2514/1.g003555.

Liu, Z., Liu, J., & Wang, L. (2018). Disturbance observer based attitude control for flexible spacecraft with input magnitude and rate constraints. *Aerospace Science and Technology, 72*, 486–492. https://doi.org/10.1016/j.ast.2017.11.036.

Sanyal, A., Fosbury, A., Chaturvedi, N., & Bernstein, D. S. (2009). Inertia-free spacecraft attitude tracking with disturbance rejection and almost global stabilization. *Journal of Guidance, Control, and Dynamics, 32*(4), 1167–1178. https://doi.org/10.2514/1.41565.

Song, S., Ge, L., Ma, S., Zhang, M., & Wang, L. (2015). Accurate measurement and detailed evaluation of static electromagnetic characteristics of switched reluctance machines. *IEEE Transactions on Instrumentation and Measurement, 64*(3), 704–714. https://doi.org/10.1109/tim.2014.2358132.

Sun, L., & Zheng, Z. (2018). Disturbance observer-based robust saturated control for spacecraft proximity maneuvers. *IEEE Transactions on Control Systems Technology, 26*(2), 684–692. https://doi.org/10.1109/tcst.2017.2669145.

Weiss, A., Kolmanovsky, I., Bernstein, D. S., & Sanyal, A. (2013). Inertia-free spacecraft attitude control using reaction wheels. *Journal of Guidance, Control, and Dynamics, 36*(5), 1425–1439. https://doi.org/10.2514/1.58363.

Wu, S. N., Sun, X. Y., Sun, Z. W., & Wu, X. D. (2010). Sliding-mode control for staring-mode spacecraft using a disturbance observer. *Proceedings of the Institution of Mechanical Engineers, Part G: Journal of Aerospace Engineering, 224*(2), 215–224. https://doi.org/10.1243/09544100jaero631.

Zhang, C., & Huang, X. L. (2016). Angular-momentum management of electromagnetic formation flight using alternating magnetic fields. *Journal of Guidance, Control, and Dynamics, 39*(6), 1292–1302. https://doi.org/10.2514/1.g001529.

Zhang, Y., Yang, L., Zhu, Y., Huang, H., & Cai, W. (2012). Nonlinear 6-DOF control of spacecraft docking with intersatellite electromagnetic force. *Acta Astronautica, 77*, 97–108. https://doi.org/10.1016/j.actaastro.2012.03.020.

Zhang, Y., Yang, L., Zhu, Y., Huang, H., & Cai, W. (2013). Angular-momentum management of spacecraft electromagnetic docking considering the Earth's magnetic field. *Journal of Guidance, Control, and Dynamics, 36*(3), 860–869. https://doi.org/10.2514/1.57723.

Zhang, Y., Yang, L., Zhu, Y., Huang, H., & Cai, W. (2015). An investigation of on-orbit release with intersatellite electromagnetic force. *Aerospace Science and Technology, 45*, 309–315. https://doi.org/10.1016/j.ast.2015.04.021.

Zhang, Y. W., Yang, L. P., Zhu, Y. W., Ren, X. H., & Huang, H. (2011). Self-docking capability and control strategy of electromagnetic docking technology. *Acta Astronautica, 69*(11–12), 1073–1081. https://doi.org/10.1016/j.actaastro.2011.07.001.

Zou, A. M., Ruiter, A. H. J., & Kumar, K. D. (2019). Disturbance observer-based attitude control for spacecraft with input MRS. *IEEE Transactions on Aerospace and Electronic Systems, 55*(1), 384–396. https://doi.org/10.1109/taes.2018.2852369.

CHAPTER 6

Model predictive control for spacecraft electromagnetic docking

6.1 Introduction

Recent years have witnessed the burgeoning demand for on-orbit servicing, satellite assembly, and active debris removal missions. With the expansion of deep space exploration missions and the surge in on-orbit service demands, spacecraft on-orbit refueling technology has become a research focus in space operations due to its core role in extending mission cycles and reducing launch costs. The realization of this technology relies on high-precision, high-safety rendezvous and docking control. However, service spacecraft carrying large-capacity propellant tanks face multiple challenges: First, the sloshing and distribution changes of liquid fuel induce liquid-solid interaction forces that are closely related to spacecraft attitude-orbit motion, leading to a dramatic increase in the complexity of coupled attitude-orbit dynamics modeling. Second, the flammable and explosive characteristics of propellants require strict avoidance of mechanical collisions and thermodynamic risks during docking. Third, traditional thruster-based docking methods have inherent defects such as plume contamination, high fuel consumption, and impact loads, making them difficult to meet long-term on-orbit service requirements.

Conventional docking methodologies predominantly rely on physical contact mechanisms such as mechanical probe-cone systems or robotic capture interfaces. These contact-based approaches, while proven in past missions, inherently introduce risks of collision, mechanical shock, plume pollution, and impingement from thruster firings, potentially jeopardizing the integrity of the involved spacecraft. Noncontact electromagnetic docking technology enables flexible pose regulation between spacecraft through controllable coils, offering advantages such as rapid response, no fuel consumption, and reversible operation, providing an innovative solution for on-orbit refueling. However, its engineering applications remain

Spacecraft Electromagnetic Docking and Separation.
DOI: https://doi.org/10.1016/B978-0-443-49334-8.00009-4

constrained by multiple technical bottlenecks: First, the electromagnetic force exhibits nonlinear attenuation characteristics with increasing distance and strong coupling with relative attitude, resulting in difficulties in high-precision modeling and leading to problems in the use of proportional integral derivative (PID) control methods in existing electromagnetic docking systems (Foust, Lupu, Nakka, Chung, & Hadaegh, 2018). Second, multi-source uncertainties, including space magnetic interference and propellant sloshing, significantly reduce controller robustness. Additionally, existing research mainly focuses on linear time-invariant systems or single disturbance compensation, lacking comprehensive consideration of elliptical orbit time-varying characteristics and multiobjective optimization requirements. Although scholars have proposed control strategies based on disturbance observers (Shi et al., 2019, 2020), extended state observers (Zhang et al., 2013), active disturbance rejection control (Liu et al., 2022), and nonfragile control (Liu et al., 2021) to improve control accuracy under multisource disturbances, these methods still have limitations in energy consumption optimization, constraint compatibility, and computational efficiency. It is, therefore, evident that the controller design is paramount in addressing the core challenges of coupling, external disturbances, and constraints in electromagnetic docking.

To address these challenges, model predictive control (MPC) offers a powerful alternative that is particularly well-suited to the electromagnetic docking problem. MPC has demonstrated unique advantages in complex system control domains such as vehicle platoons (Hu et al., 2022), UAVs (Romero et al., 2022), and spacecraft (Weiss, Baldwin, Erwin, & Kolmanovsky, 2015) due to its rolling optimization, explicit constraint handling, and multivariable coordination capabilities. Besides, MPC's optimization framework enables the design of multi-objective cost functions that can balance the docking accuracy and energy consumption. However, standard MPC is sensitive to disturbances and difficult to apply directly to scenarios with strong uncertainties. Tube MPC (TMPC), by introducing robust invariant sets and feedback compensation mechanisms, constrains system states within neighborhoods of nominal trajectories while maintaining robustness and reducing computational complexity. It has become an efficient control strategy for handling bounded uncertainties (Shi & Zhang, 2021) and has been successfully applied in UAV control (Li et al., 2023). Mayne et al. (2005) and Ghaemi et al. (2008, 2012) designed a less conservative TMPC method under bounded disturbances and provided stability proofs. However, due to the time-varying nature of electromagnetic docking systems for elliptical orbit spacecraft, these methods cannot be directly applied.

Aiming at on-orbit refueling missions for elliptical orbit spacecraft, this chapter proposes an improved TMPC strategy with the following main innovations: (1) Establishing a liquid-solid-magnetic coupled dynamics model for electromagnetic docking of on-orbit refueling spacecraft in elliptical orbits, and designing an improved TMPC controller adapted to elliptical orbits. This controller features a time-varying system adaptive constraint tightening mechanism compatible with elliptical orbit parameter variations, and proposes a multiobjective optimization index balancing energy consumption and precision to achieve dynamic equilibrium between control performance and energy consumption. (2) Verifying the method's effectiveness through rigorous proofs of closed-loop system recursive feasibility and stability, combined with numerical simulations and proportional derivative (PD) control comparisons, providing theoretical support and technical references for realizing on-orbit refueling through electromagnetic docking.

6.2 Problem formulation

Let the state vector be $\boldsymbol{x}(t) = [x \quad y \quad z \quad \dot{x} \quad \dot{y} \quad \dot{z}]^{\mathrm{T}}$, the control vector is $\boldsymbol{u}(t) = [F_{Cx} \quad F_{Cy} \quad F_{Cz}]^{\mathrm{T}}$, then the dynamic model introduced in Section 1.2, Eq. (1.78), can be converted into the following state-space form:

$$\dot{\boldsymbol{x}}(t) = (\boldsymbol{A}_{\mathrm{c}} + \Delta\boldsymbol{A}_{\mathrm{c}}(t))\boldsymbol{x}(t) + \boldsymbol{B}_{\mathrm{c}}\boldsymbol{u}(t) + \boldsymbol{w}(t) \tag{6.1}$$

where $\boldsymbol{w}(t) = \boldsymbol{B}_{\mathrm{c}}\boldsymbol{w}_0(t)$. Corresponding to the continuous-time system mentioned in Section 1.3 of Chapter 1, the system matrix $\boldsymbol{A}$, uncertainty matrix $\Delta\boldsymbol{A}$, and control matrix $\boldsymbol{B}$ are denoted here as $\boldsymbol{A}_{\mathrm{c}}$, $\Delta\boldsymbol{A}_{\mathrm{c}}(t)$ and $\boldsymbol{B}_{\mathrm{c}}$ to distinguish a continuous-time system from a discrete-time system.

Since the docking system operates in an elliptical orbit, its system matrix exhibits time-varying characteristics. To facilitate the design of the model predictive controller, the continuous electromagnetic docking system needs to be discretized. To this end, the following assumption is proposed.

Assumption 6.1 Within one sampling period T, the system matrix remains constant, that is,

$$\boldsymbol{A}_{\mathrm{c}} + \Delta\boldsymbol{A}_{\mathrm{c}}(t) = \boldsymbol{A}_{\mathrm{c}} + \Delta\boldsymbol{A}_{\mathrm{c}}(t_0),\ t \in [t_0, t_0 + T) \tag{6.2}$$

where t_0 is the sample time.

Remark 6.1 Assumption 6.1 is reasonable because $\Delta \boldsymbol{A}_c(t)$ is a periodic time-varying matrix with a period of T_O, where T_O is the orbital period. Typically, the sampling period satisfies $T \ll T_O$. Therefore, within a single sampling interval, the elements of $\Delta \boldsymbol{A}_c(t)$ vary slowly and can be approximated as constant.

Based on Assumption 6.1, the continuous-time model Eq. (6.1) is discretized using the zero-order hold (ZOH) method, yielding the discrete-time state-space model as:

$$\boldsymbol{x}(k+1) = (\boldsymbol{A} + \Delta \boldsymbol{A}(k))\boldsymbol{x}(k) + \boldsymbol{B}\boldsymbol{u}(k) + \boldsymbol{w}(k) \tag{6.3}$$

where $\boldsymbol{A} + \Delta \boldsymbol{A}(k) = \boldsymbol{\Phi}(t_{k+1}, t_k) = e^{\boldsymbol{A}_c + \Delta \boldsymbol{A}_c(t_k)}$is the state transition matrix of the docking system Eq. (6.1) at time t_k; $\boldsymbol{B} = \int_{t_k}^{t_{k+1}} \boldsymbol{\Phi}(\xi, t_k) d\xi \boldsymbol{B}_c$. Since the magnitude of $\Delta \boldsymbol{A}$ is much smaller than $\boldsymbol{A}$ and the simulation duration is short, matrix $\boldsymbol{B}$ can be approximated as a constant matrix, therefore: $\boldsymbol{B} = \int_{t_k}^{t_{k+1}} e^{\boldsymbol{A}_c} d\xi \boldsymbol{B}_c$.

For the actual system Eq. (6.3), the following prediction system is defined:

$$\boldsymbol{x}_{i+1}(k) = (\boldsymbol{A} + \Delta \boldsymbol{A}(k))\boldsymbol{x}_i(k) + \boldsymbol{B}\boldsymbol{u}_i(k) \tag{6.4}$$

where $\boldsymbol{x}_i(k)$ and $\boldsymbol{u}_i(k)$ is the state and input vectors of the prediction system at the k-th sample time.

Control input, the electromagnetic force, is calculated through the far-field model introduced in Chapter 1, Section 2. The disturbance considered in this chapter includes

1. Disturbance forces induced by the external environment

   The spacecraft's orbit around Earth is subject to multiple perturbative forces, including Earth's nonspherical gravitational perturbations, atmospheric drag perturbations, solar radiation pressure perturbations, and geomagnetic field perturbations, etc. For low-Earth-orbit (LEO) spacecraft, the dominant perturbations are the second-order zonal harmonic terms (J_2 effects) of Earth's gravitational field and atmospheric drag. The magnitude of these perturbations is approximately $O(10^{-3})$.

2. Disturbance forces induced by liquid sloshing in tanks

   To achieve on-orbit refueling, the chaser spacecraft carries propellant-filled tanks. During maneuvers, the liquid inside the tanks interacts with the tank walls, generating internal disturbance forces. The liquid-spacecraft coupling dynamics can be modeled by the following moving pulsating ball model (MPBM) (Vreeburg, 1997):

$$\boldsymbol{F}_{\mathrm{L}} = N\boldsymbol{e} + \boldsymbol{F}_{\mathrm{b}} \tag{6.5}$$

where N is the magnitude of the normal force at the contact point between the pulsating ball and the storage tank, $\boldsymbol{F}_{\mathrm{b}}$ is the tangential friction force at the contact point between the pulsating ball and the storage tank. The expression for N is given by

$$N = \frac{3m_s}{8}\left\{r_S[\boldsymbol{e}\times(\boldsymbol{\Omega}+\boldsymbol{w})]^2 - \boldsymbol{e}\frac{d\boldsymbol{V}_C}{dt}\right\} + \frac{m_s}{4}(R_t - r_S)(\boldsymbol{\omega}_S+\boldsymbol{\Omega})^2 - 5\pi\sigma(R_t - r_S) \tag{6.6}$$

where m_s is the mass of the pulsating ball; $\boldsymbol{r}_S$ is the position vector of the pulsating ball's center of mass in the spacecraft body-fixed frame, r_S is its magnitude and $\boldsymbol{e}$ is its unit vector; $\boldsymbol{\Omega}$ is the spacecraft's angular velocity in the inertial frame; $\boldsymbol{w} = \boldsymbol{e} \times (d\boldsymbol{e}/dt)$; $\boldsymbol{V}_C$ is the spacecraft's translational velocity in the inertial frame; R_t is the equivalent radius of the storage tank; $\boldsymbol{\omega}_S$ is the pulsating ball's angular velocity in the body-fixed frame; σ is the liquid surface tension.

The approximate expression for $\boldsymbol{F}_{\mathrm{b}}$ is

$$\boldsymbol{F}_{\mathrm{b}} = \frac{5\mu_l m_s}{L^2}\boldsymbol{V}_{\mathrm{u}} \tag{6.7}$$

where μ_l is the kinematic viscosity coefficient of the liquid; $\boldsymbol{V}_{\mathrm{u}}$ is the relative tangential translational velocity of the pulsating ball at the contact point, $\boldsymbol{V}_{\mathrm{u}} = -\boldsymbol{e} \times (\boldsymbol{e} \times \boldsymbol{V}_S + L\boldsymbol{\omega}_s)$; $\boldsymbol{V}_S$ is the translational velocity of the pulsating ball relative to the storage tank; L is the radius of the pulsating ball. When the radius of the pulsating ball reaches its minimum value, its radial velocity component $\boldsymbol{e} \bullet (\boldsymbol{e} \bullet \boldsymbol{V}_S)$ reverses direction. At this point, not only does an interaction force exist between the pulsating sphere and the spacecraft, but momentum exchange also occurs:

$$(m_c - m_S)\boldsymbol{V}_C^* = (m_c - m_S)\boldsymbol{V}_C + (1+\kappa)m_S\boldsymbol{e}\cdot(\boldsymbol{e}\cdot\boldsymbol{V}_S) \tag{6.8}$$

where $\boldsymbol{V}_C^*$ is the instantaneous translational velocity of the spacecraft's center of mass after collision; $\kappa \in [0, 1]$ is the coefficient of restitution, while $\kappa = 1$ in the case of a perfectly elastic collision.

At this point, under the assumption of spacecraft attitude stability, the disturbance force generated by the liquid inside the tank during the spacecraft's relative motion can be determined.

6.3 TMPC controller design

6.3.1 Design of TMPC controller

The control objective of this section is to design a TMPC controller for the electromagnetic docking spacecraft system Eq. (6.3) with time-varying control matrices and multiple disturbances, ensuring that:

$$\begin{cases} \boldsymbol{x}(k) \in X \\ \boldsymbol{u}(k) \in U \\ \boldsymbol{w}(k) \in W \end{cases} \tag{6.9}$$

where X is a closed convex set; U,W are compact convex sets containing the zero element. The system state asymptotically converges to the robust invariant set Z of the system $\boldsymbol{x}(k+1) = (\boldsymbol{A}_{\mathrm{K}} + \Delta\boldsymbol{A}(k))\boldsymbol{x}(k) + \boldsymbol{w}(k)$, which satisfies

$$(\boldsymbol{A}_{\mathrm{K}} + \Delta\boldsymbol{A})Z \oplus W \subseteq Z, \ \forall||\Delta\boldsymbol{A}||_2 \le \rho \tag{6.10}$$

where $\boldsymbol{A}_{\mathrm{K}} = \boldsymbol{A} + \boldsymbol{B}\boldsymbol{K}$; Matrix $\boldsymbol{A}_{\mathrm{K}} + \Delta\boldsymbol{A}(k)$ is Schur stable.

According to the prediction system Eq. (6.4), the state of the system Eq. (6.3) can be predicted during the prediction horizon, with state sequence and control sequence

$$\boldsymbol{X}(k) = \{\boldsymbol{x}_0(k), \boldsymbol{x}_1(k), \ldots, \boldsymbol{x}_i(k), \ldots, \boldsymbol{x}_N(k)\}$$

$$\boldsymbol{U}(k) = \{\boldsymbol{u}_0(k), \boldsymbol{u}_1(k), \ldots, \boldsymbol{u}_i(k), \ldots, \boldsymbol{u}_{N-1}(k)\}$$

where $\boldsymbol{x}_i(k), \boldsymbol{u}_i(k)$ are the predicted state and the corresponding control input at time $k+i$ at the current time k, respectively.

There are the following constraints on the prediction system.

1. Predictive state constraints

   The state and input of the prediction system satisfy the prediction system Eq. (6.4), that is,

$$\boldsymbol{x}_{i+1}(k+1) = (\boldsymbol{A} + \Delta\boldsymbol{A}(k))\boldsymbol{x}_i(k) + \boldsymbol{B}\boldsymbol{u}_i(k), i = 0, 1, \ldots, N-1 \tag{6.11}$$

The $\boldsymbol{x}_0(k)$ should satisfy the following initial constraint

$$\boldsymbol{x}_0(k) \in X \sim Z \tag{6.12}$$

where $\sim$ represents the Pontryagin difference between two sets, that is, for arbitrary sets S, T, one has $S \sim T = \{x|x+t \in S, \forall t \in T\}$.

Besides, for each predicted state, it is required to lie within the predicted state admissible set, that is,

$$\boldsymbol{x}_i(k) \in \begin{cases} \overline{X} \triangleq X \sim Z & i = 0, 1, ..., N-1 \\ X_{\mathrm{f}} & i = N \end{cases} \tag{6.13}$$

where X_{f} is the terminal constraint set satisfying

$$\begin{cases} X_{\mathrm{f}} \subset \overline{X} \\ (\boldsymbol{A}_{\mathrm{K}} + \Delta\boldsymbol{A})X_{\mathrm{f}} \subset X_{\mathrm{f}}, \forall k > 0 \\ \boldsymbol{K}X_{\mathrm{f}} \subset \overline{U} \\ J_{\mathrm{f}}((\boldsymbol{A}_{\mathrm{K}} + \Delta\boldsymbol{A})\boldsymbol{x}) + L(\boldsymbol{x}, \mathrm{K}\mathrm{x}) \leq J_{\mathrm{f}}(\boldsymbol{x}), \forall \boldsymbol{x} \in X_{\mathrm{f}} \end{cases} \tag{6.14}$$

2. Predictive input constraints

For each predicted input, it is required to lie within the predicted input admissible set, that is,

$$\boldsymbol{u}_i(k) \in \overline{U} \triangleq U \sim \boldsymbol{K}Z \quad i \in \{0, 1, ..., N-1\} \tag{6.15}$$

The constraints (6.11)-(6.15) are collectively denoted as the constraint set $C(\boldsymbol{x})$.

The following optimization function is designed for the spacecraft electromagnetic docking problem:

$$J(k, \boldsymbol{X}, \boldsymbol{U}) = \sum_{i=0}^{N-1} L(\boldsymbol{x}_i(k), \boldsymbol{u}_i(k)) + J_{\mathrm{f}}(\boldsymbol{x}_N(k)) \tag{6.16}$$

with

$$L(\boldsymbol{x}_i(k), \boldsymbol{u}_i(k)) = \frac{1}{2}\left(\boldsymbol{x}_i^{\mathrm{T}}(k)\boldsymbol{Q}\boldsymbol{x}_i(k) + \boldsymbol{u}_i^{\mathrm{T}}(k)\boldsymbol{R}\boldsymbol{u}_i(k)\right)$$

$$J_{\mathrm{f}}(\boldsymbol{x}_N(k)) = \frac{1}{2}\boldsymbol{x}_N^{\mathrm{T}}(k)\boldsymbol{P}\boldsymbol{x}_N(k)$$

where $\boldsymbol{Q}$, $\boldsymbol{R}$ are given positive-definite matrix and semi-positive-definite matrix, respectively; $\boldsymbol{P}$ is the symmetric positive-definite matrix obtained by solving Theorem 1. In the optimization function, L represents the process cost, and J_{f} denotes the terminal cost. Its domain of definition is X_N.

In summary, solving the spacecraft electromagnetic docking problem is transformed into addressing the following optimization problem:

$$\begin{cases} \min\limits_{x_0(k), \boldsymbol{U}} J(k, \boldsymbol{X}, \boldsymbol{U}) \\ \text{s.t.} \quad C(\boldsymbol{x}) \end{cases} \tag{6.17}$$

When the optimazation problem represented by Eq. (6.17) is solvable, one can get the optimal control sequence $\boldsymbol{U}^*(k)$, state sequence $\boldsymbol{X}^*(k)$ and cost function $J^*(k, \boldsymbol{X}, \boldsymbol{U})$. Define the optimal input element

$$\boldsymbol{\kappa}_N^*(k) \triangleq \boldsymbol{u}_0^*(k) \tag{6.18}$$

The model predictive control input of the electromagnetic docking system represented by Eq. (6.3) is chosen as

$$\boldsymbol{u}(k) = \boldsymbol{\kappa}_N^*(k) + \boldsymbol{K}\big(\boldsymbol{x}(k) - \boldsymbol{x}_0^*(k)\big) \tag{6.19}$$

The TMPC algorithm of the spacecraft electromagnetic docking system is presented in Algorithm 6.1.

Algorithm 6.1 : Time-varying system tube model predictive control algorithm

1. Initialization

   Given orbital semimajor axis a, eccentricity e, and spacecraft masses m_C, m_T. Set prediction horizon N, sampling time T. Initialize state $\boldsymbol{x}(0)$ and true anomaly $\theta(0)$. Set initial time step $k = 0$.

2. Control computation

   By solving Problem (6.17), the optimal control input element $\boldsymbol{\kappa}_N^*(k)$ at the current time instant is obtained. Combined with state feedback, the actual control input $\boldsymbol{u}(k)$ for the system at time k is derived.

3. State update and termination check

   Update spacecraft state $\boldsymbol{x}(k)$ and true anomaly $\theta(k)$. If docking is achieved, terminate algorithm; else set k to $k + 1$ and return to step 2.

6.3.2 Feasibility analysis of algorithm iterations

The premise of Algorithm 1 is that there exists a matrix $\boldsymbol{K}$ such that $A_K + \Delta\boldsymbol{A}(k)$ is Schur stable. The following theorem provides a method to solve for $\boldsymbol{K}$. First, it is necessary to introduce the following lemma.

Lemma 6.1 If the perturbation $\Delta\boldsymbol{A}(t)$ is sufficiently small, that is, there exists a sufficiently small positive constant υ such that $||\Delta\boldsymbol{A}(t)||_2 \leq \upsilon$, then for a finite sampling time T, the discretized matrix $||\Delta\boldsymbol{A}(k)||_2$ remains bounded and satisfies:

$$||\Delta\boldsymbol{A}(k)||_2 \leq \upsilon T e^{||\boldsymbol{A}_c||_2 T} \tag{6.20}$$

Proof: Matrixes $\boldsymbol{A}_c + \Delta\boldsymbol{A}_c(t)$ and $\boldsymbol{A} + \Delta\boldsymbol{A}(k)$ at time $t = kT$ satisfy the following equation

$$\boldsymbol{A} + \Delta\boldsymbol{A}(k) = e^{(\boldsymbol{A}_c + \Delta\boldsymbol{A}_c(t))T}$$

Expanding the right-hand side (RHS) of the above equation using Taylor series yields:

$$\mathrm{e}^{(\boldsymbol{A}_{\mathrm{c}}+\Delta\boldsymbol{A}_{\mathrm{c}})T} = \mathrm{e}^{\boldsymbol{A}_{\mathrm{c}}T} + \int_0^T \mathrm{e}^{\boldsymbol{A}_{\mathrm{c}}(T-\tau)}\Delta\boldsymbol{A}_c\mathrm{e}^{\boldsymbol{A}_{\mathrm{c}}\tau}d\tau + \mathrm{O}\left(||\Delta\boldsymbol{A}_{\mathrm{c}}||_2^2 T^2 \mathrm{e}^{2||\Delta\boldsymbol{A}_{\mathrm{c}}||_2 T}\right)$$

Therefore, while $||\Delta\boldsymbol{A}(t)||_2$ is sufficiently small, one has

$$\Delta\boldsymbol{A}(k) \approx \int_0^T \mathrm{e}^{\boldsymbol{A}_{\mathrm{c}}(T-\tau)}\Delta\boldsymbol{A}_{\mathrm{c}}\mathrm{e}^{\boldsymbol{A}_{\mathrm{c}}\tau}d\tau$$

Taking the 2-norm of the above expression and applying norm inequalities, one can obtain:

$$||\Delta\boldsymbol{A}(k)||_2 \le \int_0^T \left\|\mathrm{e}^{\boldsymbol{A}_{\mathrm{c}}(T-\tau)}\right\|_2 ||\Delta\boldsymbol{A}_{\mathrm{c}}||_2 \left\|\mathrm{e}^{\boldsymbol{A}_{\mathrm{c}}\tau}\right\|_2 d\tau$$

$$\left\|\mathrm{e}^{\boldsymbol{A}_{\mathrm{c}}\tau}\right\|_2 \le \mathrm{e}^{||\boldsymbol{A}_{\mathrm{c}}||_2 T},\ ||\Delta\boldsymbol{A}(t)||_2 \le \upsilon$$

$$||\Delta\boldsymbol{A}(k)||_2 \le \upsilon T\mathrm{e}^{||\boldsymbol{A}_{\mathrm{c}}||_2 T}$$

The proof is complete.

Based on Lemma 6.2, the following theorem can be used to solve matrix $\boldsymbol{K}$.

Theorem 6.1 If for $\exists\rho > 0$, $||\Delta\boldsymbol{A}(k)||_2 \le \rho$, there exist a symmetric positive definite matrix $\boldsymbol{P}$ and positive constants $\gamma > 0, \varepsilon > 0$, such that the following set of linear matrix inequalities (LMIs) holds:

$$\begin{bmatrix} \overline{\boldsymbol{H}} & \boldsymbol{H}^{\mathrm{T}} + \boldsymbol{A}^{\mathrm{T}}\boldsymbol{P} & \boldsymbol{H}^{\mathrm{T}} \\ \mathrm{PA} + \boldsymbol{H} & -\rho^{-1}\varepsilon\boldsymbol{I} & 0 \\ \boldsymbol{H} & 0 & -\boldsymbol{P} \end{bmatrix} < 0 \tag{6.21}$$

$$\boldsymbol{P} - \gamma\boldsymbol{I} < 0 \tag{6.22}$$

where $\overline{\boldsymbol{H}} = \boldsymbol{H}^{\mathrm{T}}\boldsymbol{A} + \boldsymbol{A}^{\mathrm{T}}\boldsymbol{H} + \boldsymbol{A}^{\mathrm{T}}\mathrm{PA} + (\rho\varepsilon + \rho^2\gamma)\boldsymbol{I} - \boldsymbol{P}$; $\boldsymbol{H} = \boldsymbol{PBK}$. Then there exists $\boldsymbol{K} = (\boldsymbol{PB})^*\boldsymbol{H} \in \mathbf{R}^{n\times m}$ such that $\boldsymbol{A}_{\mathrm{K}} + \Delta\boldsymbol{A}(k)$ is Schur stable. $(\boldsymbol{PB})^*$ is the pseudoinverse of $\boldsymbol{PB}$; $\boldsymbol{A}_{\mathrm{K}} = \boldsymbol{A} + \boldsymbol{BK}$.

Proof: Choose the Lyapunov function as

$$V(\boldsymbol{x}(k)) = \boldsymbol{x}^{\mathrm{T}}(k)\mathrm{Px}(k) \tag{6.23}$$

Its difference is

$$V(\boldsymbol{x}(k+1))-V(\boldsymbol{x}(k))=\boldsymbol{x}^{\mathrm{T}}(k)(\boldsymbol{A}_{\mathrm{K}}+\Delta\boldsymbol{A}(k))^{\mathrm{T}}\boldsymbol{P}(\boldsymbol{A}_{\mathrm{K}}+\Delta\boldsymbol{A}(k))\boldsymbol{x}(k)-\boldsymbol{x}^{\mathrm{T}}(k)\mathrm{P}\mathrm{x}(k)$$

According to Lemma 1.3 (Schur complement lemma) and letting $\boldsymbol{H} = \boldsymbol{PBK}$, inequality (6.21) is equivalent to

$$(\boldsymbol{A}+\mathrm{BK})^{\mathrm{T}}\boldsymbol{P}(\boldsymbol{A}+\mathrm{BK})+\left(\rho\varepsilon+\rho^{2}\gamma\right)\boldsymbol{I}-\boldsymbol{P}+\rho\varepsilon^{-1}(\boldsymbol{A}+\mathrm{BK})^{\mathrm{T}}\mathrm{PP}(\boldsymbol{A}+\mathrm{BK})<0 \tag{6.24}$$

Since $||\Delta\boldsymbol{A}(k)||_2 \leq \rho$, $\Delta\boldsymbol{A}(k)$ can be written into

$$\Delta\boldsymbol{A}(k) = \rho\boldsymbol{E}(k) \tag{6.25}$$

where $\boldsymbol{E}(k)$ represents the normalized uncertainty matrix, which satisfies $||\boldsymbol{E}(k)|| \leq 1$. Consequently, one has

$$\rho^{2}\boldsymbol{E}^{\mathrm{T}}(k)\mathrm{PE}(k) \leq \rho^{2}\gamma\boldsymbol{I} \tag{6.26}$$

Besides, according to Lemma 1.1, one has

$$\boldsymbol{A}_{\mathrm{K}}^{\mathrm{T}}\mathrm{PE}(k)+\boldsymbol{E}^{\mathrm{T}}(k)\boldsymbol{P}\boldsymbol{A}_{\mathrm{K}} \leq \varepsilon^{-1}\boldsymbol{A}_{\mathrm{K}}^{\mathrm{T}}\mathrm{PP}\boldsymbol{A}_{\mathrm{K}}+\varepsilon\boldsymbol{I} \tag{6.27}$$

According to inequalities (6.24), (6.26) and (6.27), one can get

$$\boldsymbol{A}_{\mathrm{K}}^{\mathrm{T}}\boldsymbol{P}\boldsymbol{A}_{\mathrm{K}}+\rho\left(\boldsymbol{A}_{\mathrm{K}}^{\mathrm{T}}\mathrm{PE}(k)+\boldsymbol{E}^{\mathrm{T}}(k)\boldsymbol{P}\boldsymbol{A}_{\mathrm{K}}\right)+\rho^{2}\boldsymbol{E}^{\mathrm{T}}(k)\mathrm{PE}(k)-\boldsymbol{P}<0 \tag{6.28}$$

Substituting Eq. (6.25) into inequality (6.28), one can finally obtain:

$$V(\boldsymbol{x}(k+1))-V(\boldsymbol{x}(k))<0,\ V(\boldsymbol{x}(k))\geq 0,\ \boldsymbol{A}_{\mathrm{K}}+\Delta\boldsymbol{A}(k)$$

is Schur stable.

The proof is complete.

The following is the lemma and the feasibility theorem for algorithm iterations.

Lemma 6.2 Let Z be a robust positively invariant set for the system $\boldsymbol{x}(k+1) = (\boldsymbol{A}_{\mathrm{K}} + \Delta\boldsymbol{A}(k))\boldsymbol{x}(k) + \boldsymbol{w}$, $\forall k \geq 0$. If the nominal system state and the actual system state satisfy $\boldsymbol{x} \in \overline{\boldsymbol{x}} \oplus Z$, with the actual control input $\boldsymbol{u}$ selected as $\boldsymbol{u} = \overline{\boldsymbol{u}} + \boldsymbol{K}(\boldsymbol{x} - \overline{\boldsymbol{x}})$ (where $\overline{\boldsymbol{u}}$ is the nominal control input), then for $\forall \boldsymbol{w} \in W$, it holds that $\boldsymbol{x}(k+1) \in \overline{\boldsymbol{x}}(k+1) \oplus Z$ (Mayne et al., 2005).

Theorem 6.2 Suppose Problem (6.17) is feasible at the initial time instant, that is, there exist initial state sequences $\boldsymbol{X}(0)$ and control sequences $\boldsymbol{U}(0)$ satisfying constraint set $C(0)$. Then for $\forall k > 0$, Problem (6.17) is solvable, with sequences $\boldsymbol{X}(k)$ and $\boldsymbol{U}(k)$ ensuring constraint set $C(\boldsymbol{x})$.

Proof By mathematical induction. Assume that the sequences $\boldsymbol{X}(k)$ and $\boldsymbol{U}(k)$ satisfy the constraint set $C(\boldsymbol{x})$. Analyze whether there exists $\boldsymbol{X}(k+1)$ and $\boldsymbol{U}(k+1)$ satisfying $C(\boldsymbol{x})$.

Suppose the optimal control sequence and state sequence at time k are

$$\boldsymbol{U}^*(k) = \left\{\boldsymbol{u}_0^*(k), \boldsymbol{u}_1^*(k), \ldots, \boldsymbol{u}_{N-1}^*(k)\right\}$$
$$\boldsymbol{X}^*(k) = \left\{\boldsymbol{x}_0^*(k), \boldsymbol{x}_1^*(k), \ldots, \boldsymbol{x}_N^*(k)\right\}$$

Since the nominal system disregards disturbances, the last N-1 elements of $\boldsymbol{U}^*(k)$ can be shifted forward to construct the first N-1 elements of the control sequence $\boldsymbol{U}^*(k+1)$ at the next time instant. Hence, its first N-1 elements satisfy constraint $C(\boldsymbol{x})$. For the N-th element, select it as $\boldsymbol{u}_N^*(k+1) = \mathrm{K}\mathrm{x}_N^*(k)$. According to the terminal state constraint (6.14), it follows that

$$\boldsymbol{u}_N^*(k+1) \in \boldsymbol{K}X_{\mathrm{f}} \subset \overline{U}$$

Thus, $\boldsymbol{U}^*(k+1)$ satisfies constraint set $C(\boldsymbol{x})$. Similarly, the first N elements of $\boldsymbol{X}(k+1)$ automatically satisfy constraint set $C(\boldsymbol{x})$, while for the $(N++1)$-th term:

$$\boldsymbol{x}_N^*(k+1) = (\boldsymbol{A}_{\mathrm{K}} + \Delta\boldsymbol{A})\boldsymbol{x}_N^*(k) \in (\boldsymbol{A}_{\mathrm{K}} + \Delta\boldsymbol{A})X_{\mathrm{f}} \subset X_{\mathrm{f}}$$

In summary, there exist $\boldsymbol{X}(k+1)$ and $\boldsymbol{U}(k+1)$ satisfying $C(\boldsymbol{x})$. In practice, the receding horizon optimization applied at each time step will yield control performance no worse than this feasible solution.

6.3.3 Algorithm stability analysis

To establish the stability theorem for the algorithm, it is necessary to introduce the following lemmas.

Lemma 6.3 For the system $\boldsymbol{x}(k+1) = (\boldsymbol{A} + \Delta\boldsymbol{A})\boldsymbol{x}(k) + \boldsymbol{Bu}(k)$, $||\Delta\boldsymbol{A}||_2 \leq \rho$, if the optimization problem

$$\begin{cases} \min\limits_{\boldsymbol{U}} J_0\left(k, \tilde{\boldsymbol{X}}, \tilde{\boldsymbol{U}}\right) \\ \text{s.t.} \;\; C(\boldsymbol{x}) \\ \qquad \boldsymbol{x}_0(k) = \boldsymbol{x}(k) \end{cases} \tag{6.29}$$

is feasible with optimal sequences $\tilde{\boldsymbol{X}}^*(k)$, $\tilde{\boldsymbol{U}}^*(k)$, and optimal cost function $J_0^*(k, \tilde{\boldsymbol{X}}^*, \tilde{\boldsymbol{U}}^*)$, and the system adopts the model predictive control law $\boldsymbol{u}(k) =$

$\widetilde{\boldsymbol{\kappa}}_N^0(k) \overset{\Delta}{=} \widetilde{\boldsymbol{u}}_0^*(k)$, then for any initial state within X_N, the optimization problem remains feasible for all subsequent times, and the following inequality holds:

$$J_0^*\left(k+1, \widetilde{\boldsymbol{X}}^*, \widetilde{\boldsymbol{U}}^*\right) \leq J_0^*\left(k, \widetilde{\boldsymbol{X}}^*, \widetilde{\boldsymbol{U}}^*\right) - L\left(\boldsymbol{x}, \boldsymbol{\kappa}_N^0\right) \tag{6.30}$$

Proof Recursive feasibility of this optimization problem can be proven following the approach in Theorem 6.2. Now prove inequality (6.30).

Consider the candidate feasible sequences at the next time step:

$$\widetilde{\boldsymbol{U}}(k+1) = \left\{\widetilde{\boldsymbol{u}}_1(k), \ldots, \widetilde{\boldsymbol{u}}_{N-1}(k), \boldsymbol{K}\widetilde{\boldsymbol{x}}_N(k)\right\}$$

$$\widetilde{\boldsymbol{X}}(k+1) = \left\{\widetilde{\boldsymbol{x}}_1(k), \ldots, \widetilde{\boldsymbol{x}}_{N-1}(k), (\boldsymbol{A}_\mathrm{K} + \Delta \boldsymbol{A})\widetilde{\boldsymbol{x}}_N(k)\right\}$$

The corresponding cost function is:

$$J_0\left(k+1, \widetilde{\boldsymbol{X}}, \widetilde{\boldsymbol{U}}\right) = J_0\left(k, \widetilde{\boldsymbol{X}}, \widetilde{\boldsymbol{U}}\right) - L(\widetilde{\boldsymbol{x}}_0(k), \widetilde{\boldsymbol{u}}_0(k)) + L(\boldsymbol{x}_N(k), \boldsymbol{K}\boldsymbol{x}_N(k)) + J_f((\boldsymbol{A}_\mathrm{K} + \Delta \boldsymbol{A})\boldsymbol{x}_N(k)) \tag{6.31}$$

Since the candidate sequences are feasible but not optimized, their cost function is no less than the optimal cost function:

$$J_0^*\left(k+1, \widetilde{\boldsymbol{X}}^*, \widetilde{\boldsymbol{U}}^*\right) \leq J_0\left(k+1, \widetilde{\boldsymbol{X}}, \widetilde{\boldsymbol{U}}\right) \tag{6.32}$$

According to constraint (6.14), for the terminal state $\widetilde{\boldsymbol{x}}_N(k)$, there holds:

$$J_f\left((\boldsymbol{A}_\mathrm{K} + \Delta \boldsymbol{A})\widetilde{\boldsymbol{x}}_N(k)\right) + L\left(\widetilde{\boldsymbol{x}}_N(k), \boldsymbol{K}\widetilde{\boldsymbol{x}}_N(k)\right) \leq J_f\left(\widetilde{\boldsymbol{x}}_N(k)\right)$$

Substituting this into Eq. (6.31) yields:

$$J_0\left(k, \widetilde{\boldsymbol{X}}, \widetilde{\boldsymbol{U}}\right) \leq J_f\left(\widetilde{\boldsymbol{x}}_N(k)\right) - L\left(\widetilde{\boldsymbol{x}}_0(k), \widetilde{\boldsymbol{u}}_0(k)\right)$$

Combining with inequality (6.32), we obtain inequality (6.30). This completes the proof.

Theorem 6.3 If for the initial state $\boldsymbol{x}(0)$, there exist control and state sequences satisfying constraint $C(\boldsymbol{x})$, then the closed-loop system formed by Eq. (6.3) and (6.19) is stable with a region of attraction X_N.

Proof. Since Problem (6.17) is feasible, choose the optimal cost function as the Lyapunov function:

$$V(\boldsymbol{x}(k)) = J^*(k, \boldsymbol{X}, \boldsymbol{U})$$

The distinction between Problem (6.17) and Problem (6.29) lies in whether the initial value of the state sequence is treated as an optimization variable. Specifically, in Problem (6.17), the initial state of the sequence does not necessarily equal the current actual system state. Therefore, the optimal cost function for the current state $\boldsymbol{x}(k)$ in Problem (6.17) equals the value function of Problem (6.29) evaluated at the optimal initial state $\boldsymbol{x}_0^*(k)$:

$$V(\boldsymbol{x}(k)) = J_0^*\left(k, \widetilde{\boldsymbol{X}}^*, \widetilde{\boldsymbol{U}}^*\right), \tilde{\boldsymbol{x}}_0(k) = \boldsymbol{x}_0^*(k) \tag{6.33}$$

The Lyapunov function at the next time step, $V(\boldsymbol{x}(k+1))$ will be no worse than the cost of the candidate's feasible sequences:

$$V(\boldsymbol{x}(k+1)) \le J(k+1, \boldsymbol{X}(k+1), \boldsymbol{U}(k+1)) = J_0^*\left(k+1, \widetilde{\boldsymbol{X}}^*, \widetilde{\boldsymbol{U}}^*\right) \tag{6.34}$$

where $\boldsymbol{X}(k+1) = \{\boldsymbol{x}_1^*(k), ..., \boldsymbol{x}_{N-1}^*(k), (\boldsymbol{A}_{\mathrm{K}} + \Delta\boldsymbol{A})\boldsymbol{x}_N(k)\}$and $\boldsymbol{U}(k+1) = \{\boldsymbol{u}_1^*(k), ..., \boldsymbol{u}_{N-1}^*(k), \boldsymbol{K}\boldsymbol{x}_N(k)\}$ are candidate feasible sequences for Problem (6.17).

From inequalities (6.30), (6.34) and Eq. (6.33), we conclude that for $\forall \boldsymbol{x} \in X_N$, when the system state and control input are non-zero:

$$V(\boldsymbol{x}(k+1)) - V(\boldsymbol{x}(k)) \le -L\left(\boldsymbol{x}(k), \boldsymbol{\kappa}_N^0(k)\right) < 0 \tag{6.35}$$

This completes the proof.

6.4 Simulation test

To verify the effectiveness of the proposed control approach for electromagnetic docking control during on-orbit refueling, numerical simulations are conducted.

Considering a chaser spacecraft with mass $m_{\mathrm{C}} = 40\mathrm{kg}$, carrying hydrogen peroxide propellant, performs refueling with a target spacecraft (mass $m_{\mathrm{T}} = 100\mathrm{kg}$). The pulsating ball (modeling liquid slosh) elastically collides with the tank walls. The target's electromagnetic docking system generates a synthetic magnetic moment $\boldsymbol{\mu}_{\mathrm{T}} = [-2.29 \quad -8.17 \quad -2.33]^{\mathrm{T}} \times 10^5\mathrm{Am}^2$. The initial state is $\boldsymbol{x}(0) = [3\mathrm{m} \quad 4\mathrm{m} \quad 5\mathrm{m} \quad 0 \quad 0 \quad 0]^{\mathrm{T}}$.

Since the two spacecraft are controlled exclusively by electromagnetic forces and reaction wheels, the center-of-mass ("CM") orbit remains invariant. In the Earth-centered inertial frame, the selected "CM" orbital parameters are: semimajor axis $a = 10,000\mathrm{km}$, eccentricity $e = 0.15$, inclination 30°, right ascension of the ascending node 200°, argument of perigee

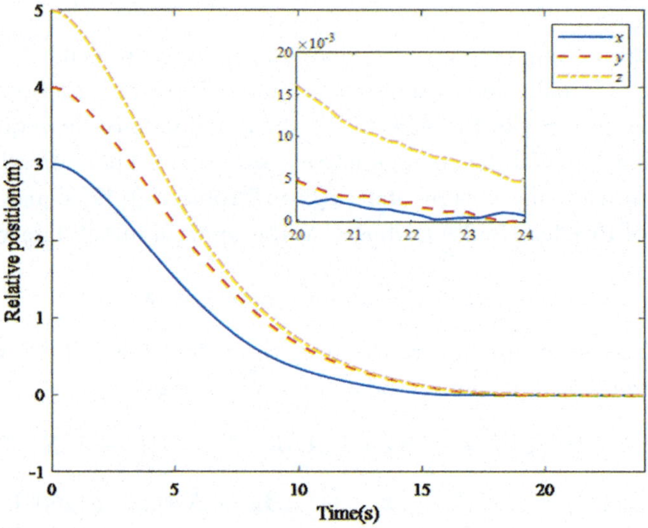

Figure 6.1 Relative position of spacecraft.

30°, initial true anomaly 0°. The sampling time is chosen as $T = 0.2$s. In Theorem 6.1, the parameters are chosen as $\varepsilon = 0.01$ and $\rho = \sqrt{6}$. Theorem 6.1 is solved via YALMIP toolbox in MATLAB, with the solver MOSEK. The controller gain matrix $\boldsymbol{K}$ is

$$\boldsymbol{K}=\begin{bmatrix} -65.4824 & 0 & 0.0008 & -242.4182 & 0 & -0.0298 \\ 0 & -65.4824 & 0 & 0 & -242.4182 & 0 \\ -0.0008 & 0 & -65.4828 & 0.0298 & 0 & -242.4181 \end{bmatrix}$$

For the TMPC, the control horizon is set to $N = 15$ and the weight matrices in the optimization function are set to $\mathbf{Q} = 10 \times \boldsymbol{I}_6$, $\boldsymbol{R} = 5 \times \boldsymbol{I}_3$. The state and control constraint sets are chosen as:

$$X = \{\boldsymbol{x}|x_{1,2,3} \leq 10\text{m}, x_{4,5,6} \leq 5\text{m/s}\} \ U = \{\boldsymbol{u}|u_{1,2,3} \leq 20\text{N}\}$$

During the entire simulation process, the spacecraft is subjected to compound disturbances consisting of random disturbances drawn from the disturbance set and fuel sloshing forces. The disturbance set is chosen as

$$W = \left\{\text{Bw}|||\boldsymbol{Bw}||_\infty \leq 10^{-3}\text{N}\right\}$$

The numerical simulation results are presented in Fig. 6.1 through Fig. 6.4. Fig. 6.1 shows the spacecraft's relative position response, demonstrating that the relative position remains within the specified state set throughout

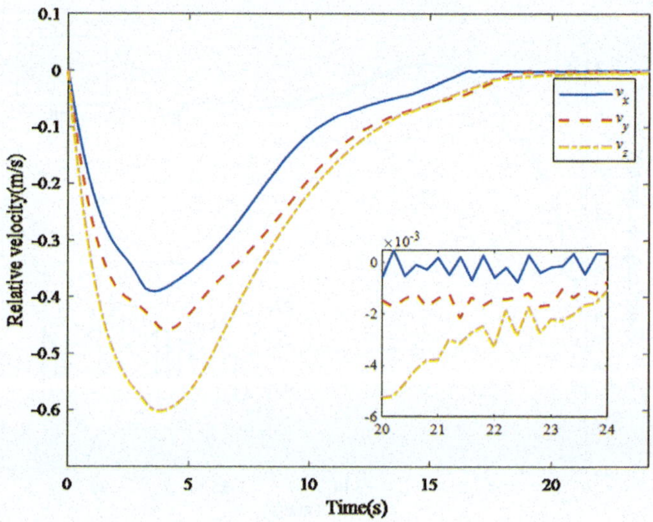

Figure 6.2 Relative velocity of spacecraft.

the process and asymptotically converges to a small neighborhood containing the origin at 20s, with steady-state errors not exceeding 10mm in all three axes. Fig. 6.2 displays the relative velocity response, revealing that the docking system first undergoes approximately uniform acceleration during the initial 3s, reaches maximum velocity, then decelerates uniformly, ultimately converging asymptotically near zero velocity at 20s, with steady-state errors within 1, 2, and 4mm/s for the respective axes. The simulation results indicate that the optimal docking trajectory for spacecraft electromagnetic docking follows an accelerate-then-decelerate profile. It can be concluded that the electromagnetic docking process is completed within 20s.

Fig. 6.3 presents the electromagnetic force profile calculated using the far-field model. This study assumes the persistent applicability of the far-field model, achieved through multiple compact coils with high current-carrying capacity to generate the required magnetic moment. During most docking phases, the relative distance exceeds six times the coil radius, maintaining far-field modeling errors within acceptable limits. The results demonstrate that electromagnetic forces saturate in the first 1s, and remain within prescribed control bounds in the rest of the simulation. Due to liquid sloshing disturbances, post-5s electromagnetic forces stabilize below 0.3N in all axes. Fig. 6.4 displays the chaser spacecraft's magnetic moment variation

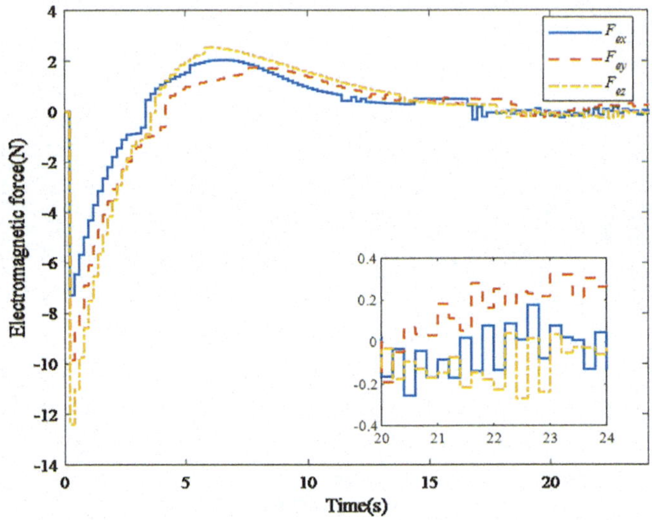

Figure 6.3 Required electromagnetic force.

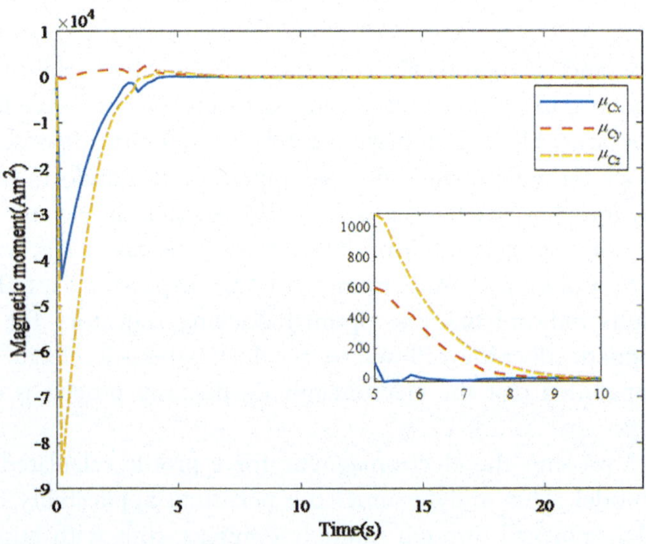

Figure 6.4 Required magnetic moment.

Table 6.1 Parameters of PD controller.

	Proportional Gain K_P	Differential Gain K_D
PD1	diag(−3,−3,−3,−10,−10,−10)	diag(−10,−10,−10,−15,−15,−15)
PD2	diag(−20,−20,−20,−40,−40,−40)	diag(−10,−10,−10,−15,−15,−15)

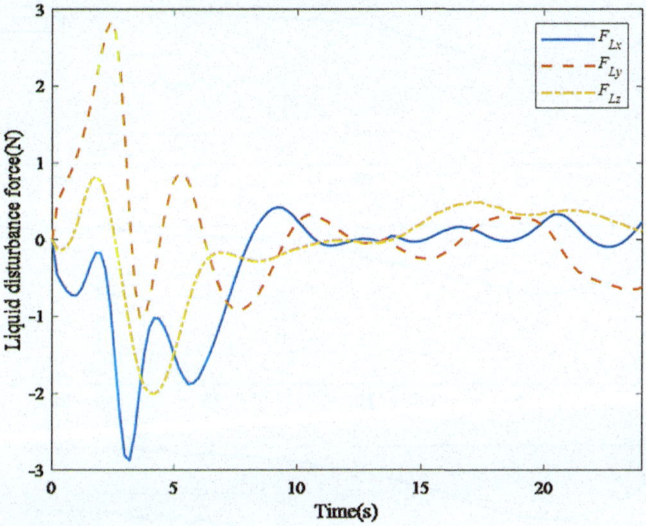

Figure 6.5 Liquid sloshing disturbance force.

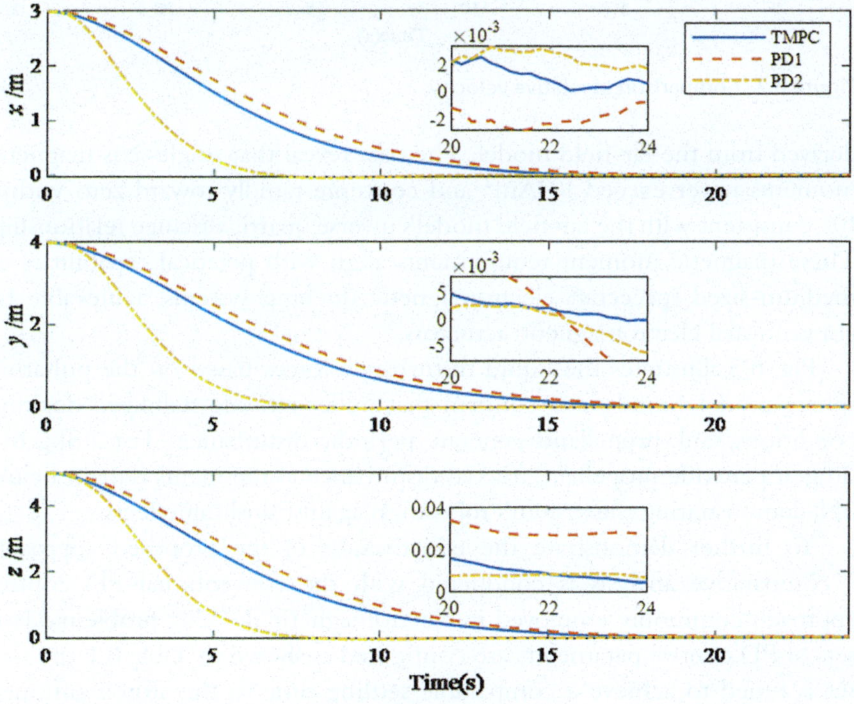

Figure 6.6 Comparison of relative position.

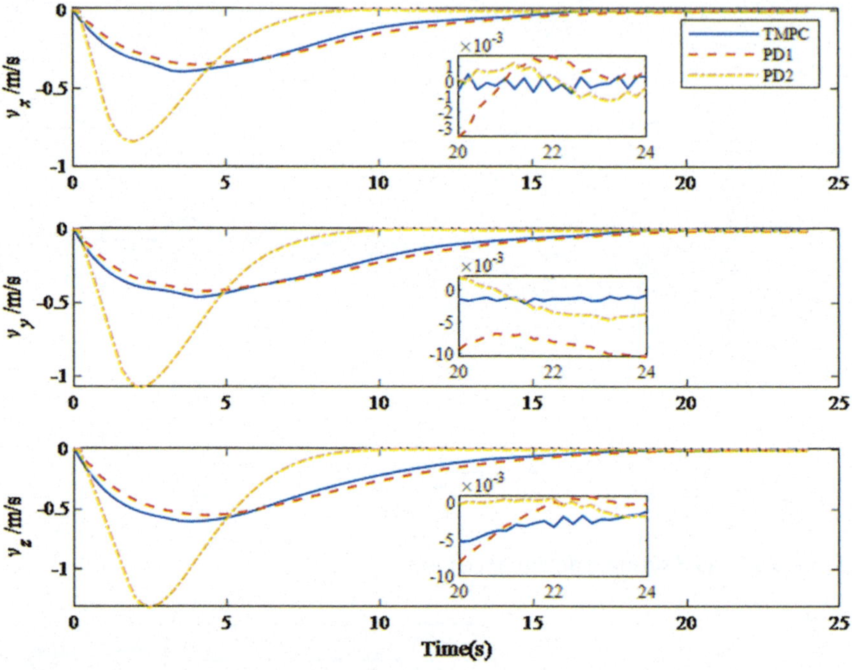

Figure 6.7 Comparison of relative velocity.

derived from the far-field model. The data reveal that single-axis magnetic moments never exceed $10^5 Am^2$ and converge rapidly toward zero within 10s, consistent with the far-field model's inverse quartic distance relationship. These magnetic moment requirements align with practical capabilities of medium-sized spacecraft electromagnetic docking systems, achievable by conventional electromagnetic actuators.

Fig. 6.5 illustrates the liquid disturbance forces based on the pulsating sphere model. Simulations confirm that under attitude-stabilized docking conditions, tank propellants generate periodic disturbances. For a 5kg hydrogen peroxide propellant, triaxial disturbance components remain below 3N, demonstrating the system's robustness against sloshing effects.

To further demonstrate the effectiveness of the proposed approach, a comparative analysis is conducted with the conventional PD control approach commonly employed in electromagnetic docking problems. Two sets of PD control parameters are configured as shown in Table 6.1: the first set is tuned to achieve a comparable settling time to the aforementioned TMPC simulation results, while the second set is optimized for higher control precision.

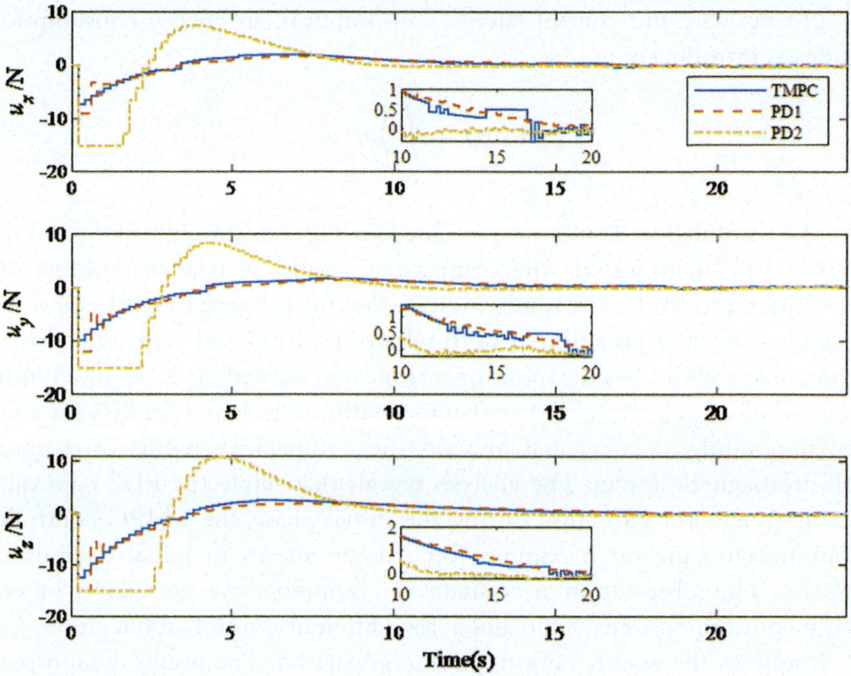

Figure 6.8 Comparison of required electromagnetic force.

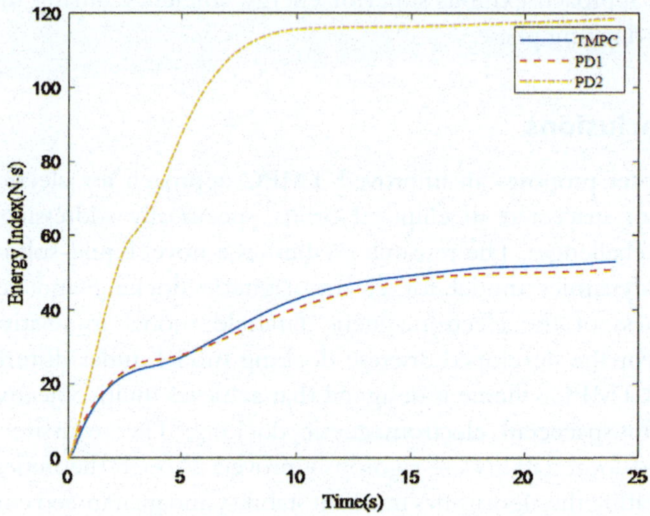

Figure 6.9 Comparison of energy index.

To evaluate the control energy consumption, an energy consumption index is introduced.

$$E_e(t_k) = \int_0^{t_k} ||\boldsymbol{u}||_2 dt$$

The simulation results are presented in Fig. 6.6 through Fig. 6.9. Figs. 6.6 and 6.7 demonstrate the comparative results of relative position and velocity, respectively. The results indicate that the proposed TMPC approach achieves superior precision in both relative position and velocity across all three axes, with absolute position errors not exceeding 2, 5, and 10mm, and absolute velocity errors remaining within 1.1, 3, and 5mm/s for each corresponding axis. Fig. 6.8 presents the comparative results of required electromagnetic forces. The analysis reveals that while the PD2 controller exhibits actuator saturation during the initial phase, the TMPC controller demonstrates greater maximum force requirements in initial operational phases. This observation necessitates a comprehensive evaluation of energy consumption characteristics across different control architectures. Fig. 6.9 presents the energy consumption comparison. The results demonstrate that the TMPC approach achieves the optimal energy consumption index during the operation period. Therefore, it can be concluded that the proposed approach exhibits superior energy efficiency during the electromagnetic docking phase.

6.5 Conclusions

This chapter proposes an improved TMPC approach for electromagnetic docking of spacecraft in elliptical orbits, specifically addressing on-orbit refueling challenges. The research establishes a novel liquid-solid-magnetic coupled dynamics model for electromagnetic docking and verifies the effectiveness of the electromagnetic far-field model in relative motion control. For the developed discrete docking model under disturbances, an enhanced TMPC scheme is designed that achieves multi-objective optimal control for spacecraft electromagnetic docking. The recursive feasibility and closed-loop stability are rigorously proved through theoretical analysis, demonstrating the algorithm's inherent stability and guaranteed convergence of the time-varying linear system states to the minimal robust invariant set. Comparative simulations with two PD controllers further validate the superior performance of the proposed approach.

References

Foust, R., Lupu, E. S., Nakka, Y. K., Chung, S., & Hadaegh, F. (2018). Ultra-soft electromagnetic docking with applications to in-orbit assembly. In *69th International Astronautical Congress (IAC): 10* (pp. 1–5). Bremen, Germany: International Astronautical Congress.

Ghaemi, R., Sun, J., & Kolmanovsky, I. (2008). Less conservative robust control of constrained linear systems with bounded disturbances. *Proceedings of the IEEE conference on decision and control* (pp. 983–988). https://doi.org/10.1109/CDC.2008.4739361.

Ghaemi, R., Sun, J., & Kolmanovsky, I. V. (2012). Robust control of constrained linear systems with bounded disturbances. *IEEE Transactions on Automatic Control, 57*(10), 2683–2688. https://doi.org/10.1109/tac.2012.2192362.

Hu, M., Li, C., Bian, Y., Zhang, H., Qin, Z., & Xu, B. (2022). Fuel economy-oriented vehicle platoon control using economic model predictive control. *IEEE Transactions on Intelligent Transportation Systems, 23*(11), 20836–20849. https://doi.org/10.1109/tits.2022.3183090.

Li, L., Su, P., & Zhang, X. (2023). Tube-based distributed model predictive control of nonlinear multi-agent systems based on additive state decomposition. In *Proceedings of 2023 IEEE 12th Data Driven Control and Learning Systems Conference (DDCLS)* (pp. 1433–1438). IEEE. https://doi.org/10.1109/DDCLS58216.2023.10166301.

Liu, C., Yue, X., & Yang, Z. (2021). Are nonfragile controllers always better than fragile controllers in attitude control performance of post-capture flexible spacecraft? *Aerospace Science and Technology, 118*, 107053. https://doi.org/10.1016/j.ast.2021.107053.

Liu, C., Yue, X., Zhang, J., & Shi, K. (2022). Active disturbance rejection control for delayed electromagnetic docking of spacecraft in elliptical orbits. *IEEE Transactions on Aerospace and Electronic Systems, 58*(3), 2257–2268. https://doi.org/10.1109/taes.2021.3130830.

Mayne, D. Q., Seron, M. M., & Raković, S. V. (2005). Robust model predictive control of constrained linear systems with bounded disturbances. *Automatica, 41*(2), 219–224. https://doi.org/10.1016/j.automatica.2004.08.019.

Romero, A., Sun, S., Foehn, P., & Scaramuzza, D. (2022). Model predictive contouring control for time-optimal quadrotor flight. *IEEE Transactions on Robotics, 38*(6), 3340–3356. https://doi.org/10.1109/tro.2022.3173711.

Shi, K., Liu, C., Biggs, J. D., Sun, Z., & Yue, X. (2020). Observer-based control for spacecraft electromagnetic docking. *Aerospace Science and Technology, 99*, 105759. https://doi.org/10.1016/j.ast.2020.105759.

Shi, K., Liu, C., & Sun, Z. (2019). Constrained fuel-free control for spacecraft electromagnetic docking in elliptical orbits. *Acta Astronautica, 162*, 14–24. https://doi.org/10.1016/j.actaastro.2019.05.016.

Shi, Y., & Zhang, K. (2021). Advanced model predictive control framework for autonomous intelligent mechatronic systems: A tutorial overview and perspectives. *Annual Reviews in Control, 52*, 170–196. https://doi.org/10.1016/j.arcontrol.2021.10.008.

Vreeburg, J. P. B. (1997). Dynamics and control of a spacecraft with a moving pulsating ball in a spherical cavity. *Acta Astronautica, 40*(2-8), 257–274. https://doi.org/10.1016/s0094-5765(97)00095-7.

Weiss, A., Baldwin, M., Erwin, R. S., & Kolmanovsky, I. (2015). Model predictive control for spacecraft rendezvous and docking: Strategies for handling constraints and case studies. *IEEE Transactions on Control Systems Technology, 23*(4), 1638–1647. https://doi.org/10.1109/TCST.2014.2379639.

Zhang, Y., Yang, L., Zhu, Y., & Huang, H. (2013). Dynamics and nonlinear control of space electromagnetic docking. *Journal of Systems Engineering and Electronics, 24*(3), 454–462. https://doi.org/10.1109/jsee.2013.00054.

CHAPTER 7

Deep reinforcement learning-based control for spacecraft electromagnetic docking

7.1 Introduction

With the increasing complexity and diversity of space exploration missions, the demand for tasks such as on-orbit service, space debris removal, and space infrastructure construction continues to grow, driving rapid development in spacecraft on-orbit assembly technologies. As a foundational technology for constructing large spacecraft and modular space structures, on-orbit assembly is becoming a critical direction for future aerospace development (Foust et al., 2020). Space docking, as the core aspect of large-scale spacecraft assembly missions, directly determines the accuracy and stability of assembly. Traditional spacecraft rendezvous and docking methods predominantly use thruster propulsion techniques, characterized by high docking impact forces, significant fuel consumption, plume contamination, thermal emissions, and associated vibration disturbances. These issues make it difficult to meet modern mission requirements for safety, accuracy, and flexibility, especially under complex space environments where stable and efficient docking control is challenging to achieve (Liu et al., 2022; Shi et al., 2022; Wei et al., 2023). Compared with traditional docking methods, electromagnetic docking effectively avoids inherent problems associated with conventional thruster methods, such as fuel consumption and plume contamination (Ren et al., 2021). It provides noncontact, continuous, reversible, and synchronous control capabilities, theoretically enabling low-impact docking with zero relative velocity between spacecraft (Gronauer & Diepold, 2022; Shi et al., 2019; Yan et al., 2022).

The technical challenges of electromagnetic on-orbit assembly are concentrated in the dynamics and control aspects. For instance, elliptical orbit parameters exhibit time-varying characteristics leading to highly nonlinear

Spacecraft Electromagnetic Docking and Separation.
DOI: https://doi.org/10.1016/B978-0-443-49334-8.00008-2

relative motion equations; electromagnetic coupling effects between space platforms and assembly modules induce high-order harmonic disturbances; and large-scale space structures demand higher precision in positioning and docking. Conventional control methods often fail to meet multiple requirements of high accuracy, antiinterference capability, and efficiency in complex on-orbit assembly tasks. Therefore a deep reinforcement learning (DRL)-based multiobjective optimization control approach must be explored to meet electromagnetic docking control requirements. Focusing on spacecraft electromagnetic assembly missions on elliptical orbits, this chapter proposes an intelligent control algorithm based on deep deterministic policy gradient (DDPG). By establishing a dynamic model of relative motion for electromagnetic docking between the spacecraft platforms and assembly modules, the proposed method designs a multiobjective reward function that integrates relative distance, velocity errors, and control energy consumption. It introduces a dynamic action constraint mechanism to optimize electromagnetic force outputs.

DRL is an artificial intelligence method combining neural networks and reinforcement learning, which acquires optimal strategies through trial-and-error feedback obtained by interacting with the environment, without prior knowledge. It has been widely employed in the design of novel controllers and has become an essential tool for solving robot control and spacecraft control problems, due to its superior performance in high-dimensional and complex spaces (Lyu, 2025; Tarekegn et al., 2024; Yu et al., 2024; Yue et al., 2024). Moon et al. (2021) proposed a trajectory control method for multiflight-route base stations using a two-stage learning strategy, autonomously adjusting multi-UAV trajectories through centralized multiagent DRL to optimize communication performance, thereby addressing challenges of maximizing ground user communication coverage and network throughput. Chai et al. (2024) proposed a hierarchical deep-learning-based control framework, using a recurrent deep neural network during the motion planning stage to predict optimal trajectories for mobile robots, investigating optimal maneuver trajectory planning and target-driven navigation in uncertain environments. Gao, Yan, Li, Wang, & Chen, 2023) introduced an improved DDPG algorithm capable of collision-free trajectory planning among multiple dynamic obstacles, helping to reduce path length and movement time, thus facilitating dynamic obstacle avoidance for mobile robots.

The DDPG algorithm combines the advantages of policy gradient methods and Q-learning, using an Actor-Critic training structure that efficiently achieves policy optimization in continuous control problems. Compared with traditional numerical optimization methods, DDPG can learn through environment interactions, adapting to unstable and uncertain environments. Existing spacecraft docking control methods, such as sliding mode control and adaptive control (Wang et al., 2023; Zhang et al., 2024), typically require precise dynamic models and time-varying parameters. However, elliptical orbit spacecraft platforms have time-varying orbital parameters, and traditional methods heavily rely on precise modeling or parameter estimation, resulting in limited robustness and adaptability. As a DRL algorithm, DDPG autonomously learns strategies through interaction with the environment, without relying on precise dynamic models or parameters. For instance, in trajectory planning of reconfigurable spacecraft, DDPG maps states directly to actions via Markov decision models, eliminating strong dependencies on reconfiguration parameters and kinematic models inherent in traditional approaches. This characteristic is especially advantageous for scenarios involving noncooperative targets or unknown parameters.

Based on the electromagnetic force far-field model and the Tschauner–Hempel (TH) equation, this chapter establishes a relative-motion dynamic model capable of characterizing the dynamics of electromagnetic docking. To address multiobjective optimization requirements comprehensively—such as relative position, relative velocity, acceleration constraints, and complex disturbances during electromagnetic actuation—this research innovatively applies the DDPG algorithm to spacecraft electromagnetic assembly control. Specifically, considering the high-precision control, antiinterference capability, optimal energy consumption, and docking time optimization required for large-scale space platforms and assembly modules conducting electromagnetic assembly in elliptical orbits, this chapter proposes a multiobjective cooperative optimization control method based on DDPG. The method ensures on-orbit assembly accuracy while balancing mission timeliness and fuel consumption.

7.2 Problem formulation

7.2.1 Description of DDPG in on-orbit electromagnetic assembly

During the on-orbit electromagnetic assembly process of spacecraft in elliptical orbits, the electromagnetic force model between the spacecraft platform and the assembly module is complex, and a far-field model is typically used as a simplified mathematical expression, which is introduced in Section 1.2. Moreover, the actuators of the control system often face saturation and nonlinear constraints. The strong nonlinearity and coupling between the spacecraft platform and the assembly module negatively impact the system.

In the on-orbit assembly process, the electromagnetic docking control performance of the spacecraft directly affects its orbit and operational status. Traditional studies typically focus on optimizing the spacecraft's optimal velocity profile or on how to accurately track the optimal trajectory. However, in actual spacecraft control, it is necessary to consider complex space environments, strong nonlinearity, strong coupling, actuator failures, and other issues. Additionally, the relative motion equations of elliptical orbits contain time-varying parameters, which must be calculated based on the platform and assembly module's orbital information. During the rendezvous process, obtaining real-time mass information is usually very difficult, and offline control methods fail to meet the requirements for real-time performance and precision, resulting in insufficient autonomy for the spacecraft. This makes the task exceptionally difficult, increasing the probability of mission failure.

As an online reinforcement learning algorithm, DDPG can collect and process environmental data in real time during the task execution, continuously updating the control strategy. Based on the Actor-Critic architecture, DDPG achieves precise control in continuous action spaces and can dynamically adjust and optimize the assembly module's state. Through online learning, DDPG intelligently adapts to complex dynamic environments and external disturbances. It has significant advantages for solving the multiobjective optimization control problems in the spacecraft electromagnetic docking process, such as control precision, fuel consumption, and time optimization. By integrating the electromagnetic docking model and considering practical constraints such as relative position, relative velocity, time, and external disturbances, the DDPG algorithm can optimize the assembly module's control strategy to achieve high-precision on-orbit assembly.

7.2.2 Dynamics model of electromagnetic docking of spacecraft in elliptical orbits

Assuming a sampling time interval of dt, the continuous state-space Eq. (1.78) can be converted into a discrete-time form. Specifically, for the time step k, the discrete state-space equation is obtained as

$$x(k+1) = A_d x(k) + B_d u(k) + B_d w(k)$$

where k is the current time index, $\boldsymbol{A}_d = e^{(\boldsymbol{A}+\Delta\boldsymbol{A})dt}$ and $\boldsymbol{A}_d \in \mathbf{R}^{6\times 6}$, $\boldsymbol{B}_d = \left(\int_0^{dt} e^{(\boldsymbol{A}+\Delta\boldsymbol{A})} d\tau\right)(\boldsymbol{B} + \Delta\boldsymbol{B})$ and $\boldsymbol{B}_d \in \mathbf{R}^{6\times 3}$.

The far-field electromagnetic force model (1.68) is accurate only when the relative distance between the two spacecraft exceeds a certain threshold. During the assembly process, however, the platform and the module operate in close proximity. Therefore the error introduced by using the far-field model at close range can be regarded as an external disturbance to the system.

Based on the mission context, the large spacecraft platform's orbit is simplified to a fixed orbit that neglects perturbations and dynamic changes. The platform's position in its elliptical orbit can be described by orbital parameters (e.g., the true anomaly), which are determined by solving the orbital dynamics using an ODE45 solver. In this chapter, we focus on constraint modeling, including maximum velocity constraints, control input saturation, relative distance constraints, and relative velocity constraints.

1) Maximum velocity constraints

Assume the assembly module's maximum velocities along the x, y and z axes are $\dot{x}_{\max}$, $\dot{y}_{\max}$, and $\dot{z}_{\max}$, respectively. For the module's state at time k, these velocities must satisfy the constraint as

$$\begin{bmatrix} \boldsymbol{I}_3 \\ -\boldsymbol{I}_3 \end{bmatrix} \boldsymbol{H}\boldsymbol{\eta}(k) \leq \boldsymbol{V}_{\max} \tag{7.1}$$

The expressions for the matrix $\boldsymbol{H}$ and vector $\boldsymbol{V}_{\max}$ are

$$\begin{cases} \boldsymbol{H} = \begin{bmatrix} 0 & 0 & 0 & 1 & 0 & 0 \\ 0 & 0 & 0 & 0 & 1 & 0 \\ 0 & 0 & 0 & 0 & 0 & 1 \end{bmatrix} \\ \boldsymbol{V}_{\max} = \begin{bmatrix} \dot{x}_{\max} & \dot{y}_{\max} & \dot{z}_{\max} & \dot{x}_{\max} & \dot{y}_{\max} & \dot{z}_{\max} \end{bmatrix}^{\mathrm{T}} \end{cases}$$

2) Control input saturation constraints

Let the assembly module's control inputs along the three axes be denoted as u_x, u_y, u_z. Each control input is limited by a saturation function, that is, $\mathrm{sat}(u_{i=x,\,y,\,z}(t)) = \mathrm{sign}(u_i(t))\min\{\,|u_i(t)|\,,u_{mi}\}$, where u_{mi} is the upper bound of the control input.

3) Relative distance and relative velocity constraints

The relative distance between the spacecraft platform and the assembly module is defined as:

$$\eta = \sqrt{x^2 + y^2 + z^2} \tag{7.2}$$

If the platform and module get too close, the electromagnetic field intensity increases significantly, making precise control of the docking process difficult and possibly causing electromagnetic interference or system failure. To ensure stable and effective docking from a safe distance, a minimum allowable separation $\eta_{\min} \leq \eta \leq \eta_{\max}$ is imposed on this relative distance. Similarly, the relative velocity magnitude between the assembly module and the platform is given by

$$\dot{\eta} = \sqrt{\dot{x}^2 + \dot{y}^2 + \dot{z}^2} \tag{7.3}$$

During the docking process, this relative velocity must remain sufficiently low; in the final docking stage, it should approach zero to ensure a smooth contact. Assuming the maximum allowable relative velocity is $v_{\max}$, this constraint can be expressed as $\dot{\eta} \leq v_{\max}$.

7.3 DDPG-based controller design

DDPG is a model-free reinforcement learning algorithm designed for continuous action spaces. It combines deep neural networks with deterministic policy gradient theory to efficiently learn optimal policies. The goal of the algorithm is to find an optimal strategy in the environment that maximizes the cumulative reward for the module.

The DDPG algorithm employs a dual neural network architecture, where both the policy function and value function utilize dual neural network models (i.e., online and target networks). This structure enhances the stability of the learning process and accelerates convergence. Additionally, the algorithm incorporates an experience replay mechanism: the interaction data between the Actor and the environment is stored in an experience replay buffer, and training is performed by sampling mini-batches of data. Similar to DQN's experience replay mechanism, this approach reduces the correlation and dependency between samples, facilitating convergence. The algorithm framework is illustrated in Fig. 7.1.

The Critic network updates its parameters by minimizing the temporal difference error of sampled mini-batches:

$$\begin{cases} y_t = r_t + \gamma Q'\left(s_{t+1}, \mu'\left(s_{t+1}|\theta^{\mu'}\right)|\theta^{Q'}\right) \\ L\left(\theta^Q\right) = \frac{1}{2N}\sum_{t=1}^{N}\left[y_t - Q\left(s_t, a_t|\theta^Q\right)\right]^2 \\ \nabla_{\theta^Q} L\left(\theta^Q\right) = -\frac{1}{N}\sum_{t=1}^{N}\left(y_t - Q(s_t, a_t)\right)\nabla_{\theta^Q} Q(s_t, a_t) \\ \theta_{t+1}^Q = \theta_t^Q - \alpha_Q \nabla_{\theta^Q} L\left(\theta^Q\right) \end{cases} \tag{7.4}$$

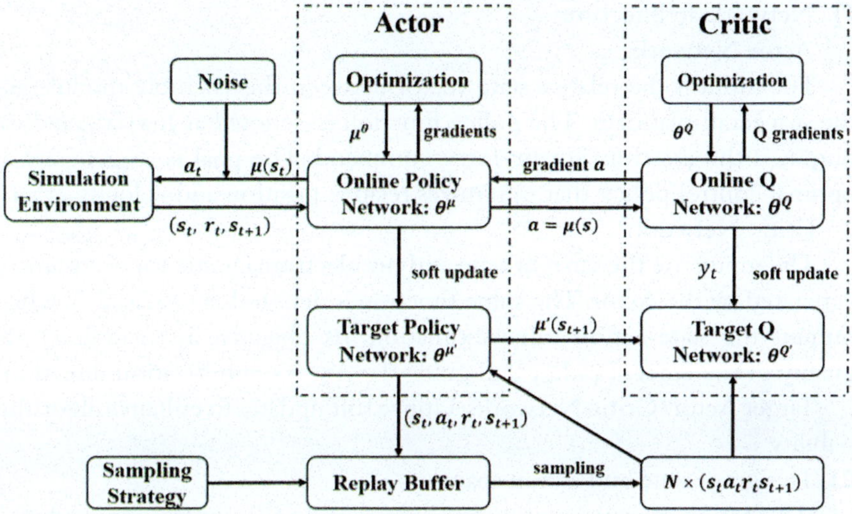

Figure 7.1 DDPG algorithm framework diagram.

where y_t guides the parameter updates for the Critic evaluation network, ensuring smoother updates, γ is the discount factor, s and s_{t+1} represent the current and next states of the environment, respectively, $\mu(s \mid \theta^{\mu}.)$ denotes the deterministic policy approximated by the Actor network with parameters θ^{μ}, $Q(s, a \mid \theta^{Q})$ is the state-action value function approximated by the Critic network, with parameters θ^{Q}, μ' and Q' are the target Actor and Critic networks, respectively, r_t is the immediate reward at time t, N is a sample from the experience replay buffer, L is the loss function, and a_Q is the learning rate of the Actor network.

The Actor network updates its parameters using the deterministic policy gradient. Let a_μ denote its learning rate:

$$\begin{cases} \nabla_{\theta^\mu} J(\theta^\mu) = \frac{1}{N} \sum_{t=1}^{N} \nabla_{\theta^\mu} \mu(s_t) \nabla_a Q(s_t, a_t)|_{a_t=\mu(s_t)} \\ \theta_{t+1}^{\mu} = \theta_t^{\mu} + \alpha_\mu \nabla_{\theta^\mu} J(\theta^\mu) \end{cases} \tag{7.5}$$

For the parameter updates of the target Actor and Critic networks, let τ denote the update rate

$$\begin{cases} \theta^{Q'} \leftarrow \tau\theta^{Q} + (1-\tau)\theta^{Q'} \\ \theta^{\mu'} \leftarrow \tau\theta^{\mu} + (1-\tau)\theta^{\mu'} \end{cases} \tag{7.6}$$

The structure of the DDPG-based intelligent control algorithm is as follows.

1) Network architecture

Actor Network:

The input is the relative state vector between the assembly module and the spacecraft platform. The policy function is denoted as $\mu(\boldsymbol{s} \mid \theta^{\mu})$, and the output is the electromagnetic force command. The goal is to generate an optimal control policy that minimizes relative position and velocity errors.

Critic Network:

The inputs are the current state and the electromagnetic force command generated by the Actor. The value function is denoted as $Q(\boldsymbol{s}, \boldsymbol{a} \mid \theta^{Q})$, which outputs the state-action value estimation. Its objective is to evaluate the quality of the current policy and guide the Actor's optimization direction.

Target Actor/Critic Networks: Utilize soft updates to enhance algorithm stability.

2) Design of state and action spaces

The action space is continuous, representing the control inputs during the electromagnetic docking process. The action is designed as a control vector $\boldsymbol{a}_t = [a_x \ a_y \ a_z]^{\mathrm{T}}$, representing the electromagnetic acceleration applied to the assembly module, which directly affects its position and velocity.

To ensure system stability and docking precision, the input to the action space is scaled during training to better adapt to the system dynamics. The state space in reinforcement learning refers to the set of observable variables by the learning agent. The state vector is defined as

$$\boldsymbol{s} = \left[\Delta x \ \Delta y \ \Delta z \ \Delta \dot{x} \ \Delta \dot{y} \ \Delta \dot{z}\right]^{\mathrm{T}} \tag{7.7}$$

where Δx denotes the relative position errors between the assembly module and the spacecraft platform along x axis, and $\Delta \dot{x}$ denotes the corresponding relative velocity errors; the same applies to the other directions.

3) Reward function design

To achieve multiobjective optimal control, a composite reward function is designed by integrating relative distance error, relative velocity error, time duration, and control energy consumption, which is

$$R = -0.2 \cdot \left(\frac{\arctan(e_{\mathrm{M}})}{\pi}\right) + \sum_{i=1}^{n} \alpha_i \cdot \left(\frac{1}{1 + \beta_i \cdot e_{\mathrm{M}}}\right) - \gamma \cdot \|\boldsymbol{v}_{\mathrm{M}} - \boldsymbol{v}_{\mathrm{P}}\|^2 \tag{7.8}$$

where $e_{\mathrm{M}} = \sqrt{\Delta x^2 + \Delta y^2 + \Delta z^2}$ denotes the relative distance error between the assembly module and the spacecraft platform, R is the reward value,

v_M and v_P respectively denote the velocity vectors of the assembly module and the spacecraft platform, α_i and β_i are adjustment factors under different conditions used to modulate the reward across various distance ranges, γ is the velocity penalty factor used to control the impact of velocity error on the reward.

In practice, when the distance between the assembly module and the spacecraft platform is large, the reward function focuses on penalizing the distance error, guiding the system to gradually reduce the error through negative reinforcement. As the distance decreases below certain thresholds, significant positive rewards are provided to encourage the module to adopt more refined control strategies during the final approach. Additionally, to prevent large variations in relative velocity during the docking process, a velocity penalty term is included in the reward function. This ensures that when the speed difference between the module and the platform is large, the reward is correspondingly reduced, effectively constraining the velocity error. This composite reward function not only balances the fine regulation of both distance and velocity but also ensures task completion within a limited time and at low control cost through penalties on time and energy. Furthermore, when the relative distance error is smaller than a specified threshold, a significant reward increment is applied to encourage precise control actions during the final docking phase.

7.4 Simulation test

7.4.1 Module training and analysis

The module adopts a multilayer fully connected neural network. The Actor network is designed as a two-hidden-layer neural network with six input neurons corresponding to the 6-dimensional state. The first hidden layer contains 400 neurons and employs the ReLU activation function. A normalization layer is applied to the output of this layer to help stabilize training. The second hidden layer contains 300 neurons, also using the ReLU function. The output layer consists of 3 neurons corresponding to the equivalent control commands in each channel. The output values are constrained within the range $[-1, 1]$ using the tanh activation function and scaled with a proportional factor to match the actual output range.

In the design of the Critic network, the state and action inputs are handled separately. The state input is 6-dimensional, and the action input is 3-dimensional. The state branch includes two fully connected layers: the

Table 7.1 Network parameters of DDPG algorithm.

Parameter	Value
Actor network learning rate α	0.0003
Critic network learning rate β	0.0003
Discount factor γ	0.99
Soft update coefficient τ	0.005
Memory size	1024

first has 400 neurons with LayerNorm and ReLU activations; the second has 300 neurons and also uses LayerNorm. The action branch has one fully connected layer with 300 neurons. The outputs of the state and action branches—each being a 300-dimensional vector—are added element-wise to generate a scalar value at the output layer.

As different tasks have varying parameter combinations, finding an appropriate learning rate is crucial. The number of layers and neurons in the neural network also significantly influences the algorithm's performance. Analysis of the training of the DRL algorithm mentioned in this chapter shows that it eventually converges under the network parameters listed in Table 7.1.

Based on a single experiment duration of 20s and a control cycle of 0.2s, along with the reward function in Eq. (7.8), the cumulative reward per experiment tends to approach 4900 once the assembly module completes the entire electromagnetic docking task. The curve of reward over experiment iterations is shown in Fig. 7.2. It is shown that the cumulative reward increases steadily and gradually converges. From around the 200th training episode, the reward increases, indicating that the algorithm is learning an effective policy, improving docking accuracy, and reducing energy consumption. After about 4000 episodes, the reward stabilizes or continues increasing with reduced fluctuation, signifying convergence and stability. The large fluctuations observed during convergence are typical for complex tasks such as electromagnetic docking control and indicate that the DDPG algorithm avoids premature convergence, thus demonstrating good stability in this application.

Analysis of the training results validates the effectiveness of the DDPG-based control approach for electromagnetic assembly tasks in elliptical orbits. As shown in Figs 7.3 and 7.4, the Actor network loss increases with training while the Critic network loss decreases, and both eventually stabilize. During each update, the Actor adjusts its parameters to maximize the Critic's output, while the Critic minimizes its output error to match the current policy value.

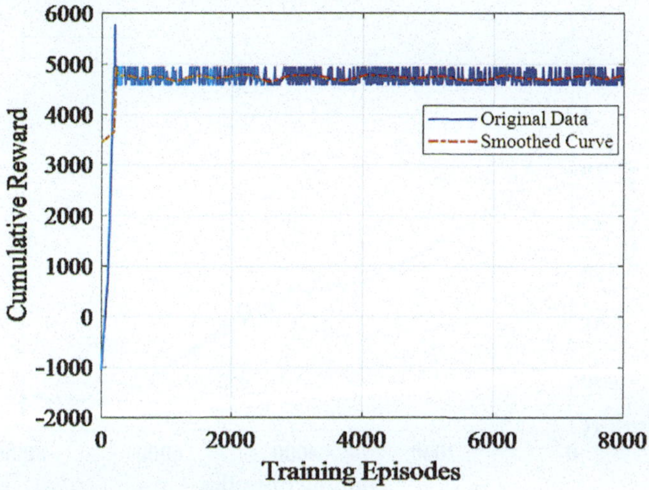

Figure 7.2 Training iteration response.

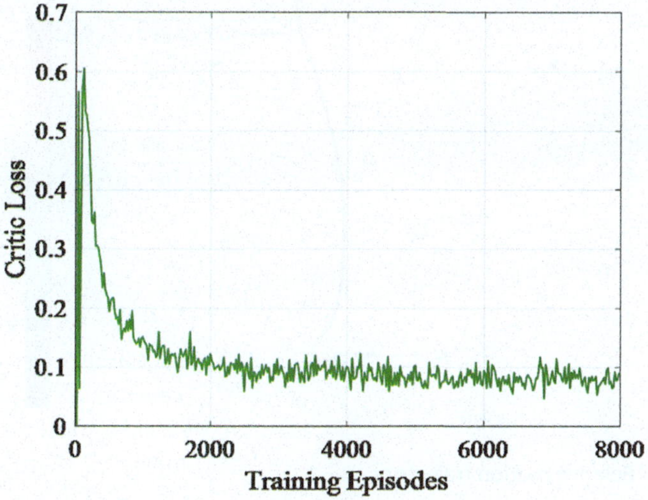

Figure 7.3 Training iteration response of critic loss.

7.4.2 Simulation result analysis

This section presents the simulation analysis of the electromagnetic docking task between the assembly module and the spacecraft platform in an elliptical orbits. A state-space model is established using the TH equation, and the

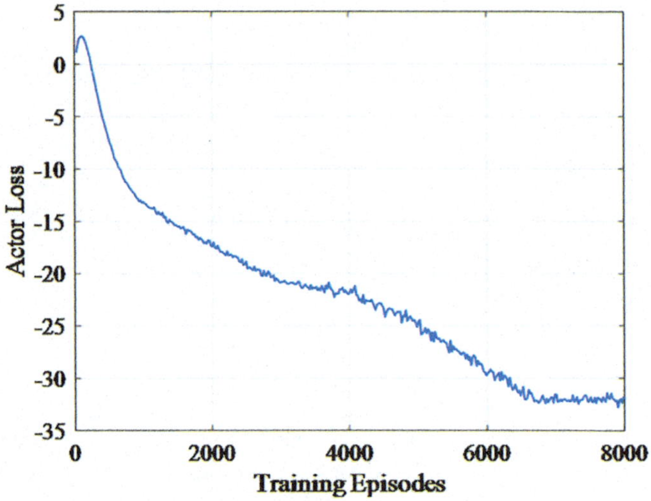

Figure 7.4 Training iteration response of actor loss.

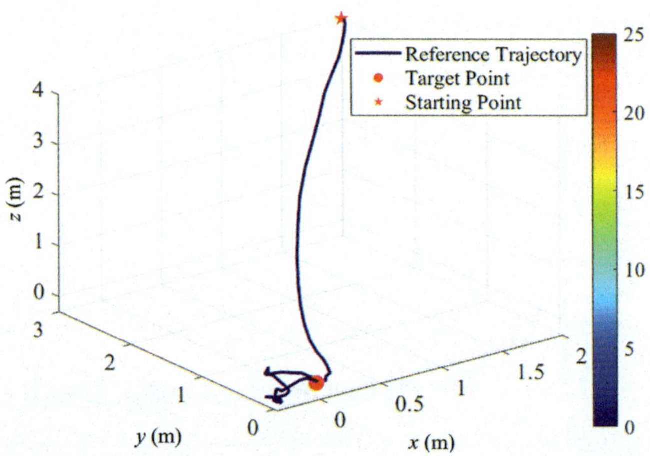

Figure 7.5 Relative motion trajectory.

DDPG algorithm is used to control the relative position and velocity of the assembly module. The simulation results are then analyzed.

Assume the spacecraft platform and the assembly module have mass that are $m_P = 100$kg, $m_M = 40$kg and the effective mass is $m_0 = 20$kg. The semimajor axis of the orbit is $a = 10,000$km, with eccentricity $e = 0.15$, and the initial true anomaly in the "CM" orbital coordinate system is $\theta_0 = 0$. The magnetic dipole moment

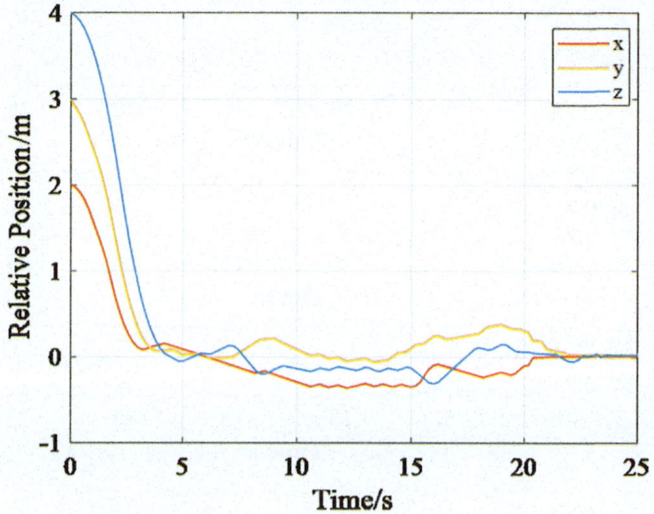

Figure 7.6 Time response of relative position.

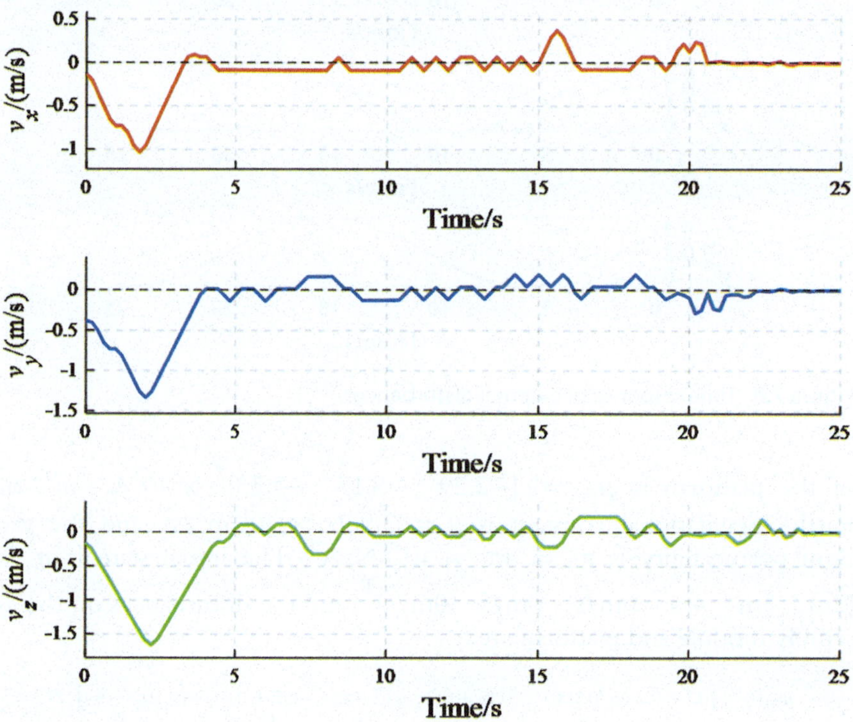

Figure 7.7 Time response of relative velocity.

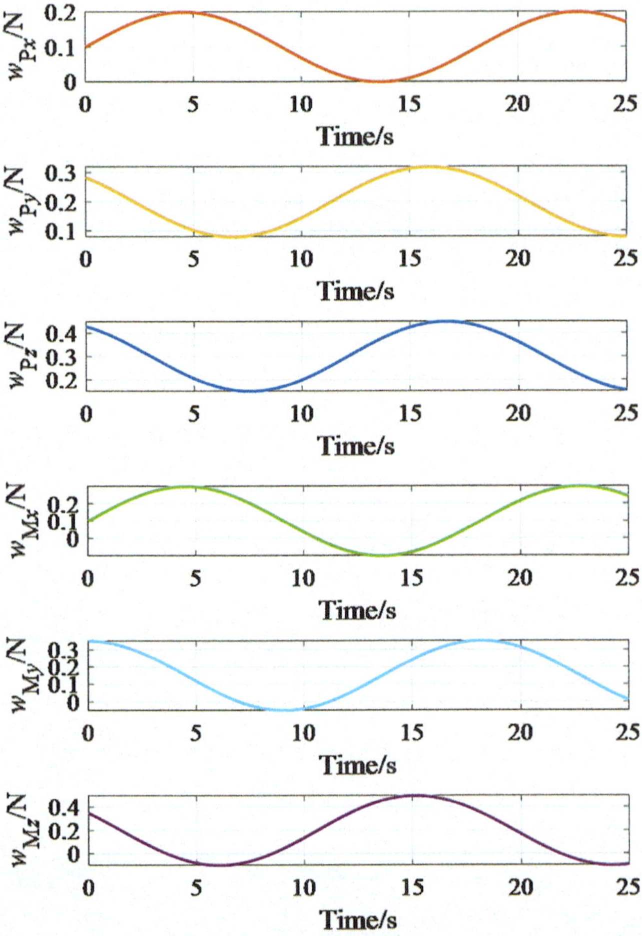

Figure 7.8 Time response of external disturbances.

of the platform is $\boldsymbol{\mu}_{\mathrm{P}} = [-2.29 \quad -8.17 \quad -2.33]^{\mathrm{T}} \times 10^5 \mathrm{A} \cdot \mathrm{m}^2$, the Earth's gravitational constant is $\mu = 3.986 \times 10^{14} \mathrm{m}^3/\mathrm{s}^2$, and the vacuum permeability is $\pi_0 = 4\pi \times 10^{-7} \mathrm{N/A}^2$. The initial state is $\boldsymbol{\eta}_0 = [2\mathrm{m} \quad 3\mathrm{m} \quad 4\mathrm{m} \quad 0\mathrm{m/s} \quad 0\mathrm{m/s} \quad 0\mathrm{m/s}]^{\mathrm{T}}$, and the disturbances on the assembly module and platform are

$$w_{\mathrm{M}}(t) = \left[0.1 + 0.2\sin(0.11\pi t) \quad 0.15 + 0.2\cos(0.11\pi t) \quad 0.2 + 0.3\cos\left(0.11\pi t + \tfrac{\pi}{3}\right)\right]^{\mathrm{T}} \mathrm{N}$$

$$w_{\mathrm{P}}(t) = \left[0.1 + 0.1\sin(0.11\pi t) \quad 0.2 + 0.12\cos(0.11\pi t + \pi/4) \quad 0.3 + 0.15\cos(0.11\pi t + \pi/6)\right]^{\mathrm{T}} \mathrm{N}$$

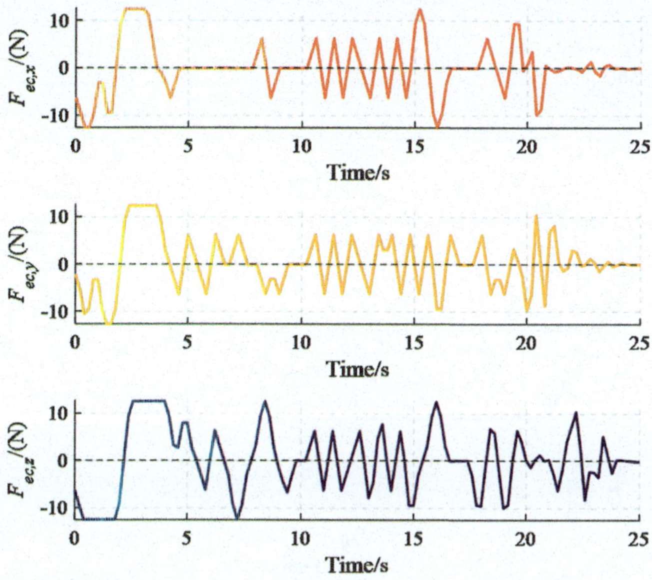

Figure 7.9 Time response of electromagnetic force.

The maximum allowable relative velocity is $v_{\max} = 1\text{m/s}$, and the distance constraint is $0 \le \eta \le 1\text{m}$.

To illustrate the relative motion between the assembly module and the spacecraft platform, a trajectory with a success rate of 0.99 is selected. The relative motion trajectory during the assembly is shown in Fig. 7.5, where the color bar represents simulation time. The target point is the initial position of the platform, and the start point is that of the module. The results show that the module successfully docks with the platform within a short time.

The module successfully completes docking within 56s. As shown in Figs 7.6 and 7.7, the module's relative velocity increases first and then decreases to zero, satisfying the soft docking requirement. This verifies the effectiveness of the DDPG algorithm in electromagnetic docking control. The external disturbances include two parts, the assembly module's and the platform's, which are shown in Fig. 7.8. The time response of the electromagnetic force is shown in Fig. 7.9, with control forces along all axes remaining below 10N. The resulting magnetic moment of the module is shown in Fig. 7.10, with values remaining below $2 \times 10^5 \text{A} \cdot \text{m}^2$, thereby satisfying practical conditions.

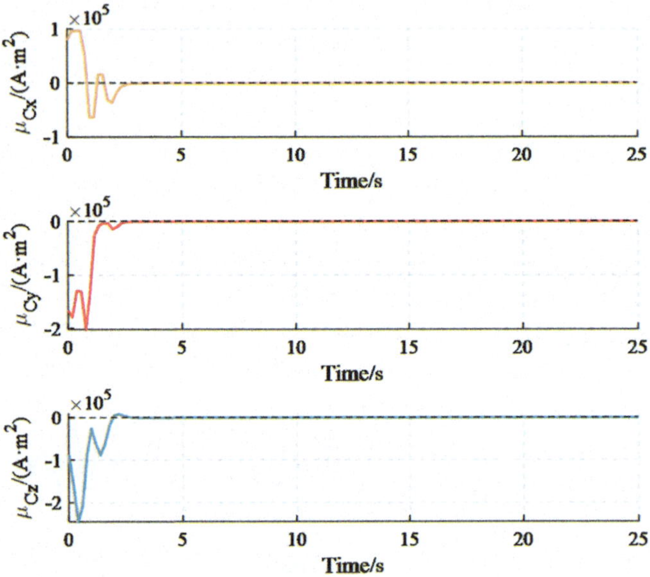

Figure 7.10 Time response of magnetic moment.

7.5 Conclusions

This section proposes a DDPG-based multiobjective optimal electromagnetic docking control method for spacecraft on elliptical orbits. Through trajectory planning and control input prediction, the method addresses the complex nonlinear constraints and dynamic environment of the docking process. First, a dynamic model for electromagnetic docking is established, incorporating multiple constraints such as relative position, velocity, disturbances, and time into the reinforcement learning environment. A comprehensive reward function and Actor-Critic network are then designed, enabling autonomous online control optimization in complex orbital environments.

This method does not rely on accurate dynamic models or strict prior parameters; instead, it learns and adjusts policies in real time through interaction with the environment. It ensures rapid and accurate docking under complex disturbances, optimizing docking time and control precision. Simulation results confirm that the DDPG algorithm demonstrates significant advantages in multiobjective electromagnetic docking control. It not only accurately guides the assembly module through the docking process but also exhibits strong robustness and adaptability, highlighting its value for on-orbit assembly applications.

References

Chai, R., Niu, H., Carrasco, J., Arvin, F., Yin, H., & Lennox, B. (2024). Design and experimental validation of deep reinforcement learning-based fast trajectory planning and control for mobile robot in unknown environment. *IEEE Transactions on Neural Networks and Learning Systems, 35*(4), 5778–5792. https://doi.org/10.1109/tnnls.2022.3209154.

Foust, R. C., Lupu, E. S., Nakka, Y. K., Chung, S. J., & Hadaegh, F. Y. (2020). Autonomous in-orbit satellite assembly from a modular heterogeneous swarm. *Acta Astronautica, 169*, 191–205. https://doi.org/10.1016/j.actaastro.2020.01.006.

Gao, X., Yan, L., Li, Z., Wang, G., & Chen, I. M. (2023). Improved deep deterministic policy gradient for dynamic obstacle avoidance of mobile robot. *IEEE Transactions on Systems, Man, and Cybernetics: Systems, 53*(6), 3675–3682. https://doi.org/10.1109/tsmc.2022.3230666.

Gronauer, S., & Diepold, K. (2022). Multiagent deep reinforcement learning: a survey. *Artificial Intelligence Review, 55*(2), 895–943. https://doi.org/10.1007/s10462-021-09996-w.

Liu, C., Yue, X., Zhang, J., & Shi, K. (2022). Active disturbance rejection control for delayed electromagnetic docking of spacecraft in elliptical orbits. *IEEE Transactions on Aerospace and Electronic Systems, 58*(3), 2257–2268. https://doi.org/10.1109/taes.2021.3130830.

Lyu, P. (2025). Robot path planning algorithm with improved DDPG algorithm. *International Journal on Interactive Design and Manufacturing, 19*(2), 1123–1133. https://doi.org/10.1007/s12008-024-01834-x.

Moon, J., Papaioannou, S., Laoudias, C., Kolios, P., & Kim, S. (2021). Deep reinforcement learning multi-UAV trajectory control for target tracking. *IEEE Internet of Things Journal, 8*(20), 15441–15455. https://doi.org/10.1109/jiot.2021.3073973.

Ren, J., Kong, N., Zhuang, Y., Zhang, J., Ma, S., Wang, B., & Liu, W. (2021). A review on the interfaces of orbital replacement unit: Great efforts for modular spacecraft. *Proceedings of the Institution of Mechanical Engineers, Part G: Journal of Aerospace Engineering, 235*(14), 1941–1967. https://doi.org/10.1177/0954410021993086.

Shi, K., Liu, C., & Sun, Z. (2019). Constrained fuel-free control for spacecraft electromagnetic docking in elliptical orbits. *Acta Astronautica, 162*, 14–24. https://doi.org/10.1016/j.actaastro.2019.05.016.

Shi, K., Liu, C., Sun, Z., & Yue, X. (2022). Coupled orbit-attitude dynamics and trajectory tracking control for spacecraft electromagnetic docking. *Applied Mathematical Modelling, 101*, 553–572. https://doi.org/10.1016/j.apm.2021.08.030.

Tarekegn, G. B., Juang, R. T., Tesfaw, B. A., Lin, H. P., Hsu, H. C., Tarekegn, R. B., & Tai, L. C. (2024). A centralized multiagent drl-based trajectory control strategy for unmanned aerial vehicle-enabled wireless communications. *IEEE Open Journal of Vehicular Technology, 5*, 1230–1241. https://doi.org/10.1109/OJVT.2024.3451143.

Wang, Y., Fang, S., Hu, J., & Huang, D. (2023). A novel active disturbance rejection control of PMSM based on deep reinforcement learning for more electric aircraft. *IEEE Transactions on Energy Conversion, 38*(2), 1461–1470. https://doi.org/10.1109/tec.2023.3235927.

Wei, B., Zhang, J., Kong, N., Ma, S., Wang, B., & Han, R. (2023). Self-docking characteristics and sliding mode control on space electromagnetic docking mechanism. *Machines., 11*(3), 332. https://doi.org/10.3390/machines11030332.

Yan, B., Ling, P., Zhou, Y., Wu, C., & Zhang, W. (2022). Shock isolation characteristics of a bistable vibration isolator with tunable magnetic controlled stiffness. *Journal of Vibration and Acoustics, 144*(2). https://doi.org/10.1115/1.4051850.

Yu, W., Liu, C., & Yue, X. (2024). Reinforcement learning-based decision-making for spacecraft pursuit-evasion game in elliptical orbits. *Control Engineering Practice, 153*, 106072. https://doi.org/10.1016/j.conengprac.2024.106072.

Yue, C., Wang, X., & Yue, X. (2024). Event-triggered model predictive controller with neural network for autonomous rendezvous and proximity on elliptical orbits. *IEEE Transactions on Aerospace and Electronic Systems, 60*(5), 6163–6180. https://doi.org/10.1109/TAES.2024.3400171.

Zhang, H., Zhang, J., Ji, M., Liu, H., & Yu, Y. (2024). Anti-unwinding adaptive robust backstepping attitude maneuver control for rigid spacecraft. *Advances in Space Research, 74*(2), 964–974. https://doi.org/10.1016/j.asr.2024.04.019.

CHAPTER 8

Sliding mode tracking control for spacecraft electromagnetic separation

8.1 Introduction

Large-scale spacecraft are major pieces of aerospace equipment for future space resource utilization, exploration of cosmic mysteries, and long-term on-orbit residence. They often require multiple launches and on-orbit assembly, and cannot be constructed through a single launch and on-orbit deployment. As the on-orbit assembly strategy can provide an attractive solution for the on-orbit construction of various large space structures, for example, space telescopes, satellite antennas, solar power systems, and a deep space exploration transit station, it has been identified as a significant research direction for future space missions (He et al., 2022; She et al., 2019). On-orbit assembly usually comes in one of two ways. One option is to assemble and operate task modules through robots related to tasks (Hou et al., 2022; Mohan, 2010), but the corresponding module cannot be too large and is only suitable for some small tasks. The other option is that each spacecraft or mission module can be self-driven for on-orbit assembly (Cao et al., 2021; Zhao et al., 2024). However, the on-orbit assembly process requires the spacecraft module to be close to the spacecraft platform to form a large-scale combined spacecraft, which makes its mass characteristics difficult to be accurately known due to significant noncooperative characteristics, the large scale, and configuration variations of the platform. It should be noted that spacecraft docking and separation are the key technologies for on-orbit assembly (Foust et al., 2018), but traditional thruster-based energy supply systems inevitably cause restrictions on the flexibility of missions and lifetime (Olivieri & Francesconi, 2016; Yang et al., 2023), which promotes the development of electromagnetic docking and separation using noncontact forces.

Spacecraft electromagnetic docking is to use the inter-craft electromagnetic force to govern the relative motion between the target and chaser

Spacecraft Electromagnetic Docking and Separation.
DOI: https://doi.org/10.1016/B978-0-443-49334-8.00003-3

spacecraft, and it has some special, less obvious internal abilities, such as self-docking ability, which can reduce the burden on the controller to a certain extent (Zhang et al., 2016). Many scholars have investigated the controller design in Shi et al. (2019), Zhang et al., 2012, and references therein. In Zhang et al., 2012, the relative dynamics model used in electromagnetic docking was derived based on Hill's model, so it is not difficult to see that this technique is only for circular orbits and does not apply to elliptical orbits in this work. Furthermore, it should be noted that the mass of the spacecraft was previously known. To deal with this, Shi et al. (2019) designed a disturbance observer-based controller, where no information on elliptical eccentricity or spacecraft mass is required, but the relative motion information is required for accurate measurements. Most of the existing technologies in the field of electromagnetic docking are aimed at circular orbits, and applying them to large-eccentric orbits may cause great errors and lead to mission failures (Ahsun et al., 2010; Shi et al., 2020). Based on differential flatness, an orbit planning algorithm for spacecraft proximity operations in a large-eccentric orbit using spacecraft electromagnetic force is studied in Zhu et al., 2018. In Shi et al. (2020), a robust electromagnetic docking problem was addressed for spacecraft in the presence of unknown mass, fault signals, external disturbances, input constraints, measurement errors, and elliptical eccentricity. However, the compliant trajectory was not accounted for, which may violate some optimality. The trajectory planning problem is generally solved by a pseudo-spectral approach (Boesso & Francesconi, 2013; Hallaj & Assadian, 2015), driven by the electromagnetic force of the spacecraft. However, as the number of spacecraft increases, the numerical difficulty of approaching operational problems for electromagnetic spacecraft increases. To this end, based on a predetermined reference trajectory, an improved sliding mode controller was designed in Shi et al. (2022), and the good capability of coupled orbit-attitude tracking was illustrated.

Similar to electromagnetic docking, electromagnetic separation can be seen as the inverse process of electromagnetic docking, which has the same distinct advantages, for example, no propellant consumption, no plume contamination, continuous, reversible, and noncontact actuation (Liu et al., 2022a). However, the dynamics of electromagnetic separation are also complex and coupled, which makes the controller design a great challenge. Thus, it deserves further attention to solve the high-precision control problem of spacecraft electromagnetic separation considering external disturbances, elliptical eccentricity, and unknown mass, which is still an open problem. Any of the above problems can greatly degrade the control performance, and if these problems are considered simultaneously, it is difficult to meet the control objectives. Much literature has largely alleviated or solved this problem by using sliding mode control techniques (Fei et al., 2022; Pukdeboon & Kumam, 2015). For example, Bayat (2019) designed a continuous nonsingular sliding mode control law to achieve finite time approximate tracking by assuming that the spacecraft inertia matrix contains a certain proportion of bounded parameter uncertainty and applying bounded unknown external disturbances midway. But tracking performance lacks universality and only depends on the uncertainty bound, mainly from external disturbances and design parameters. An adaptive sliding mode controller based on fuzzy logic was designed for the realization of satellite attitude tracking, but no compliant performance was considered in Xing et al. (2021). In addition, the above approaches cannot be applied to electromagnetic separation directly with strong couplings and high nonlinearity. An artificial-potential-function-based control strategy was proposed in Zhang et al., 2011, and an extended state observer was designed to address the inherent control issue involved in self-docking, which showed good performance. In Liu et al. (2022b), by using a second-order integral sliding mode control approach, an attitude controller with global robustness and global finite-time convergence is designed. The gain of the controller is calculated by an adaptive law, and the advanced information of uncertainty and external interference boundary is eliminated. Based on the coupling of sliding mode control and a fuzzy logic system, Liu et al. (2020) proposed a robust adaptive control strategy to reduce chattering of the traditional sliding mode approach effectively without knowing the boundaries of parameter uncertainty and external disturbance. In Guo & Chen (2019), based on adaptive fast terminal sliding mode control, a fault-tolerant control strategy was proposed to solve the fault-tolerant control problem of a spacecraft attitude stabilization system under mismatching disturbances and actuator faults.

To sum up, the above control strategies are mostly aimed at attitude control systems or circular orbits with known mass, and the electromagnetic separation process has stronger nonlinearity and coupling compared to the docking process in the initial stage of the relative motion system. Thus, the application of the above control strategies in the electromagnetic separation process may result in poor stability and robustness, and even instability. It is difficult to achieve the goal of high-precision and stable separation of two spacecraft under multisource disturbances.

Motivated by the above discussions, this chapter overcomes the propellant consumption and plume pollution in the actual separation process, as well as various complex disturbances from multiple sources, and achieves a continuously controllable, high-precision, and multidirectional separation process for two spacecraft orbits. An improved sliding mode tracking control strategy has been proposed for electromagnetic separation without prior knowledge of mass and external disturbances to separate the platform and module from a combined spacecraft with very good performance. There are two main contributions. First, compared to Liu et al. (2022b), this chapter focuses on the research of electromagnetic separation control approaches for on-orbit assembly of large-scale spacecraft, which improves fault tolerance and wide application in the on-orbit assembly process. Combined with docking, it can achieve on-orbit self-healing of large-scale spacecraft. Secondly, compared to the traditional sliding mode control approaches in Zhang et al. (2021), this chapter combines the advantages of exponential sliding mode and power sliding mode. When the error is far away from the sliding surface, the linear part of the reaching law plays a major role, so that the error can approach the sliding surface quickly. When the error is close to the sliding surface, the fractional power term in the reaching law has much more influence than the linear part, which can drive the error to reach the sliding surface in finite time to achieve a smoother and more efficient separation process. The error system is globally asymptotically stable, and the error state can reach the sliding surface from any initial position in a finite time.

The remaining structures in this work are as follows. Section 8.2 describes the nonlinear model for spacecraft electromagnetic separation, which has a predesigned reference trajectory, and puts forward the corresponding control problem. Section 8.3 presents a sliding mode control scheme with a finite-time part and an asymptotical-stability part, and the corresponding stability analysis is carried out by the Lyapunov stability theorem. Finally, numerical simulations are carried out to verify the performance of the developed controller, and some conclusions are drawn.

8.2 Problem formulation

8.2.1 Dynamic modeling

According to the dynamic model introduced in Section 1.2 and considering actuator fault, Eq. (1.78) can be converted into the following state-space form:

$$\dot{\boldsymbol{\eta}}(t) = \boldsymbol{A}\boldsymbol{\eta}(t) + \boldsymbol{B}\boldsymbol{u}(t) + \boldsymbol{B}\boldsymbol{w}(t) \tag{8.1}$$

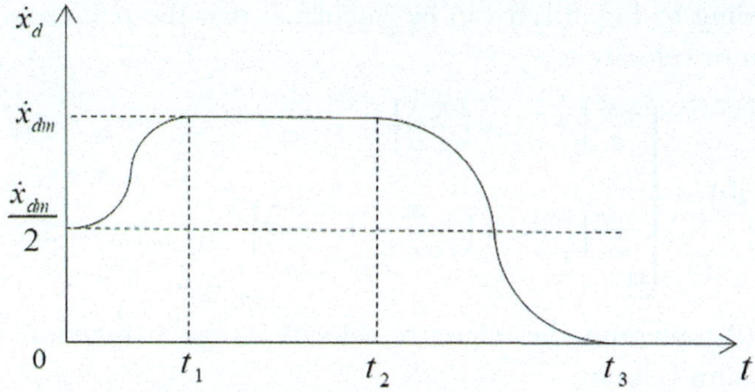

Figure 8.1 Velocity reference trajectory in the *x* direction.

with

$$\boldsymbol{w}(t) = \boldsymbol{B}^{\mathrm{T}}\left(\boldsymbol{B}\boldsymbol{B}^{\mathrm{T}}\right)^{*}(\Delta \boldsymbol{A}\boldsymbol{\eta}(t) + \Delta \boldsymbol{B}\boldsymbol{u}(t) + (\boldsymbol{B} + \Delta \boldsymbol{B})\boldsymbol{w}_0(t))$$

which can be regarded as a comprehensive interference, where $(\boldsymbol{B}\boldsymbol{B}^{\mathrm{T}})^{*}$ is the pseudoinverse of $\boldsymbol{B}\boldsymbol{B}^{\mathrm{T}}$.

Assumption 8.1 The synthetic disturbance $\boldsymbol{w}(t)$ and its first derivative is piecewise continuous and bounded, and the upper bound is $\boldsymbol{D}$.

Remark 8.1 Assumption 8.1 is reasonable. First, the three components of the control input are continuous and bounded in practice, due to limited control capabilities. Second, the external disturbances are continuous and bounded, such as solar radiation, residual atmosphere, and magnetic effects. In particular, relative position and velocity in motion are also continuous and bounded in practice. The disturbance of composition is decomposed into $w_x(t)$, $w_y(t)$ and $w_z(t)$ in the x, y and z directions. Although the exact value of the disturbance in each direction is unknown, $|w_x(t)| \leq D_x$, $|w_y(t)| \leq D_y$ and $|w_z(t)| \leq D_z$. Thus, in the subsequent simulation, the upper bound $\boldsymbol{D} = [D_x\ D_y\ D_z]^{\mathrm{T}}$ is used to replace the actual comprehensive disturbance value. To sum up, the synthetic disturbance $\boldsymbol{w}(t)$ and its first derivative is piecewise continuous and bounded, and the upper bound is $\boldsymbol{D}$.

8.2.2 Reference trajectory design

As shown in Fig. 8.1, assuming that the separation process is sequentially divided into three consecutive stages of acceleration, uniform, and deceleration, and the reference trajectory of relative velocity is $\dot{x}_d$.

According to Fig. 8.1, it can be calculated that the reference trajectory expression of velocity is

$$\dot{x}_d(t) = \begin{cases} \dfrac{\dot{x}_{dm}}{4}\left[3 - \cos\left(\dfrac{\pi}{t_1}t\right)\right] & 0 \le t \le t_1 \\ \dot{x}_{dm} & t_1 < t \le t_2 \\ \dfrac{\dot{x}_{dm}}{2}\left[1 + \cos\left(\dfrac{\pi}{t_3 - t_2}(t - t_2)\right)\right] & t_2 < t \le t_3 \\ 0 & t_3 < t \end{cases} \tag{8.2}$$

where $\dot{x}_d(t)$ represents the reference velocity in the x direction, and $\dot{x}_{dm}$ is the maximum velocity.

Based on Eq. (8.2), one has

$$x_d(t) = \begin{cases} -\dfrac{\dot{x}_{dm}t_1}{4\pi} + \dfrac{3\dot{x}_{dm}}{4}t - \dfrac{\dot{x}_{dm}t_1}{4\pi}\left(\sin\dfrac{\pi t}{t_1} - 1\right) & 0 \le t \le t_1 \\ \dfrac{3\dot{x}_{dm}}{4}t_1 + \dot{x}_{dm}(t - t_1) \quad t_1 < t \le t_2 & \\ \dfrac{3\dot{x}_{dm}}{4}t_1 + \dot{x}_{dm}(t_2 - t_1) + \dfrac{\dot{x}_{dm}}{2}(t - t_2) & \\ \qquad + \dfrac{\dot{x}_{dm}(t_3 - t_2)}{2\pi}\sin\left(\dfrac{\pi}{t_3 - t_2}(t - t_2)\right) & t_2 < t \le t_3 \\ \dfrac{3\dot{x}_{dm}}{4}t_1 + \dot{x}_{dm}(t_2 - t_1) + \dfrac{\dot{x}_{dm}}{2}(t_3 - t_2) & t_3 < t \end{cases} \tag{8.3}$$

where $x_d(t)$ represents the reference position along the x direction.

$$\ddot{x}_d(t) = \begin{cases} \dfrac{\dot{x}_{dm}\pi}{4t_1}\sin\left(\dfrac{\pi}{t_1}t\right) & 0 \le t \le t_1 \\ 0 & t_1 < t \le t_2 \\ -\dfrac{\dot{x}_{dm}\pi}{2(t_3 - t_2)}\sin\left(\dfrac{\pi}{t_3 - t_2}(t - t_2)\right) & t_2 < t \le t_3 \\ 0 & t_3 < t \end{cases} \tag{8.4}$$

where $\ddot{x}_d(t)$ represents the reference acceleration along x direction.

8.2.3 Control objectives

The control scheme should achieve the following two objectives:

1) The closed-loop control system is asymptotically stable.
2) The error state of any initial position will reach the sliding surface within a finite time.

8.3 Improved sliding mode controller design

The mathematical equations of reference position, velocity, and acceleration in the y and z directions are similar to Eqs. (8.2)– (8.4). Defining the reference state $\boldsymbol{\eta}_d = [x_d \; y_d \; z_d \; \dot{x}_d \; \dot{y}_d \; \dot{z}_d]^{\mathrm{T}}$, the error $\boldsymbol{e} = \boldsymbol{\eta} - \boldsymbol{\eta}_d$, and the first derivative of the error $\dot{\boldsymbol{e}} = \dot{\boldsymbol{\eta}} - \dot{\boldsymbol{\eta}}_d$, then a sliding surface is chosen as

$$\boldsymbol{s} = c\boldsymbol{e} \tag{8.5}$$

where c is a given positive constant.

The sliding mode control law consists of an equivalent control term $\boldsymbol{u}_{eq}$ and a switching control term $\boldsymbol{u}_{sw}$. The former ensures the state of the system remains on the sliding surface, while the latter drives the state to cross continuously on both sides of the sliding surface. When $\dot{\boldsymbol{s}} = 0$, the equivalent control term can be obtained, then the reaching law $\dot{\boldsymbol{s}}$ is introduced and $\boldsymbol{u} = \boldsymbol{u}_{eq} + \boldsymbol{u}_{sw}$.

Based on Eq. (8.5), the first derivative of $\boldsymbol{s}$ is

$$\dot{\boldsymbol{s}} = c\left(\boldsymbol{A}\boldsymbol{\eta}(t) + \boldsymbol{B}\boldsymbol{u}(t) + \boldsymbol{B}\boldsymbol{w}(t) - \dot{\boldsymbol{\eta}}_d(t)\right) \tag{8.6}$$

Let $\dot{\boldsymbol{s}} = 0$, $\boldsymbol{u}_{eq}$ is

$$\boldsymbol{u}_{eq} = \boldsymbol{B}^{\mathrm{T}}\left(\boldsymbol{B}\boldsymbol{B}^{\mathrm{T}}\right)^{*}\left[\dot{\boldsymbol{\eta}}_d(t) - \boldsymbol{A}\boldsymbol{\eta}(t) - \boldsymbol{B}\boldsymbol{w}(t)\right] \tag{8.7}$$

where $(\boldsymbol{B}\boldsymbol{B}^{\mathrm{T}})^{*}$ is the pseudoinverse of $\boldsymbol{B}\boldsymbol{B}^{\mathrm{T}}$.

Lemma 8.1 The reaching law is designed as

$$\dot{\boldsymbol{s}} = -k\boldsymbol{s} - \varepsilon|\boldsymbol{s}|^{\overline{\alpha}}\mathrm{sgn}(\boldsymbol{s}),\ 0 < \overline{\alpha} < 1,\ k > 0,\ \varepsilon > 0 \tag{8.8}$$

Then, the error $\boldsymbol{e}$ will reach the sliding surface within finite time t_r.

$$t_r = \frac{1}{k\left(\overline{\alpha} - 1\right)}\ln\frac{\varepsilon}{k|\boldsymbol{s}(0)|^{1-\overline{\alpha}} + \varepsilon} \tag{8.9}$$

Remark 8.2 The innovation of sliding mode control in this chapter is that the reaching law $\dot{\boldsymbol{s}} = -k\boldsymbol{s} - \varepsilon|\boldsymbol{s}|^{\overline{\alpha}}\mathrm{sgn}(\boldsymbol{s})$ has the advantages of both exponential reaching law $\dot{\boldsymbol{s}} = -k\boldsymbol{s} - \varepsilon\mathrm{sgn}(\boldsymbol{s})$ and power reaching law $\dot{\boldsymbol{s}} = -\varepsilon|\boldsymbol{s}|^{\overline{\alpha}}\mathrm{sgn}(\boldsymbol{s})$, which is used to drive the error $\boldsymbol{e}$ to reach the sliding mode surface in finite time. When the error $\boldsymbol{e}$ is far away from the sliding surface, the linear part $-k\boldsymbol{s}$ of the reaching law plays a major role, and has a similar structure to the exponential reaching law, and satisfies the exponential convergence condition, so that the error $\boldsymbol{e}$ can approach the sliding surface quickly. When the error $\boldsymbol{e}$ is close to the sliding surface, the fractional power term $-\varepsilon|\boldsymbol{s}|^{\overline{\alpha}}\mathrm{sgn}(\boldsymbol{s})$ in the reaching law has much more influence than the linear part $-k\boldsymbol{s}$, and has similar structure with the power reaching law, which can drive the error $\boldsymbol{e}$ to reach the sliding surface in finite time. Thus, compared with other sliding mode control, it has the advantages of both exponential reaching law and power reaching law, and has better performance.

Proof. Regarding Eq. (8.8), one has

$$\begin{aligned} \boldsymbol{s}^{\mathrm{T}}\dot{\boldsymbol{s}} &= \boldsymbol{s}^{\mathrm{T}}\left(-k\boldsymbol{s} - \varepsilon|\boldsymbol{s}|^{\overline{\alpha}}\mathrm{sgn}(\boldsymbol{s})\right) \\ &= -k|\boldsymbol{s}|^2 - \varepsilon\boldsymbol{s}^{\mathrm{T}}|\boldsymbol{s}|^{\overline{\alpha}-1}|\boldsymbol{s}|\mathrm{sgn}(\boldsymbol{s}) \\ &= -k|\boldsymbol{s}|^2 - \varepsilon|\boldsymbol{s}|^{\overline{\alpha}+1} \end{aligned} \tag{8.10}$$

then, Eq. (8.10) can perform the following transformations:

$$\begin{aligned} &-k|\boldsymbol{s}|^2 - \varepsilon|\boldsymbol{s}|^{\overline{\alpha}+1} = \frac{1}{2}\frac{d(\boldsymbol{s}^{\mathrm{T}}\boldsymbol{s})}{dt} = |\boldsymbol{s}|\frac{d|\boldsymbol{s}|}{dt} \\ &\Rightarrow -k|\boldsymbol{s}| - \varepsilon|\boldsymbol{s}|^{\overline{\alpha}} = \frac{d|\boldsymbol{s}|}{dt} \\ &\Rightarrow dt = -\frac{d|\boldsymbol{s}|}{k|\boldsymbol{s}| + \varepsilon|\boldsymbol{s}|^{\overline{\alpha}}} = -\frac{|\boldsymbol{s}|^{-\overline{\alpha}}d|\boldsymbol{s}|}{k|\boldsymbol{s}|^{1-\overline{\alpha}} + \varepsilon} \\ &= -\frac{1}{1-\overline{\alpha}}\frac{d|\boldsymbol{s}|^{1-\overline{\alpha}}}{k|\boldsymbol{s}|^{1-\overline{\alpha}} + \varepsilon} = \frac{1}{\overline{\alpha}-1}\frac{d|\boldsymbol{s}|^{1-\overline{\alpha}}}{k|\boldsymbol{s}|^{1-\overline{\alpha}} + \varepsilon} \end{aligned} \tag{8.11}$$

Integrating both sides of Eq. (8.11) to get

$$t_r = \frac{1}{k(\overline{\alpha}-1)}\ln\frac{\varepsilon}{k|\boldsymbol{s}(0)|^{1-\overline{\alpha}} + \varepsilon} \tag{8.12}$$

Thus, based on Eq. (8.12), it is obvious that when $t > t_r$, $\dot{\boldsymbol{s}} = 0$.

This completes the proof.

Theorem 8.1 The improved sliding mode control law is designed as

$$\boldsymbol{u} = \boldsymbol{B}^{\mathrm{T}}\left(\boldsymbol{B}\boldsymbol{B}^{\mathrm{T}}\right)^{*}\left[\dot{\boldsymbol{\eta}}_d(t) - \boldsymbol{A}\boldsymbol{\eta}(t) - \boldsymbol{S}\boldsymbol{B}\boldsymbol{D} - \frac{k}{c}\boldsymbol{s} - \frac{\varepsilon}{c}|\boldsymbol{s}|^{\overline{\alpha}}\mathrm{sgn}(\boldsymbol{s})\right] \tag{8.13}$$

where

$$\boldsymbol{S} = \mathrm{diag}\left(\left[\mathrm{sgn}(s_1) \quad \mathrm{sgn}(s_2) \quad \mathrm{sgn}(s_3) \quad \mathrm{sgn}(s_4) \quad \mathrm{sgn}(s_5) \quad \mathrm{sgn}(s_6)\right]\right)$$

Then, the error state can reach the sliding surface in a finite time from any initial position.

Proof. Design a Lyapunov function candidate as

$$V = \frac{1}{2}\boldsymbol{s}^{\mathrm{T}}\boldsymbol{s} \tag{8.14}$$

Substituing Eq. (8.1) and Eq. (8.13) into Eq. (8.14),the derivation of V is

$$\begin{aligned}
\dot{V} &= \boldsymbol{s}^{\mathrm{T}}\dot{\boldsymbol{s}} = \boldsymbol{s}^{\mathrm{T}}\left[c\left(\boldsymbol{A}\boldsymbol{\eta}(t) + \boldsymbol{B}\boldsymbol{u}(t) + \boldsymbol{B}\boldsymbol{w}(t) - \dot{\boldsymbol{\eta}}_d(t)\right)\right] \\
&= \boldsymbol{s}^{\mathrm{T}}\left[c\left(\boldsymbol{A}\boldsymbol{\eta}(t) + \dot{\boldsymbol{\eta}}_d(t) - \boldsymbol{A}\boldsymbol{\eta}(t) - \boldsymbol{SBD} - \frac{k}{c}\boldsymbol{s} - \frac{\varepsilon}{c}|\boldsymbol{s}|^{\overline{\alpha}}\mathrm{sgn}(\boldsymbol{s}) + \boldsymbol{B}\boldsymbol{w}(t) - \dot{\boldsymbol{\eta}}_d(t)\right)\right] \\
&= \boldsymbol{s}^{\mathrm{T}}\left(-k\boldsymbol{s} - \varepsilon|\boldsymbol{s}|^{\overline{\alpha}}\mathrm{sgn}(\boldsymbol{s}) - c\boldsymbol{SB}(\boldsymbol{D} - \boldsymbol{w}(t))\right) \\
&= -k|\boldsymbol{s}|^2 - \varepsilon\boldsymbol{s}^{\mathrm{T}}|\boldsymbol{s}|^{\overline{\alpha}-1}|\boldsymbol{s}|\mathrm{sgn}(\boldsymbol{s}) - c\boldsymbol{s}^{\mathrm{T}}\boldsymbol{SB}(\boldsymbol{D} - \boldsymbol{w}(t)) \\
&\leq -k|\boldsymbol{s}|^2 - \varepsilon|\boldsymbol{s}|^{\overline{\alpha}+1} < 0
\end{aligned} \tag{8.15}$$

The error system is globally asymptotically stable. Thus, under this control law, in finite time the error will reach the sliding surface and remain on the sliding surface. This completes the proof.

To reduce chattering, one can replace the sign function with the following saturation function, defined as,

$$\mathrm{sat}\left(\frac{\boldsymbol{s}}{\kappa}\right) = \left[\mathrm{sat}\left(\frac{s_1}{\kappa}\right)\mathrm{sat}\left(\frac{s_2}{\kappa}\right)\mathrm{sat}\left(\frac{s_3}{\kappa}\right)\mathrm{sat}\left(\frac{s_4}{\kappa}\right)\mathrm{sat}\left(\frac{s_5}{\kappa}\right)\mathrm{sat}\left(\frac{s_6}{\kappa}\right)\right]^{\mathrm{T}},$$

$$\mathrm{sat}\left(\frac{s_i}{\kappa}\right) = \begin{cases} 1, & \frac{s_i}{\kappa} > 1 \\ \frac{s_i}{\kappa}, & \left|\frac{s_i}{\kappa}\right| \leq 1 \quad (i = 1, 2, .., 6) \\ -1, & \frac{s_i}{\kappa} < -1 \end{cases} \tag{8.16}$$

where $s_1 = c(x - x_d)$, $s_2 = c(y - y_d)$, $s_3 = c(z - z_d)$, $s_4 = c(\dot{x} - \dot{x}_d)$, $s_5 = c(\dot{y} - \dot{y}_d)$, $s_6 = c(\dot{z} - \dot{z}_d)$, and κ is a constant.

8.4 Simulation test

The simulation of the spacecraft electromagnetic separation system is performed, where the relative position of the two spacecraft is separated from (0.1, 0, 0)m to (2, 3, 4)m in this section, to illustrate the effectiveness of the controller. In the simulation example below, set the eccentricity e to 0.15, the initial true anomaly $\theta(0)$ to 0, the semimajor axis a to 10,000km, and a valid mass m_0 to 20kg. The masses of the chaser and target spacecraft are assumed to be $m_C = 40$kg and $m_T = 100$kg respectively. Set the initial states $\boldsymbol{\eta}(0)$ to $[0.1\text{m}\ 0\text{m}\ 0\text{m}\ 0\text{m/s}\ 0\text{m/s}\ 0\text{m/s}]^{\mathrm{T}}$, the gravitational parameter μ to $3.986\times10^5\text{km}^3/\text{s}^2$, the permeability of free space π_0 to $4\pi\times10^{-7}\text{N}\cdot\text{A}^{-2}$, and the magnetic moment of the target spacecraft $\boldsymbol{\pi}_T$ to $[-2.29\ -8.17\ -2.33]^{\mathrm{T}}\times10^5\text{Am}^2$. The control force and its first derivative are limited to less than 30N and 300N/s, respectively. In the control law, the parameters are $\kappa = 20$, $c = 10$, $\varepsilon = 10^{-6}$, $k = 5$, $\bar{\alpha} = 4$, and t_1, t_2 and t_3 in the reference trajectory are 15, 20 and 30s respectively. The external disturbances $\boldsymbol{w}(t)$ of the two spacecraft are assumed to be:

$$\boldsymbol{w}_{\mathrm{C}}(t) = \begin{bmatrix} 0.1 + 0.2 \sin(0.11\pi t) \\ 0.15 + 0.2 \cos(0.11\pi t) \\ 0.2 + 0.3 \cos\left(0.11\pi t + \frac{\pi}{3}\right) \end{bmatrix} \mathrm{N}$$

$$\boldsymbol{w}_{\mathrm{T}}(t) = \begin{bmatrix} 0.1 + 0.1 \sin(0.11\pi t) \\ 0.2 + 0.12 \cos\left(0.11\pi t + \pi/4\right) \\ 0.3 + 0.15 \cos\left(0.11\pi t + \pi/6\right) \end{bmatrix} \mathrm{N}$$

Remark 8.3 The selection of parameters c, k, ε, and $\bar{\alpha}$ for improving sliding mode control, follows constraints such as control force and steady-state error. By balancing various constraint conditions, the optimal parameters are selected. Among them, within limits, the larger the parameter c is, the smaller the steady-state error of the system is, but a more severe chattering phenomenon may also occur and more control force may also increase at the same time. The larger the parameter k is, the faster the system approaches the sliding surface, but the control force and its rate also increase correspondingly within limits. The larger the parameter ε is, the faster the system approaches, but it also causes greater chattering phenomenon. Parameter $\bar{\alpha}$ can enable the system state to approach at a higher velocity when it is far away from the sliding surface, and at a lower velocity when it is close to the sliding surface. When selecting parameters, appropriate adjustments can be made according to current performance requirements.

Substituting $\boldsymbol{w}_{\mathrm{C}}(t)$ and $\boldsymbol{w}_{\mathrm{T}}(t)$ into the simulation, it can be seen from Fig. 8.2 that the upper bound of external disturbance $\boldsymbol{D}$ is $[0.27\ 0.22\ 0.24]^{\mathrm{T}}$N. The upper bound is used to replace the external disturbance in the controller design, and the simulation results are as follows.

As shown in Fig. 8.3, the relative positions x, y, and z gradually converge to 2m, 3m, and 4m, respectively. The steady-state error is very small and can be ignored. Thus, the final relative position between the two spacecraft remains constant; that is, the separated spacecraft completes hovering. As shown in Fig. 8.4, the velocity converges to 0m/s within 30.08s. In addition, the relative velocity meets the requirements of separation. As shown in Fig. 8.5, the tracking errors of relative positions e_x, e_y, and e_z are very small, which can be ignored. As shown in Fig. 8.6, the actual trajectory keeps up with the reference trajectory at 1.715s.

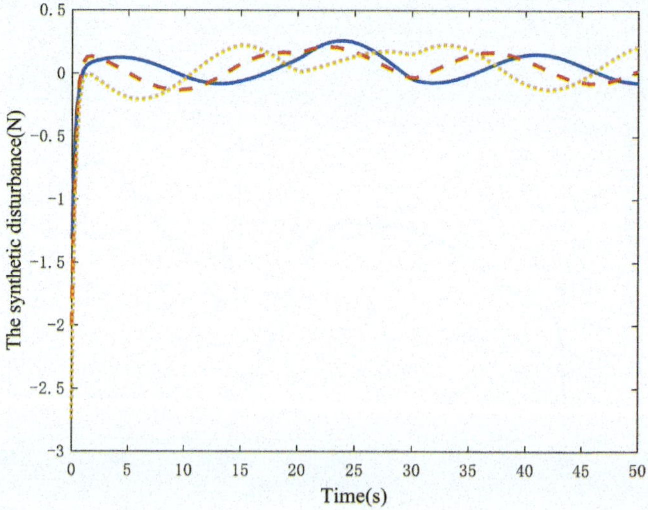

Figure 8.2 Time response of synthetic disturbance.

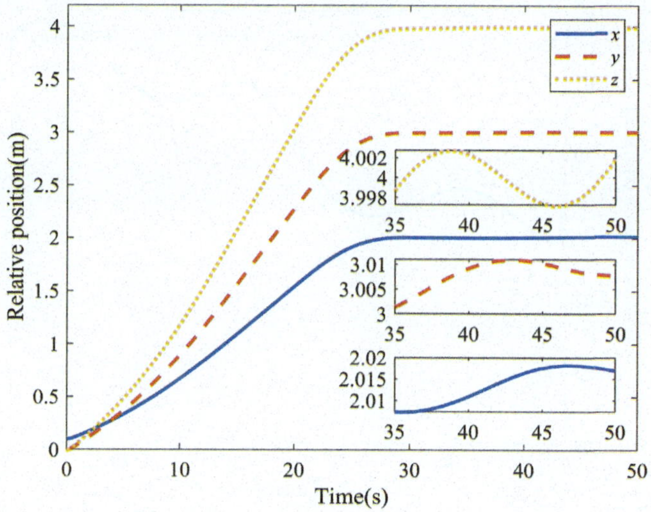

Figure 8.3 Time response of relative position.

As can be seen from the figures of control force and its rate over time in Figs. 8.7 and 8.8, the control force and its rate along any axis of x, y, and z are less than 14N and 70N/s, respectively, which satisfy their limit constraints. Given that the magnetic moment of the target spacecraft is known, Fig. 8.9 shows the response of the magnetic moment of the chaser

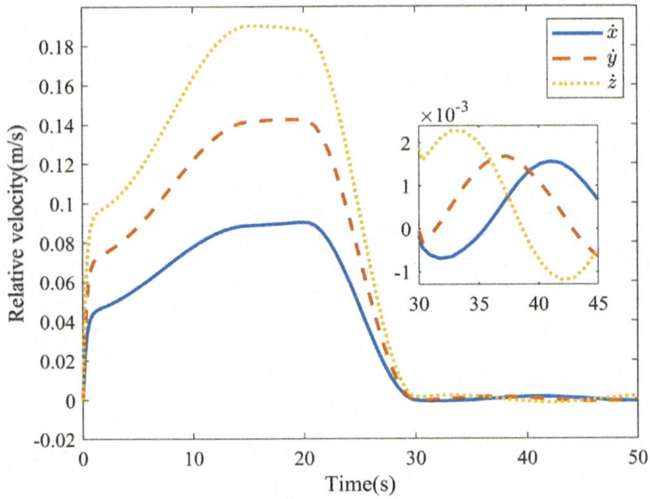

Figure 8.4 Time response of relative velocity.

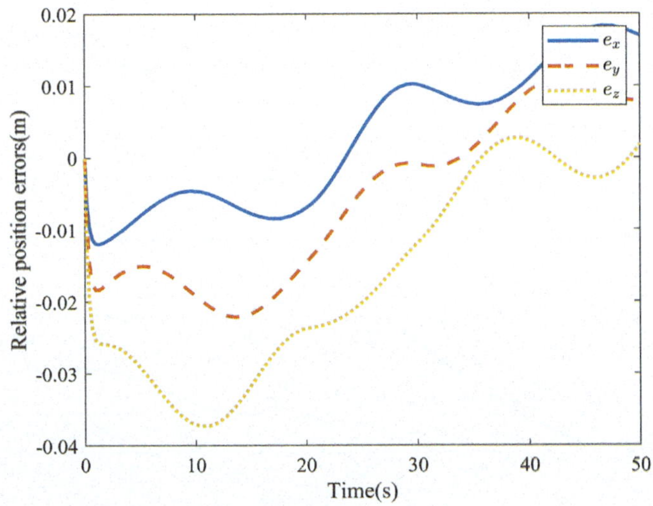

Figure 8.5 Tracking errors of relative position.

spacecraft over time after calculation. As can be seen from this figure, the magnetic moment reaches the maximum of 1837Am^2 at 25.8s. Thus, the common electromagnetic mechanism can provide sufficient magnetic moment for electromagnetic separation, which can be realized and meets the actual engineering requirements. Fig. 8.10 shows the motion of the two

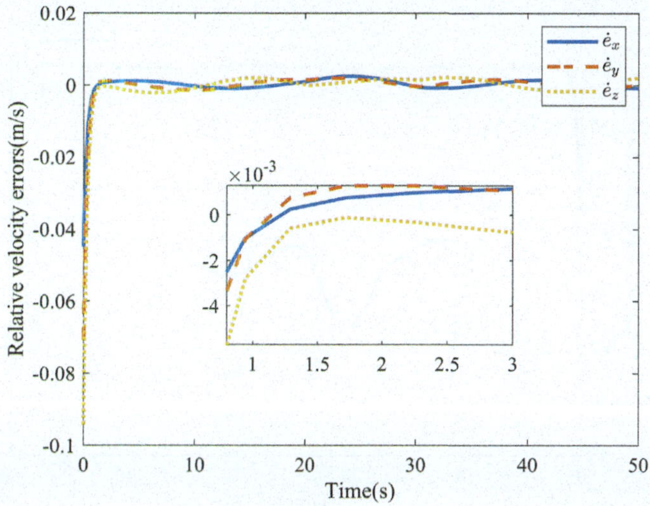

Figure 8.6 Tracking errors of relative velocity.

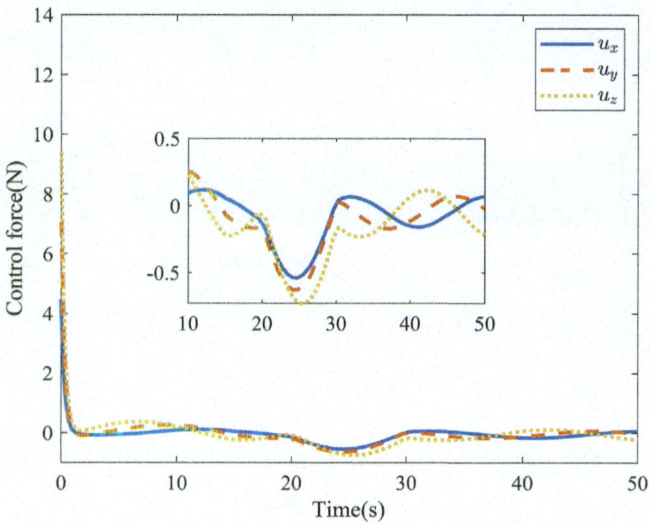

Figure 8.7 Time response of control force.

spacecraft with respect to the center of mass "CM" in a three-dimensional coordinate system, and the colored bar on the right represents the simulation time. Since the center of mass of the two spacecraft is 0.1m apart before separation, the starting points of the two motion trajectories are not the origin of the coordinate system, but (0.07143, 0, 0)m and (−0.02857, 0,

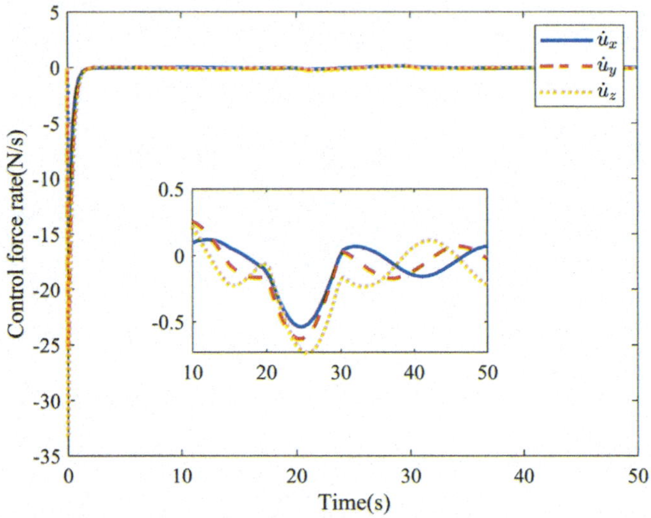

Figure 8.8 Time response of control force rate.

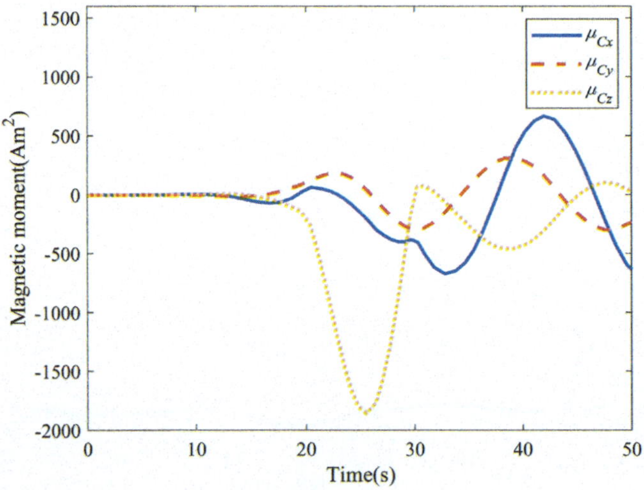

Figure 8.9 Magnetic moment of chaser spacecraft.

0)m. It can be clearly seen from the figure that the center of mass "CM" is closer to the target spacecraft, since the mass of the target spacecraft is 100kg, greater than the mass of the chaser spacecraft, which is 40kg. Thus, the relative movement distance of the chaser spacecraft is longer than the relative movement distance of the target spacecraft.

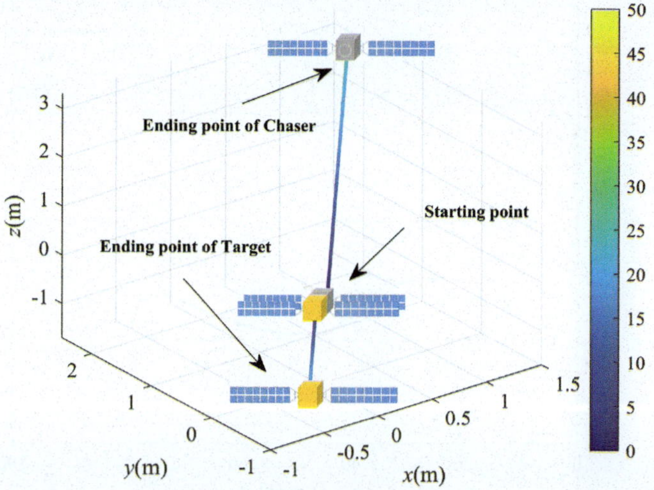

Figure 8.10 Motion of the chaser and target spacecraft relative to the center of mass "CM."

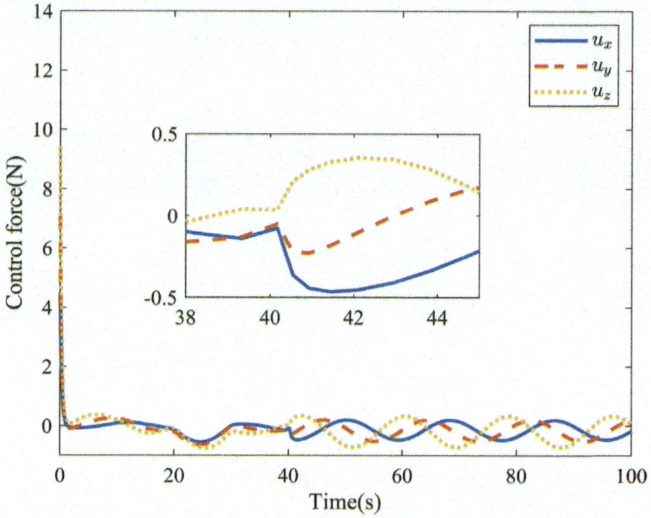

Figure 8.11 Time response of control force when adding disturbance.

In order to further demonstrate the antidisturbance ability of the control system, a triple initial disturbance is added in 40s after the system reaches its steady state. From Figs. 8.11 and 8.12, it can be seen that the system can still quickly reach its steady state. From Fig. 8.13, it can be seen that the

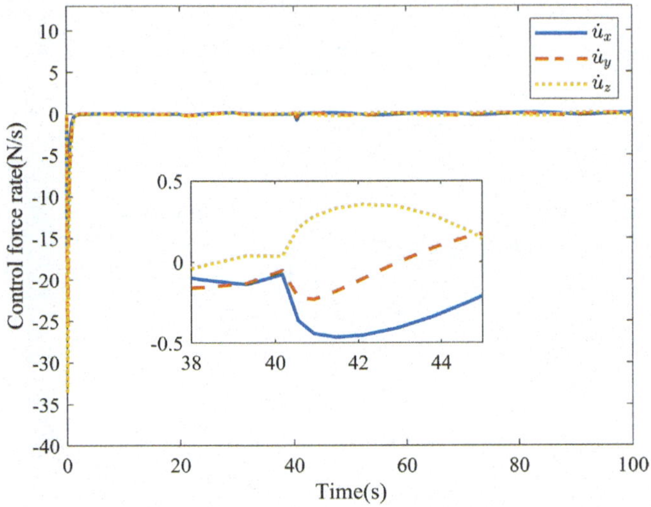

Figure 8.12 Time response of control force rate when adding disturbance.

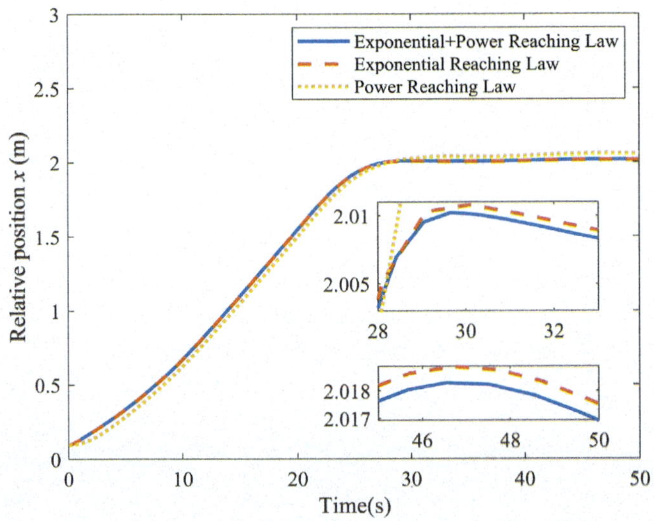

Figure 8.13 Relative positions in the *x*-direction under different reaching laws.

improved reaching law is more efficient than the power reaching law, and the steady-state error is much smaller than that of the power reaching law. In addition, compared with the exponential reaching law, the approaching velocity is faster, and the steady-state error is smaller near the sliding surface. However, the improvement effect is not very obvious. Because the initial distance of electromagnetic separation in this work is very small, the initial error of the sliding surface is determined by the initial separation distance. Thus, to a certain extent, it weakens the influence of the fractional power term. If the initial distance applied in electromagnetic separation processes is longer, the improvement effect compared to the exponential reaching law will be much more significant.

8.5 Conclusions

The electromagnetic separation problem of the two spacecraft under the background of on-orbit assembled spacecraft is investigated. Firstly, the dynamic model of the spacecraft electromagnetic separation system is established by using the electromagnetic force far-field model and taking the electromagnetic force as the control input. Secondly, the separation trajectories of relative position/velocity/acceleration are designed, and the relative position tracking error is defined. On this basis, a sliding mode controller is designed, combining the advantages of exponential reaching law and power reaching law, so that the actual trajectory in three directions quickly tracks the upper reference trajectory and the system can still stabilize under external disturbances. The improvement of the control approach in this chapter effectively improves the stability and robustness of the electromagnetic separation system, but the tracked separation trajectory still needs to be improved. In the follow-up study, the reference trajectories will be further optimized through intelligent trajectory planning algorithms. In the future, the emergence of electromagnetic docking and separation pairs will also provide a new solution for the on-orbit maintenance of spacecraft to prolong the life of spacecraft, and promote the rapid development of on-orbit service.

References

Ahsun, U., Miller, D. W., & Ramirez, J. L. (2010). Control of electromagnetic satellite formations in near-Earth orbits. *Journal of Guidance, Control, and Dynamics, 33*(6), 1883–1891. https://doi.org/10.2514/1.47637.

Bayat, F. (2019). Model predictive sliding control for finite-time three-axis spacecraft attitude tracking. *IEEE Transactions on Industrial Electronics, 66*(10), 7986–7996. https://doi.org/10.1109/tie.2018.2881936.

Boesso, A., & Francesconi, A. (2013). ARCADE small-scale docking mechanism for micro-satellites. *Acta Astronautica, 86*, 77–87. https://doi.org/10.1016/j.actaastro.2013.01.006.

Cao, K., Li, S., She, Y., Biggs, J. D., Liu, Y., & Bian, L. (2021). Dynamics and on-orbit assembly strategies for an orb-shaped solar array. *Acta Astronautica, 178*, 881–893. https://doi.org/10.1016/j.actaastro.2020.10.030.

Fei, J., Wang, H., & Fang, Y. (2022). Novel neural network fractional-order sliding-mode control with application to active power filter. *IEEE Transactions on Systems, Man, and Cybernetics: Systems, 52*(6), 3508–3518. https://doi.org/10.1109/TSMC.2021.3071360.

Foust, R., Lupu, E., Nakk, S., & Kumar, Y. (2018). Ultra-soft electromagnetic docking with applications to in-orbit assembly. *69th international astronautical congress*. Bremen, Germany: International Astronautical Congress.

Guo, B., & Chen, Y. (2019). Adaptive fast sliding mode fault tolerant control integrated with disturbance observer for spacecraft attitude stabilization system. *ISA Transactions, 94*, 1–9. https://doi.org/10.1016/j.isatra.2019.04.014.

Hallaj, M. A. A., & Assadian, N. (2015). Tethered satellite system control using electromagnetic forces and reaction wheels. *Acta Astronautica, 117*, 390–401. https://doi.org/10.1016/j.actaastro.2015.08.015.

He, G., Cao, D., Cao, Y., & Huang, W. (2022). Dynamic modeling and orbit maneuvering response analysis for a three-axis attitude stabilized large scale flexible spacecraft installed with hinged solar arrays. *Mechanical Systems and Signal Processing, 162*, 108083. https://doi.org/10.1016/j.ymssp.2021.108083.

Hou, X., Zhu, M., Sun, L., Ding, T., Huang, Z., Shi, Y., Su, Y., Li, L., Chen, T., & Lee, C. (2022). Scalable self-attaching/assembling robotic cluster (S2A2RC) system enabled by triboelectric sensors for in-orbit spacecraft application. *Nano Energy, 93*, 106894. https://doi.org/10.1016/j.nanoen.2021.106894.

Liu, C., Yue, X., Shi, K., & Sun, Z. (2022a). *Spacecraft attitude control: A linear matrix inequality approach* (pp. 1–370). Elsevier. https://doi.org/10.1016/B978-0-323-99005-9.00012-2.

Liu, C., Yue, X., Zhang, J., & Shi, K. (2022b). Active disturbance rejection control for delayed electromagnetic docking of spacecraft in elliptical orbits. *IEEE Transactions on Aerospace and Electronic Systems, 58*(3), 2257–2268. https://doi.org/10.1109/taes.2021.3130830.

Liu, E., Yang, Y., & Yan, Y. (2020). Spacecraft attitude tracking for space debris removal using adaptive fuzzy sliding mode control. *Aerospace Science and Technology, 107*, 106310. https://doi.org/10.1016/j.ast.2020.106310.

Mohan, S. (2010). *Quantative selection and design of model generation architectures for on-orbit autonomous assembly [Thesis]*. [Massachusetts Institute of Technology].

Olivieri, L., & Francesconi, A. (2016). Design and test of a semiandrogynous docking mechanism for small satellites. *Acta Astronautica, 122*, 219–230. https://doi.org/10.1016/j.actaastro.2016.02.004.

Pukdeboon, C., & Kumam, P. (2015). Robust optimal sliding mode control for spacecraft position and attitude maneuvers. *Aerospace Science and Technology, 43*, 329–342. https://doi.org/10.1016/j.ast.2015.03.012.

She, Y., Li, S., & Wang, Z. (2019). Constructing a large antenna reflector via spacecraft formation flying and reconfiguration control. *Journal of Guidance, Control, and Dynamics, 42*(6), 1372–1382. https://doi.org/10.2514/1.g004116.

Shi, K., Liu, C., Biggs, J. D., Sun, Z., & Yue, X. (2020). Observer-based control for spacecraft electromagnetic docking. *Aerospace Science and Technology, 99*, 105759. https://doi.org/10.1016/j.ast.2020.105759.

Shi, K., Liu, C., & Sun, Z. (2019). Constrained fuel-free control for spacecraft electromagnetic docking in elliptical orbits. *Acta Astronautica, 162*, 14–24. https://doi.org/10.1016/j.actaastro.2019.05.016.

Shi, K., Liu, C., Sun, Z., & Yue, X. (2022). Coupled orbit-attitude dynamics and trajectory tracking control for spacecraft electromagnetic docking. *Applied Mathematical Modelling, 101*, 553–572. https://doi.org/10.1016/j.apm.2021.08.030.

Xing, L., Zhang, J., Liu, C., & Zhang, X. (2021). Fuzzy-logic-based adaptive event-triggered sliding mode control for spacecraft attitude tracking. *Aerospace Science and Technology, 108*, 106394. https://doi.org/10.1016/j.ast.2020.106394.

Yang, Z., Liu, C., & Yue, X. (2023). NI-based static output feedback control for attitude stabilization of post-capture flexible spacecraft. *Advances in Space Research, 71*(9), 3681–3695. https://doi.org/10.1016/j.asr.2022.06.021.

Zhang, J., Chen, D., Shen, G., Sun, Z., & Xia, Y. (2021). Disturbance observer based adaptive fuzzy sliding mode control: A dynamic sliding surface approach. *Automatica, 129*, 109606. https://doi.org/10.1016/j.automatica.2021.109606.

Zhang, Y. W., Yang, L. P., Zhu, Y. W., Ao, H. J., & Qi, D. W. (2016). Self-docking analysis and velocity-aimed control for spacecraft electromagnetic docking. *Advances in Space Research, 57*(11), 2314–2325. https://doi.org/10.1016/j.asr.2016.03.022.

Zhang, Y. W., Yang, L. P., Zhu, Y. W., Huang, H., & Cai, W. W. (2012). Nonlinear 6-DOF control of spacecraft docking with inter-satellite electromagnetic force. *Acta Astronautica, 77*, 97–108. https://doi.org/10.1016/j.actaastro.2012.03.020.

Zhang, Y. W., Yang, L. P., Zhu, Y. W., Ren, X. H., & Huang, H. (2011). Self-docking capability and control strategy of electromagnetic docking technology. *Acta Astronautica, 69*(11–12), 1073–1081. https://doi.org/10.1016/j.actaastro.2011.07.001.

Zhao, C., Guo, W., Chen, M., Hu, J., Han, Y., & Lin, R. (2024). Space truss construction modeling based on on-orbit assembly motion feature. *Chinese Journal of Aeronautics, 37*(3), 365–379. https://doi.org/10.1016/j.cja.2023.07.002.

Zhu, Y. W., Cai, W. W., Yang, L. P., & Huang, H. (2018). Flatness-based trajectory planning for electromagnetic spacecraft proximity operations in elliptical orbits. *Acta Astronautica, 152*, 342–351. https://doi.org/10.1016/j.actaastro.2018.08.031.

CHAPTER 9

Optimizations-based impedance control for spacecraft electromagnetic separation

9.1 Introduction

In recent years, the high-precision electromagnetic separation control technology for spacecraft has attracted many scholars, which is the foundation of space operations in the future. The on-orbit repair and reuse of large-scale failed spacecraft has become increasingly crucial for expanding space activities and ensuring the continuity of space missions. Due to the complex operation and high precision requirement of large-scale spacecraft, component failures can lead to significant resource wastage and impede the progress of space programs (Cao et al., 2021; He et al., 2022; She et al., 2019). However, the conventional thruster-based separation strategy presents several notable drawbacks. On the one hand, the continuous consumption of propellant not only substantially increases mission costs but also limits the payload capacity of spacecraft, directly affecting their ability to perform complex tasks. On the other hand, plume contamination can damage delicate instruments, optical devices, and sensitive electronic components on spacecraft, disrupting their normal operation and diminishing the lifespan and accuracy of the equipment (Liu et al., 2022a; Olivieri & Francesconi, 2016; Zhu et al., 2018).

In contrast, electromagnetic separation technology has emerged as a focal point in the space field due to its unique advantages, e.g., no propellant consumption, plume pollution avoidance, reversible and continuous control, and noncontact operation (Liu et al., 2022a; Zhang, Yang, Zhu, Hai Ren, & Huang, 2011, Zhang, Yang, Zhu, Ao, & Qi, 2016). However, the control technology for spacecraft electromagnetic separation continues to face significant challenges. The sliding mode controller proposed by Shi et al. (2022) can effectively complete the electromagnetic docking mission of spacecraft

Spacecraft Electromagnetic Docking and Separation.
DOI: https://doi.org/10.1016/B978-0-443-49334-8.00010-0

under coupled orbit-attitude conditions. However, the desired trajectory can be further improved. The enhanced sliding mode controller proposed by Liu et al. (2024) can effectively accomplish the electromagnetic separation task for spacecraft; nevertheless, there remains potential improvement in control accuracy.

Impedance control, through the precise adjustment of system inertia, damping, and stiffness characteristics, can effectively establish a precise relationship between force and position, thereby offering a novel solution for the electromagnetic separation control of spacecraft (Ivanov et al., 2022; Wang et al., 2024). The control challenge for space robots capturing noncooperative targets was concluded by Wang et al. (2018), which elucidated the potential performance of impedance control in space robotics. An impedance control of free-flying space robots for noncooperative satellite capture operations was studied by Uyama et al. (2012), who introduced an open-loop impedance control law based on the contact dynamics model, and conducted experimental verification. The impedance controller for the capture of spacecraft using free-flying robots was developed by Gu, Liu, Wang, Mao, & Guo, 2025, effectively addressing the dynamic coupling issues inherent in such systems. However, the effectiveness of impedance control is highly sensitive to parameter selection. Traditional approaches for determining these parameters often depend on empirical formulas or extensive trial-and-error experiments, which complicates the identification of optimal parameter combinations under the complex and variable conditions associated with spacecraft electromagnetic separation. Consequently, this limitation hinders the full realization of the advantages of impedance control (Annamraju et al., 2020; Ding et al., 2023; Yang et al., 2017), making it challenging to achieve the desired ideal control performance.

Intelligent algorithms, with their unique advantages, exhibit a strong capability to optimize parameters, allowing for precise adjustments and enhancements of various control parameters. This provides innovative ideas and methodologies for addressing the control challenges associated with spacecraft electromagnetic separation control. As a novel intelligent optimization technique, the hippopotamus optimization (HO) strategy emulates the survival behaviors of hippopotamus in their natural habitat, showcasing robust global search capabilities and rapid convergence speeds within the solution space (Amiri et al., 2024). In addition, HO has demonstrated significant success in power system optimization scheduling, machine learning feature selection, and other domains (Han et al., 2025; Wang et al., 2024), yet its application in spacecraft electromagnetic separation control remains largely exploratory. Similarly, the whale optimization algorithm (WOA) is an effective intelligent optimization algorithm that mimics the distinctive hunting strategies of whales, exhibiting commendable optimization performance and an ability to escape local optima in various optimization challenges (Bentouati et al., 2017; Mirjalili & Lewis, 2016). The limitation of WOA during the exploitation phase was addressed by Heidari et al. (2020), enhancing the algorithm's exploration and exploitation capabilities. This advancement offers new perspectives and methodologies for improving WOA performance, thereby facilitating its broader application across diverse scenarios. In light of this, this chapter proposes an integrated HO and WOA strategy, merging the local fine-tuning optimization strengths of HO with the global optimization efficiency of WOA to precisely optimize impedance control parameters.

Inspired by the preceding discussions, this chapter investigates an HO-WOA-based impedance control strategy, combining the local fine-tuning optimization capabilities of HO with the global optimization efficiency of WOA, to achieve high-precision electromagnetic separation. The main contributions of this work are summarized as follows. First, compared with Liu et al. (2022b) and Shi et al. (2022), this chapter constructs a performance index function with the aim of minimizing the separation time and derives a time-optimal trajectory accordingly, which effectively shortens the time required for the electromagnetic separation and provides certain support for enhancing the execution efficiency of space missions. Second, compared with Liu et al. (2024), the optimal parameters of the impedance controller are accurately selected via integrated hippopotamus and whale optimizations, which successfully overcome the problem of poor control accuracy of the improved sliding mode control, and significantly reduce the relative position and velocity errors.

The remainder of this chapter is structured as follows: Section 9.2 describes the dynamics model of spacecraft electromagnetic separation and designs a time-optimal trajectory. Section 9.3 presents a nonlinear disturbance observer (NDO)-based impedance controller, and the global asymptotic stability of the closed-loop system is proved via Lyapunov stability analysis. The impedance control parameters are optimized by employing the integrated HO and WOA algorithms. Section 9.4 performs numerical simulations. Finally, some conclusions are drawn.

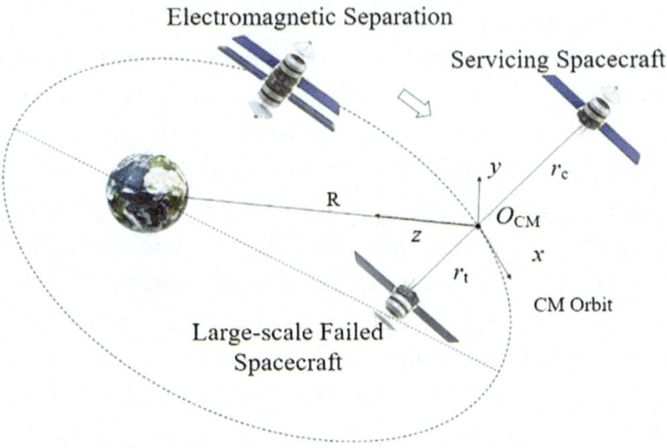

Figure 9.1 Spacecraft electromagnetic separation system.

9.2 Problem formulation

9.2.1 Dynamics modeling

As can be seen in Fig. 9.1, the center of mass (CM) orbital coordinate system is adopted, where CM is the origin of the coordinate system and also the CM of the spacecraft electromagnetic separation system. The z-axis points toward the center of the Earth, the x-axis is perpendicular to the nominal flight direction of the z-axis, and the y-axis follows the right-hand rule. The dynamics model of the spacecraft electromagnetic separation system is described as follows (Shi et al., 2019):

$$\ddot{\tilde{\boldsymbol{\eta}}} = \left(\boldsymbol{A}_{\tilde{\eta}_1} + \Delta\boldsymbol{A}_{\tilde{\eta}_1}\right)\dot{\tilde{\boldsymbol{\eta}}} + \left(\boldsymbol{A}_{\tilde{\eta}_2} + \Delta\boldsymbol{A}_{\tilde{\eta}_2}\right)\tilde{\boldsymbol{\eta}} + (\boldsymbol{B} + \Delta\boldsymbol{B})(\boldsymbol{u}(t) + \boldsymbol{w}_0(t)) + \boldsymbol{E}\boldsymbol{f}(t) \tag{9.1}$$

where, $\tilde{\boldsymbol{\eta}} = [x \quad y \quad z]^{\mathrm{T}}$ is the relative position vector from the failed spacecraft to the servicing spacecraft, $\boldsymbol{u} = [F_{eCx} \quad F_{eCy} \quad F_{eCz}]^{\mathrm{T}}$ is the electromagnetic force acting on the two spacecraft, $\boldsymbol{w}_0 = [d_x \quad d_y \quad d_z]^{\mathrm{T}}$ is the external disturbance, $\boldsymbol{f}(t)$ is the fault signal, and $\boldsymbol{E}$ is the fault distribution matrix.

The coefficient matrices are as follows:

$$\boldsymbol{A}_{\tilde{\eta}_1} = \begin{bmatrix} 0 & 0 & 2\omega_0 \\ 0 & 0 & 0 \\ -2\omega_0 & 0 & 0 \end{bmatrix}, \boldsymbol{A}_{\tilde{\eta}_2} = \begin{bmatrix} 0 & 0 & 0 \\ 0 & -\omega_0^2 & 0 \\ 0 & 0 & 3\omega_0^2 \end{bmatrix}$$

$$\Delta A_{\tilde{\eta}_1} = \begin{bmatrix} 0 & 0 & 2(\omega - \omega_0) \\ 0 & 0 & 0 \\ 2(\omega_0 - \omega) & 0 & 0 \end{bmatrix}$$

$$B = \mathrm{diag}\left(\left[m_0^{-1} \quad m_0^{-1} \quad m_0^{-1}\right]\right)$$

$$\Delta B = \mathrm{diag}\left(\left[m_{\mathrm{CT}}^{-1} \quad -m_0^{-1} \quad m_{\mathrm{CT}}^{-1} \quad -m_0^{-1} \quad m_{\mathrm{CT}}^{-1} \quad -m_0^{-1}\right]\right)$$

where, $\omega_0 = \sqrt{\mu/a^3}$, m_0 represents an effective mass, m_{C} represents the mass of the servicing spacecraft, m_{T} represents the mass of the large-scale failed spacecraft, and $m_{\mathrm{CT}} = m_{\mathrm{C}} m_{\mathrm{T}}/(m_{\mathrm{C}} + m_{\mathrm{T}})$, $d_{i=x,y,z} = (m_{\mathrm{T}} w_{\mathrm{C}i} \ m_{\mathrm{C}} w_{\mathrm{T}i})/(m_{\mathrm{C}} + m_{\mathrm{T}})$.

By approximating the coils on each spacecraft as steerable dipoles, the electromagnetic force between the two spacecraft can be represented as

$$F_{\mathrm{eC}} = \frac{3\mu_0}{4\pi} \Xi(\boldsymbol{\mu}_{\mathrm{T}}, \tilde{\boldsymbol{\eta}}) \boldsymbol{\mu}_{\mathrm{C}} \tag{9.2}$$

where, F_{eC} represents the electromagnetic force between large-scale failed spacecraft and servicing spacecraft; $\boldsymbol{\mu}_{\mathrm{C}}$ represents the magnetic moment of the servicing spacecraft; $\boldsymbol{\mu}_{\mathrm{T}}$ represents the magnetic moment of the large-scale failed spacecraft, and $\Xi(\boldsymbol{\mu}_{\mathrm{T}}, \tilde{\boldsymbol{\eta}}) = (\tilde{\boldsymbol{\eta}} \boldsymbol{\mu}_{\mathrm{T}}^{\mathrm{T}} + \boldsymbol{\mu}_{\mathrm{T}}^{\mathrm{T}} \tilde{\boldsymbol{\eta}} \boldsymbol{I}_3 + \boldsymbol{\mu}_{\mathrm{T}} \tilde{\boldsymbol{\eta}}^{\mathrm{T}})/\|\tilde{\boldsymbol{\eta}}\|^5 - 5\tilde{\boldsymbol{\eta}} \boldsymbol{\mu}_{\mathrm{T}}^{\mathrm{T}} \tilde{\boldsymbol{\eta}} \tilde{\boldsymbol{\eta}}^{\mathrm{T}}/\|\tilde{\boldsymbol{\eta}}\|^7$.

Assuming that the magnetic moment vector of the failed space platform is known, and the relative motion information $\tilde{\boldsymbol{\eta}}$ can be obtained at each moment, the matrix $\Xi(\boldsymbol{\mu}_{\mathrm{T}}, \tilde{\boldsymbol{\eta}})$ can be calculated. Then, according to Eq. (9.2), the magnetic moment vector of the servicing spacecraft can be solved using the electromagnetic force F_{eC} and the matrix $\Xi(\boldsymbol{\mu}_{\mathrm{T}}, \tilde{\boldsymbol{\eta}})$.

Thus, Eq. (9.1) can be transformed into the following form:

$$\ddot{\tilde{\boldsymbol{\eta}}} = A_{\tilde{\eta}_1} \dot{\tilde{\boldsymbol{\eta}}} + A_{\tilde{\eta}_2} \tilde{\boldsymbol{\eta}} + B u(t) + B w(t) \tag{9.3}$$

where, $E = B$ and $w(t) = B^{\mathrm{T}} (BB^{\mathrm{T}})^{*} (\Delta A_{\tilde{\eta}_1} \dot{\tilde{\boldsymbol{\eta}}}(t) + \Delta A_{\tilde{\eta}_2} \tilde{\boldsymbol{\eta}} + \Delta B u(t) + (B + \Delta B) w_0(t) + E f(t)) \in L_2[0, \infty)$ can be regarded as a lumped disturbance, which includes external disturbances, fault signals, unknown mass, and the eccentricity of the CM orbit. The matrix $(BB^{\mathrm{T}})^{*}$ is the pseudo-inverse of BB^{T}.

Assumption 9.1 (Liu et al., 2022a): The lumped disturbances $w(t)$ and its first-order derivatives are assumed to be piecewise continuous and bounded, with an upper bound of W.

9.2.2 Time-optimal trajectory design

Define a performance index function $J = t_f$ with the objective of minimizing the separation time, where t_f represents the end time of the separation process. Assuming that the upper bound of the electromagnetic force is $F_{\max}$ and the maximum force rate $\dot{F}_{\max}$, i.e., $||\boldsymbol{F}_{eC}|| \le F_{\max}$, $||\dot{\boldsymbol{F}}_{eC}|| \le \dot{F}_{\max}$. The terminal constraints are denoted by $\widetilde{\boldsymbol{\eta}}(t_f) = \boldsymbol{\eta}_d$ and $\dot{\widetilde{\boldsymbol{\eta}}}(t_f) = 0$.

The Hamiltonian function is defined as

$$\boldsymbol{H}(\widetilde{\boldsymbol{\eta}}, \dot{\widetilde{\boldsymbol{\eta}}}, \boldsymbol{u}, \boldsymbol{\lambda}) = \boldsymbol{\lambda}_1^{\mathrm{T}} \dot{\widetilde{\boldsymbol{\eta}}} + \boldsymbol{\lambda}_2^{\mathrm{T}} \left(\boldsymbol{A}_{\widetilde{\eta}_1} \dot{\widetilde{\boldsymbol{\eta}}} + \boldsymbol{A}_{\widetilde{\eta}_2} \widetilde{\boldsymbol{\eta}} + \boldsymbol{B}\boldsymbol{u} + \boldsymbol{B}\boldsymbol{w} \right) \tag{9.4}$$

where, $\boldsymbol{\lambda}_1$ and $\boldsymbol{\lambda}_2$ are skew vectors. According to Pontryagin's minimum principle, the optimal control force $\boldsymbol{u}^*$ satisfies $\partial \boldsymbol{H} / \partial \boldsymbol{u} = 0$, thus one has

$$\boldsymbol{\lambda}_2^{\mathrm{T}} \boldsymbol{B} = 0 \tag{9.5}$$

Combined with $||\boldsymbol{u}_i|| \le u_{\max}$, one can obtain $\boldsymbol{u}^*$ by judging the positive and negative of $\boldsymbol{\lambda}_2^{\mathrm{T}} \boldsymbol{B}$.

The skew vectors λ_1 and λ_2 satisfy the skew equation, i.e.,

$$\dot{\lambda}_1 = -\frac{\partial \boldsymbol{H}}{\partial \widetilde{\boldsymbol{\eta}}}, \ \dot{\lambda}_2 = -\frac{\partial \boldsymbol{H}}{\partial \dot{\widetilde{\boldsymbol{\eta}}}} \tag{9.6}$$

It yields

$$\dot{\boldsymbol{\lambda}}_1 = -\boldsymbol{\lambda}_2^{\mathrm{T}} \boldsymbol{A}_{\widetilde{\eta}_2}, \ \dot{\boldsymbol{\lambda}}_2 = -\boldsymbol{\lambda}_1^{\mathrm{T}} - \boldsymbol{\lambda}_2^{\mathrm{T}} \boldsymbol{A}_{\widetilde{\eta}_1} \tag{9.7}$$

Considering the constraints of the electromagnetic force and its rate, solving Eqs. (9.4), (9.5), and (9.7) yield the optimal control force $\boldsymbol{u}^*$.

Taking the x-direction as an example, with the action of $\boldsymbol{u}^*$, one has the initial condition $x(0) = x_0$, $\dot{x}(0) = \dot{x}_0$, the terminal condition $x(t_f) = x_f$, and the velocity trajectory $\dot{x}(t_f) = \dot{x}_f$, so that

$$\dot{x}(t) = \dot{x}_0 + \int_0^t u_x^*(s) ds \tag{9.8}$$

It indicates that the velocity is continuously accumulated and changed by the integral of the optimal control force based on the initial velocity. Thus the relative position trajectory is as follows:

$$x(t) = x_0 + \int_0^t \left(\dot{x}_0 + \int_0^s u_x^*(r) dr \right) ds \tag{9.9}$$

9.3 Integrated HO-WOA-based impedance controller design

9.3.1 Nonlinear disturbance observer-based impedance control

In this subsection, an NDO (Shi et al., 2020) is introduced to estimate the lumped disturbance, which is designed as follows:

$$\begin{aligned} \hat{\boldsymbol{w}} &= \boldsymbol{z} + \boldsymbol{L}\dot{\tilde{\boldsymbol{\eta}}} \\ \dot{\boldsymbol{z}} &= -\boldsymbol{L}\boldsymbol{B}\hat{\boldsymbol{w}} - \boldsymbol{L}\left(\boldsymbol{A}_{\tilde{\eta}_1}\dot{\tilde{\boldsymbol{\eta}}} + \boldsymbol{A}_{\tilde{\eta}_2}\tilde{\boldsymbol{\eta}} + \boldsymbol{B}\boldsymbol{u}\right) \end{aligned} \tag{9.10}$$

where, $\boldsymbol{z}$ represents the internal state vector of the observer, $\boldsymbol{L}$ denotes the observer gain matrix, and represents the estimation of the lumped disturbance.

The estimation error of the lumped disturbance can be expressed by

$$\boldsymbol{e}_{\mathrm{w}} = \hat{\boldsymbol{w}} - \boldsymbol{w} \tag{9.11}$$

Under the condition that the disturbance varies slowly, it is reasonable to assume that the lumped disturbance is a slowly time-varying bounded, i.e., $\dot{\boldsymbol{w}} = 0$ is reasonable.

The first derivative of the estimation error $\boldsymbol{e}_{\mathrm{w}}$ along time is

$$\begin{aligned} \dot{\boldsymbol{e}}_{\mathrm{w}} &= \dot{\hat{\boldsymbol{w}}} - \dot{\boldsymbol{w}} \\ &= \dot{\boldsymbol{z}} + \boldsymbol{L}\ddot{\tilde{\boldsymbol{\eta}}} - \dot{\boldsymbol{w}} \\ &= -\boldsymbol{L}\boldsymbol{B}\hat{\boldsymbol{w}} - \boldsymbol{L}\left(\boldsymbol{A}_{\tilde{\eta}_1}\dot{\tilde{\boldsymbol{\eta}}} + \boldsymbol{A}_{\tilde{\eta}_2}\tilde{\boldsymbol{\eta}} + \boldsymbol{B}\boldsymbol{u}\right) \\ &\quad + \boldsymbol{L}\left(\boldsymbol{A}_{\tilde{\eta}_1}\dot{\tilde{\boldsymbol{\eta}}} + \boldsymbol{A}_{\tilde{\eta}_2}\tilde{\boldsymbol{\eta}} + \boldsymbol{B}(\boldsymbol{u} + \boldsymbol{w})\right) - \dot{\boldsymbol{w}} \\ &= -\boldsymbol{L}\boldsymbol{B}\hat{\boldsymbol{w}} + \boldsymbol{L}\boldsymbol{B}\boldsymbol{w} - \dot{\boldsymbol{w}} \\ &= -\boldsymbol{L}\boldsymbol{B}\boldsymbol{e}_{\mathrm{w}} \end{aligned} \tag{9.12}$$

Design the impedance controller as

$$\boldsymbol{M}_{\mathrm{d}}\ddot{\boldsymbol{e}} + \boldsymbol{B}_{\mathrm{d}}(t)\dot{\boldsymbol{e}} + \boldsymbol{K}_{\mathrm{d}}(t)\boldsymbol{e} = \boldsymbol{u} - \boldsymbol{F}_{\mathrm{d}} \tag{9.13}$$

where, e is the error between the actual trajectory $\tilde{\boldsymbol{\eta}} = [x \quad y \quad z]^{\mathrm{T}}$ and the reference trajectory $\boldsymbol{\eta}_{\mathrm{d}} = [x_{\mathrm{d}} \quad y_{\mathrm{d}} \quad z_{\mathrm{d}}]^{\mathrm{T}}$; $\boldsymbol{F}_{\mathrm{d}}$ is the desired electromagnetic force; $\boldsymbol{M}_{\mathrm{d}}$, $\boldsymbol{B}_{\mathrm{d}}$, and $\boldsymbol{K}_{\mathrm{d}}$ represent the inertia matrix, damping matrix, and stiffness matrix, respectively, which are positive definite, $\boldsymbol{M}_{\mathrm{d}}$ and $\boldsymbol{K}_{\mathrm{d}}$ are symmetric. If these parameters are constant, the system will be asymptotically stable for any $\boldsymbol{M}_{\mathrm{d}}$, $\boldsymbol{B}_{\mathrm{d}}$, and $\boldsymbol{K}_{\mathrm{d}}$. In this subsection, a time-varying impedance control law is considered. It is assumed that the matrix $\boldsymbol{M}_{\mathrm{d}}$ is constant, while $\boldsymbol{B}_{\mathrm{d}}$ and $\boldsymbol{K}_{\mathrm{d}}$ are time-varying matrices. The control input $\boldsymbol{u}$ is designed as

$$\boldsymbol{u} = \boldsymbol{M}_{\mathrm{d}}\ddot{\boldsymbol{e}} + \boldsymbol{B}_{\mathrm{d}}(t)\dot{\boldsymbol{e}} + \boldsymbol{K}_{\mathrm{d}}(t)\boldsymbol{e} + \boldsymbol{F}_{\mathrm{d}}$$

$$= \boldsymbol{M}_\mathrm{d}\left(\ddot{\tilde{\boldsymbol{\eta}}} - \boldsymbol{\eta}_\mathrm{d}'\right) + \boldsymbol{B}_\mathrm{d}(t)\left(\dot{\tilde{\boldsymbol{\eta}}} - \dot{\boldsymbol{\eta}}_\mathrm{d}\right) + \boldsymbol{K}_\mathrm{d}(t)\left(\tilde{\boldsymbol{\eta}} - \boldsymbol{\eta}_\mathrm{d}\right) + \boldsymbol{F}_\mathrm{d} \tag{9.14}$$

Remark 9.1 Let $\boldsymbol{M}_\mathrm{d}$ and $\boldsymbol{L}$ be symmetric positive definite constant matrices, $\boldsymbol{B}_\mathrm{d}$ and $\boldsymbol{K}_\mathrm{d}$ be symmetric positive definite and continuously differentiable matrices. If there exists $\alpha > 0$, $\boldsymbol{u} - \boldsymbol{F}_\mathrm{d} = 0$, $\forall t \geq 0$, and the following conditions hold:

- $\alpha \boldsymbol{M}_\mathrm{d} - \boldsymbol{B}_\mathrm{d}(t)$ is a negative definite matrix;
- $\boldsymbol{K}_\mathrm{d}(t) + \alpha \boldsymbol{B}_\mathrm{d}(t) - 2\alpha \boldsymbol{K}_\mathrm{d}(t)$ is a negative definite matrix.

Then, the system is globally asymptotically stable.

Proof: Consider the following Lyapunov function:

$$V(t) = \frac{1}{2}(\dot{e} + \alpha e)^\mathrm{T} \boldsymbol{M}_\mathrm{d}(\dot{e} + \alpha e) + \frac{1}{2}\boldsymbol{e}^\mathrm{T}\boldsymbol{\beta}(t)\boldsymbol{e} + \frac{1}{2}\boldsymbol{e}_\mathrm{w}{}^\mathrm{T}\boldsymbol{e}_\mathrm{w} \tag{9.15}$$

where

$$\boldsymbol{\beta}(t) = \boldsymbol{K}_\mathrm{d}(t) + \alpha \boldsymbol{B}_\mathrm{d}(t) - \alpha^2 \boldsymbol{M}_\mathrm{d}$$

$$\dot{\boldsymbol{\beta}}(t) = \dot{\boldsymbol{K}}_\mathrm{d}(t) + \alpha \dot{\boldsymbol{B}}_\mathrm{d}(t)$$

The first derivative of $V(t)$ is

$$\begin{aligned}
\dot{V}(t) &= (\dot{e} + \alpha e)^\mathrm{T} \boldsymbol{M}_\mathrm{d}(\ddot{e} + \alpha \dot{\boldsymbol{e}}) + \boldsymbol{e}^\mathrm{T}\boldsymbol{\beta}(t)\dot{e} + \frac{1}{2}\boldsymbol{e}^\mathrm{T}\dot{\boldsymbol{\beta}}(t)\boldsymbol{e} + \boldsymbol{e}_\mathrm{w}{}^\mathrm{T}\dot{\boldsymbol{e}}_\mathrm{w} \\
&= (\dot{e} + \alpha e)^\mathrm{T}(-\boldsymbol{B}_\mathrm{d}(t)\dot{e} - \boldsymbol{K}_\mathrm{d}(t)\boldsymbol{e} + \alpha \boldsymbol{M}_\mathrm{d}\dot{e}) \\
&\quad + \boldsymbol{e}^\mathrm{T}\boldsymbol{\beta}(t)\dot{e} + \frac{1}{2}\boldsymbol{e}^\mathrm{T}\dot{\boldsymbol{\beta}}(t)\boldsymbol{e} + \boldsymbol{e}_\mathrm{w}{}^\mathrm{T}(-\boldsymbol{L}\boldsymbol{B}\boldsymbol{e}_\mathrm{w}) \\
&= \dot{\boldsymbol{e}}^\mathrm{T}(\alpha \boldsymbol{M}_\mathrm{d} - \boldsymbol{B}_\mathrm{d}(t))\dot{e} + \dot{\boldsymbol{e}}^\mathrm{T}\left(\boldsymbol{\beta}(t) + \alpha^2 \boldsymbol{M}_\mathrm{d} - \alpha \boldsymbol{B}_\mathrm{d}(t) - \boldsymbol{K}_\mathrm{d}(t)\right)\boldsymbol{e} \\
&\quad + \boldsymbol{e}^\mathrm{T}\left(\frac{1}{2}\dot{\boldsymbol{\beta}}(t) - \alpha \boldsymbol{K}_\mathrm{d}(t)\right)\boldsymbol{e} - \boldsymbol{e}_\mathrm{w}{}^\mathrm{T}\boldsymbol{L}\boldsymbol{B}\boldsymbol{e}_\mathrm{w}
\end{aligned} \tag{9.16}$$

Substituting $\beta(t)$ and $\dot{\beta}(t)$ into Eq. (9.16) yields

$$\begin{aligned}
\dot{V}(t) &= \dot{\boldsymbol{e}}^\mathrm{T}(\alpha \boldsymbol{M}_\mathrm{d} - \boldsymbol{B}_\mathrm{d}(t))\dot{e} \\
&\quad + \boldsymbol{e}^\mathrm{T}\left(\frac{1}{2}\dot{\boldsymbol{K}}_\mathrm{d}(t) + \frac{1}{2}\alpha \dot{\boldsymbol{B}}_\mathrm{d}(t) - \alpha \boldsymbol{K}_\mathrm{d}(t)\right)\boldsymbol{e} - \boldsymbol{e}_\mathrm{w}{}^\mathrm{T}\boldsymbol{L}\boldsymbol{B}\boldsymbol{e}_\mathrm{w} < 0
\end{aligned} \tag{9.17}$$

This completes the proof.

9.3.2 Integrated HO-WOA design

The HO imitates the territory contention and foraging behaviors of the hippopotamus. In the territory contention stage, to update the individual's position, one has

$$\boldsymbol{X}_i(t+1) = \begin{cases} \boldsymbol{X}_i(t) + r_1(\boldsymbol{X}_{r1}(t) - \boldsymbol{X}_{r2}(t)), & \text{if } r_2 < 0.5 \\ \boldsymbol{X}_i(t) + r_1(\boldsymbol{X}_\mathrm{best}(t) - \boldsymbol{X}_i(t)), & \text{otherwise} \end{cases} \tag{9.18}$$

where, $\boldsymbol{X}_i(t)$ represents the position of the i-th individual, $\boldsymbol{X}_{r1}(t)$ and $\boldsymbol{X}_{r2}(t)$ are the positions of two randomly selected individuals, $\boldsymbol{X}_{\text{best}}(t)$ is the position of the current global optimal individual, r_1 is a random scalar within [−1,1], and r_2 is a random scalar within [0,1].

In the foraging stage, hippopotamus individuals search for food within their territories, and the formula for updating their positions is

$$\boldsymbol{X}_i(t+1) = \boldsymbol{X}_i(t) + r_3(\boldsymbol{X}_{r3}(t) - \boldsymbol{X}_i(t)) \tag{9.19}$$

where, $\boldsymbol{X}_{r3}(t)$ is the position of a randomly selected individual, and r_3 is a random scalar within [−1,1], which further enhances the local search ability.

The WOA mimics the predatory behavior of whales, mainly including encircling prey, spiral position updating, and the contraction encirclement mechanism. In the stage of encircling prey, the position of an individual is updated following Eq. (9.20):

$$\begin{aligned} \boldsymbol{X}(t+1) &= \boldsymbol{X}^*(t) - A\boldsymbol{D} \\ \boldsymbol{D} &= \left|C\boldsymbol{X}^*(t) - \boldsymbol{X}(t)\right| \end{aligned} \tag{9.20}$$

where, $\boldsymbol{X}(t)$ is the current individual position, corresponding to a set of impedance control parameters; $\boldsymbol{X}^*(t)$ is the current optimal individual position; A and C are coefficient scalars, with $A = 2ar_4 - a$, $C = 2r_5$, a is used to control the contraction degree of the search range; r_4 and r_5 are random scalars within [−1,1], and the randomness is introduced to prevent the algorithm from falling into a local optimum.

When the whale discovers the prey, it will adopt the spiral position updating mechanism to approach the prey with Eq. (9.21).

$$\boldsymbol{X}(t+1) = \boldsymbol{D}e^{bl}\cos(2\pi l) + \boldsymbol{X}^*(t) \tag{9.21}$$

where, b is a constant defining the shape of the logarithmic spiral, and l is a random scalar within [0,1].

Remark 9.2 In the control algorithm for spacecraft electromagnetic separation, k is typically used as a phase transition threshold parameter to distinguish the global search and local development phases of the optimization algorithm. In the integrated HO-WOA algorithm, during the first k iterations, global exploration is performed through the territory contention behavior of HO to locate potential regions of better solutions. After exceeding k iterations, the algorithm switches to the spiral position-updating mechanism of WOA to adjust the parameters within the better regions. The value of k directly affects the convergence speed of the algorithm and the accuracy of the solution. It is determined by the trial-and-error method or response surface analysis. Generally, the value is between 10% and 50% of the total number of iterations. In the scenario of actuator failure, appropriately increasing the value of k can enhance the global search ability and improve the robustness of the controller.

9.3.3 Impedance parameter optimization with integrated HO-WOA

- *Initialization of parameters:* The parameters of the integrated HO-WOA algorithm are set first. The population size is N, and the maximum number of iterations is T. Initialize the positions of the hippopotamus individuals and set the parameters in the inertia matrix $\boldsymbol{M}_\mathrm{d}$, damping matrix $\boldsymbol{B}_\mathrm{d}$, and stiffness matrix $\boldsymbol{K}_\mathrm{d}$. Meanwhile, set the constant b in WOA and initialize a. For HO, set the relevant weight coefficients in the territory competition and foraging behaviors.
- *Calculation of fitness values:* Substitute each set of impedance control parameters into the electromagnetic separation system, and based on the predesigned reference trajectory, the position error $e$and velocity error $\dot{e}$ are calculated.

$$\boldsymbol{e} = \sum_{j=1}^{n}\left(\boldsymbol{\eta}_{\mathrm{d}j}\tilde{\boldsymbol{\eta}}_j\right)^2 \tag{9.22}$$

$$\dot{\boldsymbol{e}} = \sum_{j=1}^{n}\left(\dot{\boldsymbol{\eta}}_{\mathrm{d}j}\dot{\tilde{\boldsymbol{\eta}}}_j\right)^2 \tag{9.23}$$

where, $\tilde{\boldsymbol{\eta}}_j$ and $\dot{\tilde{\boldsymbol{\eta}}}_j$ represents the actual position and velocity at the moment j, $\boldsymbol{\eta}_{\mathrm{d}j}$, and $\dot{\boldsymbol{\eta}}_{\mathrm{d}j}$ represents the reference position and velocity at the moment j, and n is the number of sampling points.

Define a fitness function as:

$$F = w_1\frac{e_x + e_y + e_z}{3} + w_2\frac{\dot{e}_x + \dot{e}_y + \dot{e}_z}{3} \tag{9.24}$$

where, w_1 and w_2 are the weight coefficients.

- *Update of individual positions:* In the early stages of optimization, the HO is primarily utilized for global search. During the territory competition phase, individual positions are updated according to Eq. (9.18). In the foraging phase, positions are updated according to Eq. (9.19). As the optimization progresses, WOA is introduced for local exploitation. In the encircling prey phase, Eq. (9.20) is employed, and in the spiral position update phase, Eq. (9.21) is used. Through these updates, the new combinations of impedance control parameters can be obtained.

- *Check of termination conditions:* It is necessary to verify whether the maximum number of iterations T has been reached or if the fitness values have converged. If not, return to *Step* 2 to continue iterating; otherwise, output the optimal parameter combination.
- *Application of optimal parameters:* The optimized impedance control parameters, i.e., $\boldsymbol{M}_\mathrm{d}$, $\boldsymbol{B}_\mathrm{d}$ and $\boldsymbol{K}_\mathrm{d}$, are applied to the impedance controller of the electromagnetic separation system.

Remark 9.3 : The convergence of the fitness value is determined by judging whether the change in the amount ΔF of the fitness value in m consecutive iterations are less than a certain threshold ε, i.e., $|F_t - F_{t-m}| < \varepsilon$.

9.4 Simulation test

Considering that the mass of a large-scale failed space platform is much greater than that of the servicing spacecraft, a reasonable assumption can be made, i.e., the position variation of the servicing spacecraft does not affect the pose of the failed platform. The relative position between the two spacecraft is separated from $[0.1 \quad 0 \quad 0]^\mathrm{T}$ m to $[2 \quad 3 \quad 4]^\mathrm{T}$ m. For the CM orbit, the eccentricity $e = 0.15$, the semimajor axis $a = 10,000$km, the initial true anomaly $\theta(0) = 0$. The involved mass $m_0 = 20$kg, $m_\mathrm{C} = 40$kg, $m_\mathrm{T} = 100$kg. The initial states are chosen as $\boldsymbol{\eta}(0) = [0.1 \quad 0 \quad 0]^\mathrm{T}$ m and $\boldsymbol{e}(0) = [0 \quad 0 \quad 0]^\mathrm{T}$ N. The gravitational constant is $\mu = 3.986 \times 10^{14}\mathrm{m}^3/\mathrm{s}^2$, and the magnetic permeability of vacuum is $\mu_0 = 4\pi \times 10^{-7}\mathrm{N/A}^2$.

For comparison, two cases are considered, i.e., NDO-based impedance control and integrated HO-WOA-based impedance control, and the comparison results are provided to illustrate the superiority of the developed algorithm. In the following two simulation examples, the upper bounds of the electromagnetic force and its rate are 30N and 300N/s, respectively. The magnetic moment $\boldsymbol{\mu}_\mathrm{T}$ of the failed spacecraft is $[-2.29\ -8.17\ -2.33]\times 10^5\mathrm{Am}^2$, and the desired electromagnetic force $\boldsymbol{F}_d$ is $[2 \quad 2 \quad 2]^\mathrm{T}$N.

The fault signal is $\boldsymbol{f}(t) = [8 \quad 9 \quad 4]^\mathrm{T} \times 10^{-2}\bar{f}(t)$, and the form of $\bar{f}(t)$ is shown in Table 9.1.

The external disturbances for the servicing spacecraft and the large-scale failed spacecraft are

$$\boldsymbol{w}_\mathrm{C}(t) = \begin{bmatrix} 0.1 + 0.2\sin(0.11\pi t) \\ 0.15 + 0.2\cos(0.11\pi t) \\ 0.2 + 0.3\cos\left(0.11\pi t + \frac{\pi}{3}\right) \end{bmatrix} \mathrm{N}$$

Table 9.1 The form of fault signal.

t/s	[0,10]	(10,45]	(45,60]	(60,75]	(75,120]
$\bar{f}(t)$/N	0	$\sin(0.1\pi t)$	1	$\cos(0.1\pi t)$	0

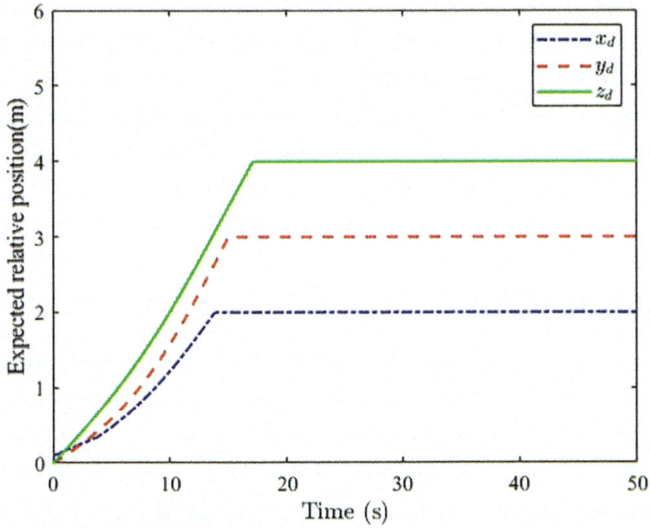

Figure 9.2 Time response of desired relative position.

$$w_{\mathrm{T}}(t) = \begin{bmatrix} 0.1 + 0.1\sin(0.11\pi t) \\ 0.2 + 0.12\cos\left(0.11\pi t + \pi/4\right) \\ 0.3 + 0.15\cos\left(0.11\pi t + \pi/6\right) \end{bmatrix} \mathrm{N}$$

9.4.1 Nonlinear disturbance observer-based impedance control

The following simulation values are chosen, i.e., $\boldsymbol{L} = \mathrm{diag}([0.1\ 0.1\ 0.1])$, $\boldsymbol{M}_{\mathrm{d}} = \mathrm{diag}([5\ 5\ 5])$, $\boldsymbol{B}_{\mathrm{d}} = \mathrm{diag}([b\ b\ b])$, $\boldsymbol{K}_{\mathrm{d}} = \mathrm{diag}([k\ k\ k])$, $b = 100 + 10\sin(0.1t)$, $k = 200 + 10\sin(0.1t)$. The adjoint equations, transversality conditions, and the optimal controller are combined to form a two-point boundary-value problem. The shooting method is used to solve this problem, and the desired relative position and the relative velocity trajectory are shown in Figs. 9.2 and 9.3. The desired relative positions x_{d}, y_{d}, and z_{d} converge to 2m, 3m, and 4m, respectively, and the desired relative velocity converges to 0 at 18s.

Applying the NDO-based impedance controller to the electromagnetic separation system yields Figs. 9.4– 9.9. As shown in Fig. 9.4, the upper bound of the lumped disturbance $\boldsymbol{w}$ is $[-0.36\ -0.42\ -0.63]^{\mathrm{T}}$N. The time response of relative position error is shown in Fig. 9.5, which is always less

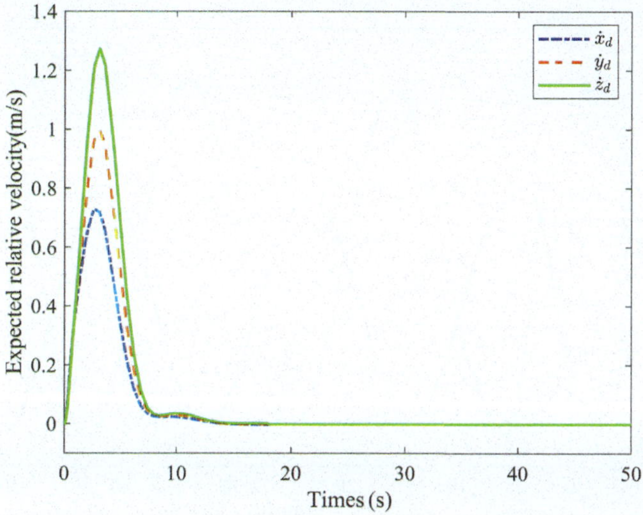

Figure 9.3 Time response of desired relative velocity.

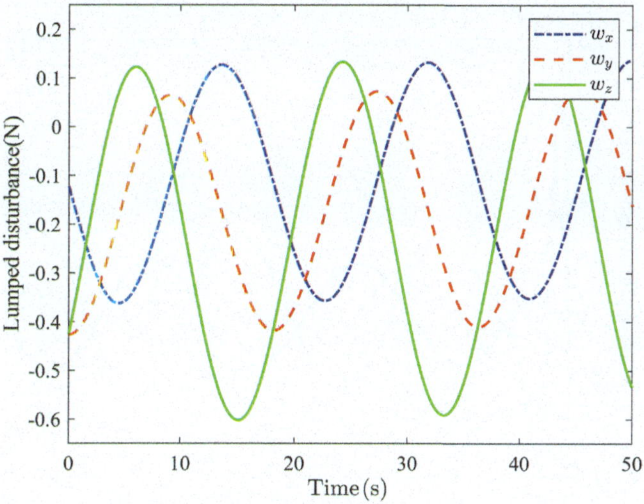

Figure 9.4 Time response of lumped disturbance.

than 2.2×10^{-3}m. The time response of relative velocity error is shown in Fig. 9.6, which is always less than 6.5×10^{-4}m/s. It should be noted that a certain range of chattering phenomenon appears in the initial stage of separation, and the actual trajectory follows the desired trajectory. The electromagnetic force and its rate are shown in Figs. 9.7 and 9.8, which are less than 16N and 61N/s, respectively, satisfying the constraint requirements. As shown in Fig. 9.9, the magnetic moment reaches the maximum value

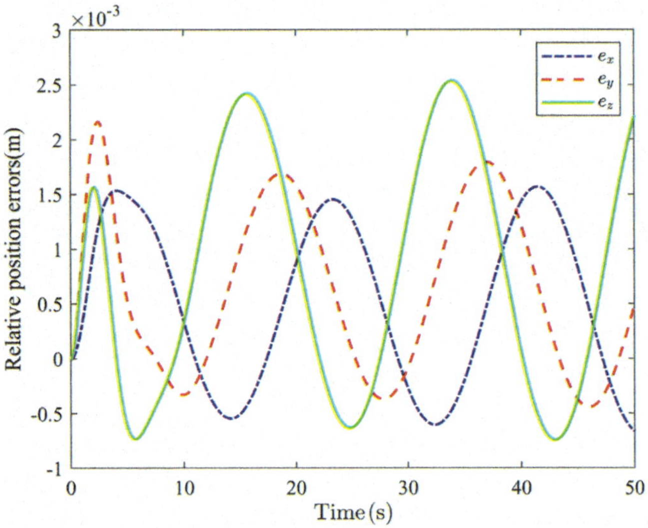

Figure 9.5 Tracking errors of relative position.

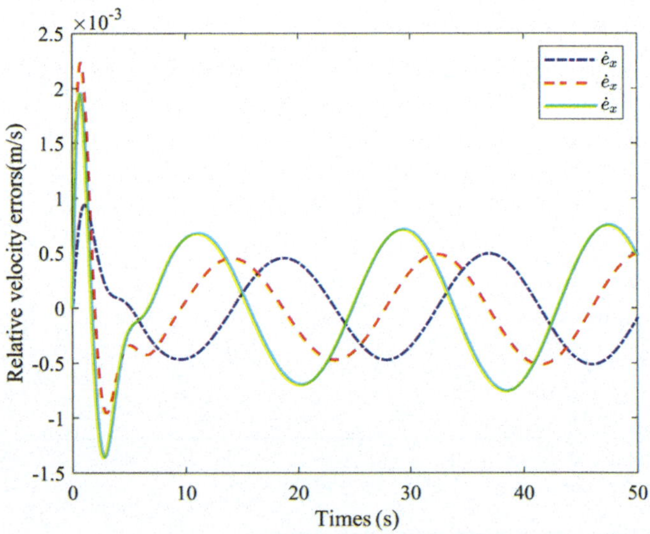

Figure 9.6 Tracking errors of relative velocity.

of 490Am^2 at 24s. Thus, the common electromagnetic mechanism can provide sufficient magnetic moment for electromagnetic separation, which can be realized and satisfy the engineering requirements. In light of this, the separation between the servicing spacecraft and the failed platform is achieved.

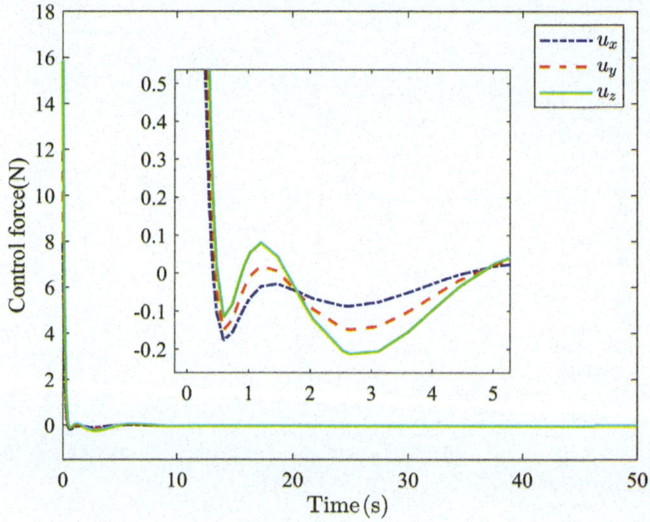

Figure 9.7 Time response of control force.

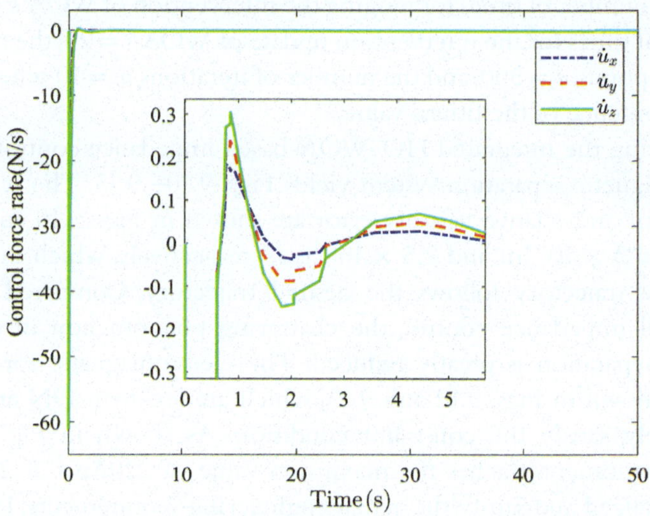

Figure 9.8 Time response of control force rate.

9.4.2 Integrated HO-WOA-based impedance control

In the integrated HO-WOA algorithm, the population size $N = 100$, the maximum number of iterations $T = 500$, the weight coefficient $w_1 = 0.7$, $w_2 = 0.3$, the step size of territorial competition $step_1 = 0.5$, the foraging

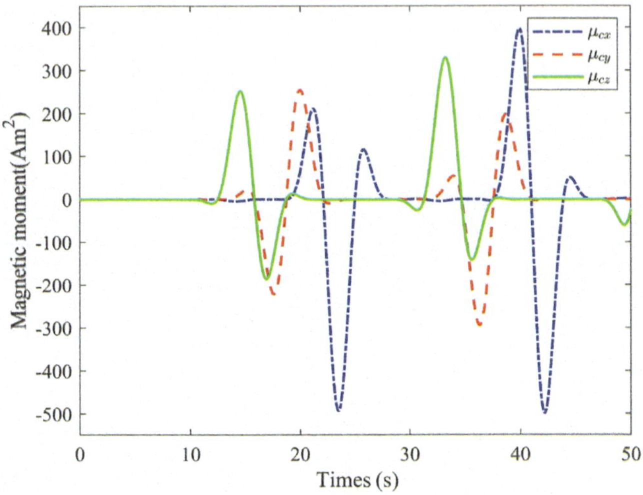

Figure 9.9 Magnetic moment of service spacecraft.

step size $step_2$ = 0.5, the contraction factor decreases linearly from 2 to 0, $b = 1$, the number of times for judging the intervention of WOA $k = 150$, the number of times for the intervention update of WOA $s = 20$, the number of sampling points $n = 500$, and the number of iterations $m = 10$ when judging the convergence of the fitness value.

Applying the integrated HO-WOA-based impedance controller to the electromagnetic separation system yields Figs. 9.10– 9.15. The relative position error and relative velocity error are shown in Figs. 9.10 and 9.11 are less than 6.3×10^{-4}m and 2.5×10^{-4}m/s, respectively, which implies that the actual trajectory follows the desired trajectory. Compared with the traditional impedance control, the chattering phenomenon in the initial stage of separation is greatly reduced. The electromagnetic force and its rate are shown in Figs. 9.12 and 9.13, which are less than 19N and 37N/s, respectively, satisfy the constraint condition. As shown in Fig. 9.14, the magnetic moment reaches the maximum value of 220Am^2 at 22s, which can be realized and satisfy the actual engineering requirements. In order to further illustrate the motion of the servicing spacecraft and the large-scale failed spacecraft relative to CM, Fig. 9.15 is plotted, where the color bar represents the simulation time. One can see that the relative motion of the servicing spacecraft is much longer than that of the failed platform. This is because the mass of the failed platform is greater than that of the servicing spacecraft, resulting in the CM being much closer to the CM of the large-scale failed space platform.

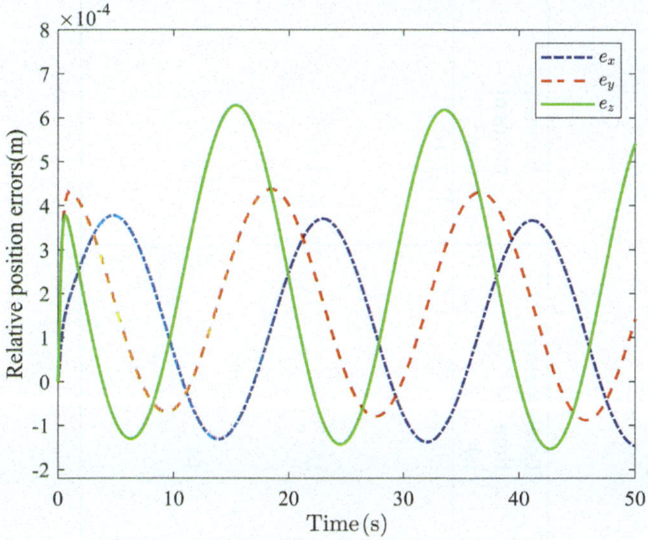

Figure 9.10 Time response of external disturbance.

9.4.3 Comparison of the two control approaches

Based on the aforementioned simulation results, Figs. 9.16– 9.19 show the comparison between the NDO-based impedance controller and the integrated HO-WOA-based impedance controller, which is shown in Table 9.2.

By combining Figs. 9.16– 9.19 with Table 9.2, it can be concluded that the integrated HO-WOA-based impedance controller has a smaller steady-state error in relative position, which is reduced by 98.3% and 76.7%, respectively. In addition, it has a smaller steady-state error in relative velocity, which is reduced by 86.4% and 66.67%, illustrated in Fig. 9.17. Although the required control force increased, illustrated in Fig. 9.18, the magnetic moment generated by the servicing spacecraft is smaller, reduced by 70.7% and 55.1%, illustrated in Fig. 9.19. Thus, the developed integrated HO-WOA-based impedance controller can effectively reduce the dependence on the actuator and mitigate the impact of faults. In the initial stage of separation, the required control force of the developed algorithm is relatively high, but it complies with the constraint conditions. The simulation results show that the integrated HO-WOA-based impedance control approach can achieve high-precision, high-efficiency, and high-stability electromagnetic separation of spacecraft.

Table 9.2 Comparison of control effects.

Control method	Relative position errors (m)	Relative velocity errors (m/s)	Convergence time (s)	Magnetic moment (Am^2)
Improved sliding mode control (Liu et al., 2024)	2×10^{-2}	1×10^{-3}	30	750
Traditional NDO-based impedance control	1.5×10^{-3}	3×10^{-4}	18	490
Integrated HO-WOA-based impedance control	3.5×10^{-4}	1×10^{-4}	18	220

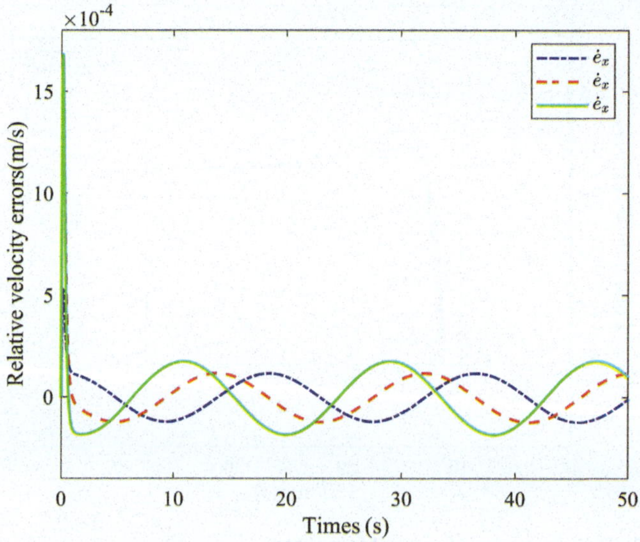

Figure 9.11 Estimation errors of relative velocity.

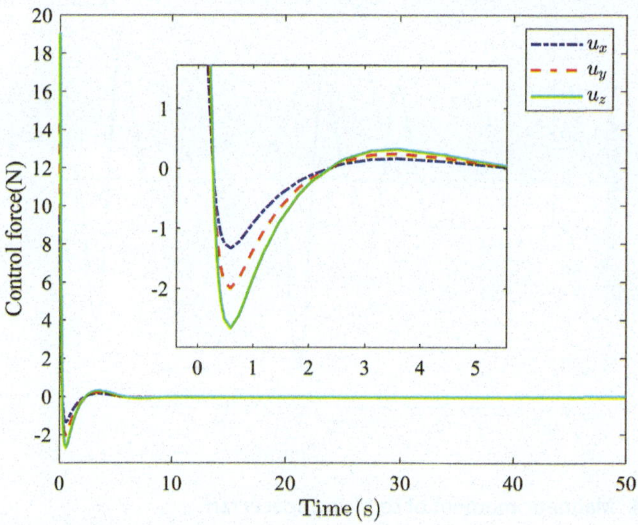

Figure 9.12 Time response of control force.

9.5 Conclusions

This chapter addresses the electromagnetic separation problem of large-scale failed spacecraft by optimizing impedance control through the integration

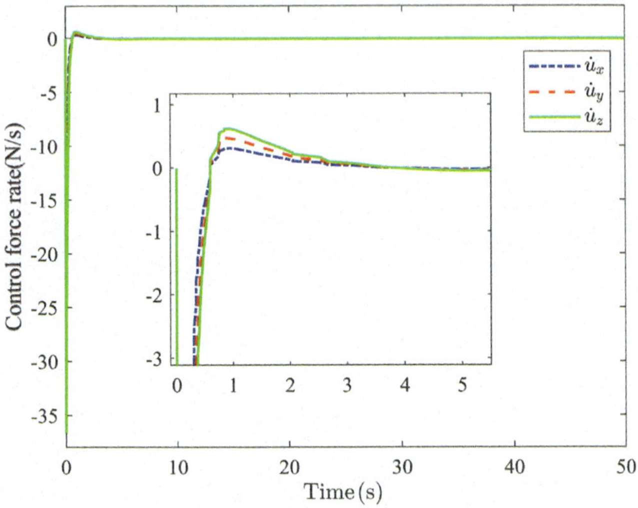

Figure 9.13 Time response of control force rate.

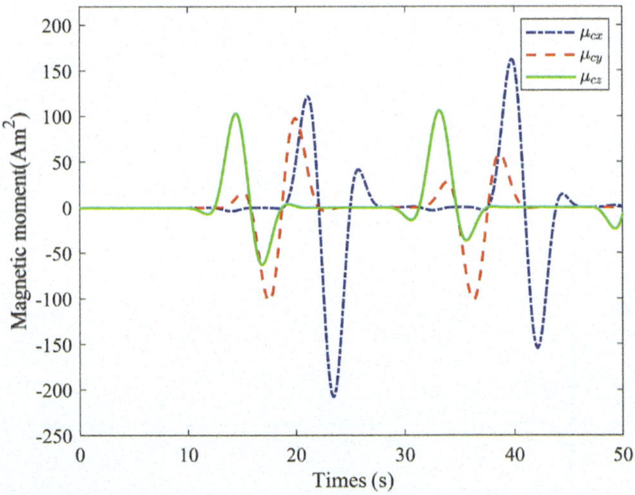

Figure 9.14 Magnetic moment of servicing spacecraft.

of the HO and the WOA. Traditional control approaches are inadequate for managing multisource disturbances and fail to satisfy the high-precision requirements essential for electromagnetic separation control. Based on the dynamics model of spacecraft electromagnetic separation, an impedance controller is developed utilizing an NDO. Subsequently, the impedance

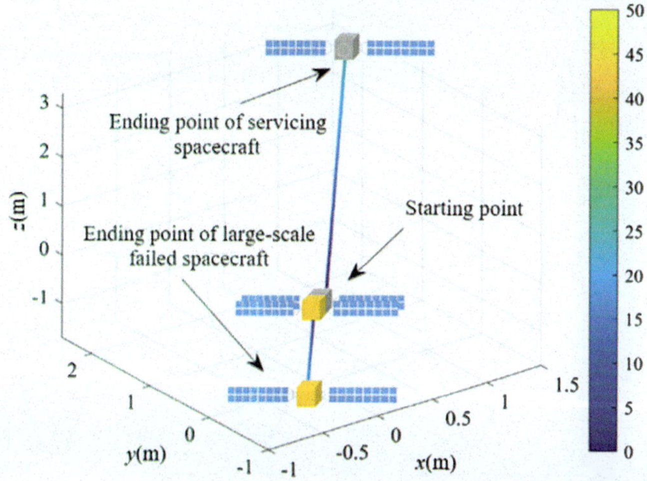

Figure 9.15 Motion of the two-spacecraft relative to "CM".

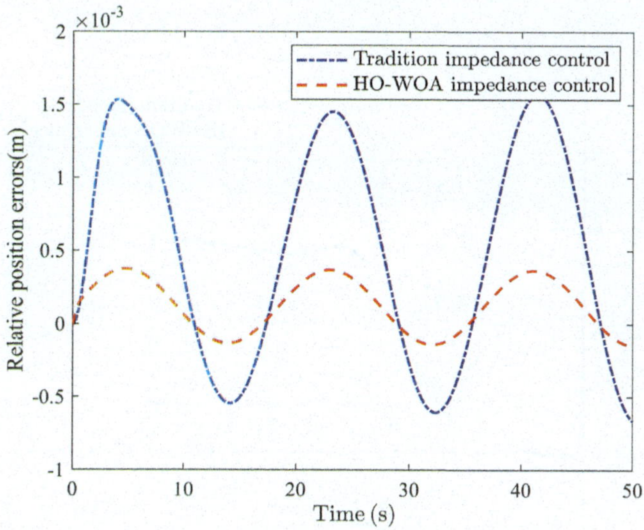

Figure 9.16 Comparison of relative position errors.

control parameters are optimized using an integrated HO-WOA tailored for spacecraft electromagnetic separation. Simulation results demonstrate that the integrated algorithm effectively reduces the relative tracking error of the spacecraft and minimizes convergence time. In future work, the

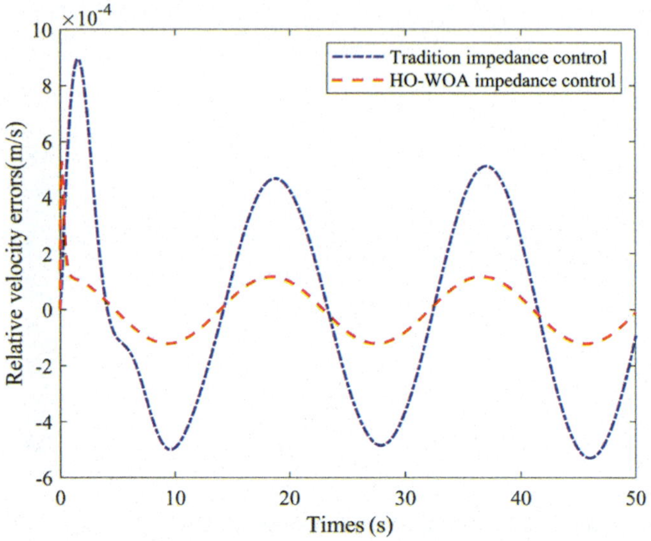

Figure 9.17 Comparison of relative velocity errors.

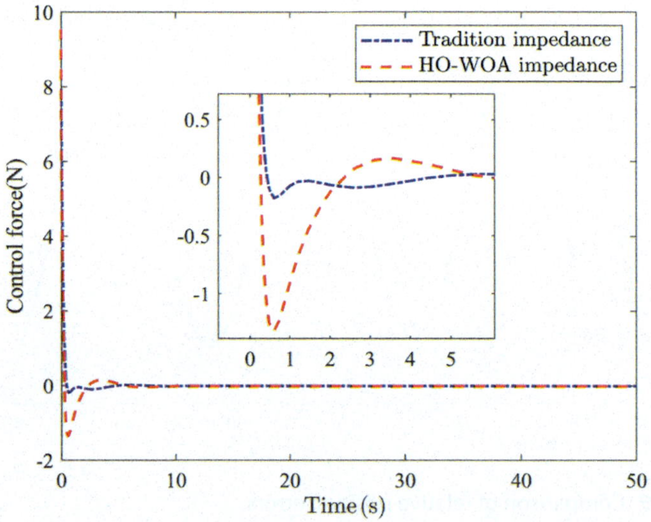

Figure 9.18 Comparison of control force.

multiple couplings between attitude, orbit, and electromagnetic force during spacecraft separation should be further explored.

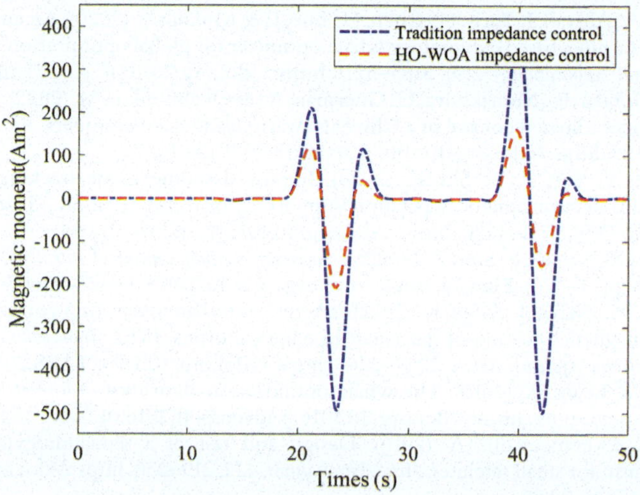

Figure 9.19 Comparison of the magnetic moment.

References

Amiri, M. H., Hashjin, N. M., Montazeri, M., Mirjalili, S., & Khodadadi, N. (2024). Hippopotamus optimization algorithm: A novel nature-inspired optimization algorithm. *Scientific Reports, 14*(1). https://doi.org/10.1038/s41598-024-54910-3.

Annamraju, S., Raj, S., Pediredla, V. K., & Thondiyath, A. (2020). Parameter determination technique for impedance control of interactive robots using transformation matrices. *IFAC-PapersOnLine, 53*, 201–205. https://doi.org/10.1016/j.ifacol.2020.06.034.

Bentouati, B., Chaib, L., & Chettih, S. (2017). A hybrid whale algorithm and pattern search technique for optimal power flow problem. *Proceedings of 2016 8th international conference on modelling, identification and control, ICMIC 2016* (pp. 1048–1053). IEEE. https://doi.org/10.1109/ICMIC.2016.7804267.

Cao, K., Li, S., She, Y., Biggs, J. D., Liu, Y., & Bian, L. (2021). Dynamics and on-orbit assembly strategies for an orb-shaped solar array. *Acta Astronautica, 178*, 881–893. https://doi.org/10.1016/j.actaastro.2020.10.030.

Ding, S., Peng, J., Xin, J., Zhang, H., & Wang, Y. (2023). Vision-based virtual impedance control for robotic system without prespecified task trajectory. *IEEE Transactions on Industrial Electronics, 70*(6), 6046–6056. https://doi.org/10.1109/tie.2022.3199917.

Gu, X., Liu, L., Wang, L., Mao, J., & Guo, Y. (2025). Predefined-time impedance control of free-flying flexible-joint space robots for force sensor-less target capturing with prescribed performance. *International Journal of Robust and Nonlinear Control, 35*(2), 452–478. https://doi.org/10.1002/rnc.7657.

Han, T., Wang, H., Li, T., Liu, Q., & Huang, Y. (2025). MHO: A modified Hippopotamus optimization algorithm for global optimization and engineering design problems. *Biomimetics, 10*(2), 90. https://doi.org/10.3390/biomimetics10020090.

He, G., Cao, D., Cao, Y., & Huang, W. (2022). Dynamic modeling and orbit maneuvering response analysis for a three-axis attitude stabilized large scale flexible spacecraft installed with hinged solar arrays. *Mechanical Systems and Signal Processing, 162*, 108083. https://doi.org/10.1016/j.ymssp.2021.108083.

Heidari, A. A., Aljarah, I., Faris, H., Chen, H., Luo, J., & Mirjalili, S. (2020). An enhanced associative learning-based exploratory whale optimizer for global optimization. *Neural Computing and Applications, 32*(9), 5185–5211. https://doi.org/10.1007/s00521-019-04015-0.

Ivanov, D., Gondar, R., Monakhova, U., Guerman, A., & Ovchinnikov, M. (2022). Electromagnetic uncoordinated control of a ChipSats swarm using magnetorquers. *Acta Astronautica, 192*, 15–29. https://doi.org/10.1016/j.actaastro.2021.12.014.

Liu, C., Ma, Y., Yue, X., & Shi, K. (2024). Improved sliding mode tracking control for spacecraft electromagnetic separation supporting on-orbit assembly. *Advances in Space Research, 73*(9), 4711–4720. https://doi.org/10.1016/j.asr.2024.01.034.

Liu, C., Yue, X., Shi, K., & Sun, Z. (2022a). *Spacecraft attitude control: A linear matrix inequality approach* (pp. 1–370). Elsevier. https://doi.org/10.1016/B978-0-323-99005-9.00012-2.

Liu, C., Yue, X., Zhang, J., & Shi, K. (2022b). Active disturbance rejection control for delayed electromagnetic docking of spacecraft in elliptical orbits. *IEEE Transactions on Aerospace and Electronic Systems, 58*(3), 2257–2268. https://doi.org/10.1109/TAES.2021.3130830.

Mirjalili, S., & Lewis, A. (2016). The whale optimization algorithm. *Advances in Engineering Software, 95*, 51–67. https://doi.org/10.1016/j.advengsoft.2016.01.008.

Olivieri, L., & Francesconi, A. (2016). Design and test of a semiandrogynous docking mechanism for small satellites. *Acta Astronautica, 122*, 219–230. https://doi.org/10.1016/j.actaastro.2016.02.004.

She, Y., Li, S., & Wang, Z. (2019). Constructing a large antenna reflector via spacecraft formation flying and reconfiguration control. *Journal of Guidance, Control, and Dynamics, 42*(6), 1372–1382. https://doi.org/10.2514/1.g004116.

Shi, K., Liu, C., & Sun, Z. (2019). Constrained fuel-free control for spacecraft electromagnetic docking in elliptical orbits. *Acta Astronautica, 162*, 14–24. https://doi.org/10.1016/j.actaastro.2019.05.016.

Shi, K., Liu, C., Sun, Z., & Yue, X. (2022). Coupled orbit-attitude dynamics and trajectory tracking control for spacecraft electromagnetic docking. *Applied Mathematical Modelling, 101*, 553–572. https://doi.org/10.1016/j.apm.2021.08.030.

Shi, K., Liu, C., Sun, Z., & Yue, X. (2020). Disturbance observer-based attitude stabilization for rigid spacecraft with input MRCs. *Advances in Space Research, 66*(3), 689–701. https://doi.org/10.1016/j.asr.2020.04.016.

Uyama, N., Nakanishi, H., Nagaoka, K., & Yoshida, K. (2012). Impedance-based contact control of a free-flying space robot with a compliant wrist for non-cooperative satellite capture. *IEEE international conference on intelligent robots and systems*. Japan: IEEE https://doi.org/10.1109/IROS.2012.6386082.

Wang, J., Zhou, Q., Zheng, W., & Shao, J. (2024). Passivity-based control with disturbance observer of electromagnetic formation flight spacecraft in the port-hamiltonian framework. *Applied Sciences, 14*(10), 4248. https://doi.org/10.3390/app14104248.

Wang, M., Luo, J., Yuan, J., & Walter, U. (2018). An integrated control scheme for space robot after capturing non-cooperative target. *Acta Astronautica, 147*, 350–363. https://doi.org/10.1016/j.actaastro.2018.04.016.

Yang, R., Yang, C., Chen, M., & Na, J. (2017). Adaptive impedance control of robot manipulators based on Q-learning and disturbance observer. *Systems Science and Control Engineering, 5*(1), 287–300. https://doi.org/10.1080/21642583.2017.1347532.

Zhang, Y. W., Yang, L. P., Zhu, Y. W., Ao, H. J., & Qi, D. W. (2016). Self-docking analysis and velocity-aimed control for spacecraft electromagnetic docking. *Advances in Space Research, 57*(11), 2314–2325. https://doi.org/10.1016/j.asr.2016.03.022.

Zhang, Y. W., Yang, L. P., Zhu, Y. W., Hai Ren, X., & Huang, H. (2011). Self-docking capability and control strategy of electromagnetic docking technology. *Acta Astronautica, 69*(11–12), 1073–1081. https://doi.org/10.1016/j.actaastro.2011.07.001.

Zhu, Y., Cai, W., Yang, L., & Huang, H. (2018). Flatness-based trajectory planning for electromagnetic spacecraft proximity operations in elliptical orbits. *Acta Astronautica, 152*, 342–351. https://doi.org/10.1016/j.actaastro.2018.08.031.

CHAPTER 10

Integrated control for spacecraft electromagnetic docking and separation

10.1 Introduction

Spacecraft docking and separation is one of the core technologies and plays pivotal roles in complex space missions, e.g., on-orbit assembly, refueling, and debris management (Li et al., 2019). Thrusters traditionally serve as the primary actuators in these operations (Fauré et al., 2023; Xia & Zou, 2021), which bring inherent disadvantages including plume contamination, vibration, propellant consumption, and significant impact forces during docking. In recent years, electromagnetic docking and separation technology has garnered attention due to its potential for soft docking, extended operational lifespan, and precise control accuracy. This approach leverages electromagnetic forces and moments as control inputs, offering continuous, reversible, noncontact, and self-docking capabilities, which reduce the reliance on relative motion measurements and enhances control accuracy (Zhang and Chen, 2023; Zhang et al., 2016). However, the precise model of electromagnetic forces introduces significant nonlinearity, posing substantial challenge in modeling and controlling. To address this problem, a far-field model based on the magnetic dipole hypothesis of electromagnetic forces and moments was established in "Schiweighart (2005) and then various electromagnetic docking and separation control methodologies have been proposed, including extended state observer-based control (Zhang et al., 2013), sliding mode control (Liu et al., 2024; Shi et al., 2022), and piecewise control using evolving relative motion laws (Zhang, 2021), etc.

Aiming at mitigating disturbances encountered during docking and separation processes, e.g., external perturbations, input constraints, elliptical eccentricities, fault signals, and unknown mass, disturbance observer-based control has been widely employed as a robust approach for estimating and compensating for disturbances in spacecraft systems. Many scholars have

Spacecraft Electromagnetic Docking and Separation.
DOI: https://doi.org/10.1016/B978-0-443-49334-8.00002-1

successfully applied disturbance observer-based strategies to handle disturbances from external sources (Wang et al., 2024), parameter uncertainties (Wu, Yang, Li, Li, & Guo, 2018), input constraints (Zhang et al., 2024), and noncooperative behaviors (Zhu, Chen, & Zhu, 2021). In electromagnetic docking and separation, many disturbance-observer-based controllers have been proposed. Shi et al. (2019) and Shi et al. (2020a) investigated the observer-based control problem for spacecraft electromagnetic docking and proposed a disturbance observer-based control strategy with state feedback to address comprehensive disturbances. In these works, a linear matrix inequalities (LMIs)-based approach is utilized to determine the controller and observer gains, which has been proven effectively in managing electromagnetic docking and separation challenges amidst diverse disturbances.

Despite these disturbances, challenges persist due to perturbations in controller and observer gains arising from factors like computer truncation errors and signal conversion between digital and analog formats (Keel & Bhattacharyya, 1997). To achieve high-precision control amidst these challenges, nonfragile control strategies have been explored. There are two widely recognized and researched types of gain perturbations, i.e., additive perturbation and multiplicative perturbation. The former typically occurs during the initial operational period of the system, and the latter is more common during the decay period (Liu et al., 2017), while both types coexist in the transition period. Significant researches have addressed nonfragile control in other domains such as network control (Dong et al.,2020), neural networks control (Rajavel et al., 2017), multiagent system control (Kaviarasan et al., 2021), uncertain system control (Hafezi & Khandani, 2022) and repeated scalar nonlinear system control (Safana et al., 2024), with usually only the additive perturbation considered, and its application to spacecraft systems remains relatively nascent. In recent years, several studies have been done to deal with the spacecraft nonfragile control problems concerning the above two types of gain perturbation.

In spacecraft attitude control, Han and Kim (2023) proposed a disturbance observer-based fuzzy attitude tracking control approach that considers additive perturbations in controller gain, where an LMI condition is established to ensure the asymptotic stability of the closed-loop system. Liu et al. (2017) introduced a nonfragile control strategy featuring robust dynamic output feedback, which accounted for model uncertainty, input saturation, external disturbances, and both additive and multiplicative controller gain perturbations. This approach achieved faster convergence and improved control performance. In another study, Liu et al. (2021b), a novel attitude stabilization and vibration suppression control strategy was developed under various disturbances, taking into account additive and multiplicative perturbations. A detailed comparison of the performance between fragile and nonfragile controllers demonstrated the advantages of the nonfragile controller. Similarly, another research (Liu et al., 2021a) presented a hybrid nonfragile control strategy for postcapture spacecraft attitude stabilization, also considering additive and multiplicative perturbations. Lyu et al. (2022) introduced an intermediate observer into the nonfragile attitude stabilization problem for spacecraft, where both additive and multiplicative observer gain perturbations were considered, while the controller gain perturbations were neglected. In general, some of the aforementioned works did not address observer gain perturbations, while others focused solely on observer gain perturbations, overlooking concurrent perturbations in controller gains. Moreover, few studies have targeted the nonfragile control of spacecraft relative position. Fang et al. (2018) provided a brief introduction to disturbance observer-based nonfragile position control of a spacecraft equipped with two solar panels, considering additive gain perturbations in both the controller and observer.

However, to our knowledge, the dual hybrid nonfragile strategy has not been previously explored. Compared to existing nonfragile control strategies used in spacecraft control, this chapter aims to comprehensively address the nonfragile problem in both the controller and disturbance observer. The main theoretical contributions of this chapter are summarized as follows:

1. Compared with Wu, Yang, Li, Li, & Guo, 2018 and Liu et al. (2021a), the coexisting perturbations of additive and multiplicative types in observer and controller are dealt with simultaneously, which results in a novel dual hybrid nonfragile strategy, i.e., a hybrid nonfragile disturbance observer-based hybrid nonfragile control scheme.
2. An integrated electromagnetic docking and separation control approach is proposed and the influence of input saturation is separated into the comprehensive disturbance, which is beneficial to stability analysis. Compared with Shi et al. (2020b), this approach improves the accuracy of relative position and velocity with only a slight increase in input force and energy.
3. Compared with existing fragile controller in electromagnetic docking (Shi et al., 2019) and nonfragile controller in attitude stabilization (Liu et al., 2021a; Lyu et al., 2022), this approach shows improved overall performance in terms of position and velocity accuracy, control input requirements, and energy consumption.

The chapter is structured as follows. Section 10.2 introduces a hybrid nonfragile control strategy and proves the stability of the closed-loop system via Lyapunov stability analysis. Section 10.3 gives some simulation results to demonstrate the feasibility and superiority of the developed hybrid nonfragile strategy. Finally, Section 10.4 draws some conclusions.

10.2 Hybrid nonfragile disturbance observer and controller design

The control input $\boldsymbol{u}$ for the spacecraft is the electromagnetic force, which is subject to the current and power constraints of its coils. The maximum magnetic moment of the coils is finite, thus resulting in a bounded electromagnetic interaction force between spacecraft. Although this force is highly dependent on the relative attitude and position between spacecraft, it is modeled here as a saturated control input sat($\boldsymbol{u}$) for simplicity. Thus, the dynamic model introduced in Section 1.2, the Eq. (1.78) can be converted into the following state-space form:

$$\dot{\boldsymbol{\eta}}(t) = (\boldsymbol{A} + \Delta\boldsymbol{A})\boldsymbol{\eta}(t) + (\boldsymbol{B} + \Delta\boldsymbol{B})(\mathrm{sat}(\boldsymbol{u}) + \boldsymbol{w}_0(t)) \tag{10.1}$$

with

$$\mathrm{sat}(\boldsymbol{u}) = \left[\mathrm{sat}(u_x)\mathrm{sat}(u_y)\mathrm{sat}(u_z)\ \right]^{\mathrm{T}},$$

$$\mathrm{sat}(u_i) = \begin{cases} U_i(u_i > U_i) \\ u_i(-U_i \le u_i \le U_i)\left(i = x, y, z\right) \\ -U_i(u_i < -U_i) \end{cases}$$

$$\boldsymbol{w}_0 = \left[d_x \quad d_y \quad d_z\right]^{\mathrm{T}}$$

$$d_{i=x,y,z} = \frac{m_{\mathrm{T}} w_{\mathrm{C}i} - m_{\mathrm{C}} w_{\mathrm{T}i}}{m_{\mathrm{C}} + m_{\mathrm{T}}}$$

where U_i is the upper bound of u_i.

In order to facilitate the subsequent proof of stability, one can easily transform the electromagnetic docking system into the state-space form of Eq. (10.2).

$$\dot{\boldsymbol{\eta}}(t) = \boldsymbol{A}\boldsymbol{\eta}(t) + \boldsymbol{B}\boldsymbol{u}(t) + \boldsymbol{B}\boldsymbol{w}(t) \tag{10.2}$$

with

$$\boldsymbol{w}(t) = \boldsymbol{B}^{\mathrm{T}}\left(\boldsymbol{B}\boldsymbol{B}^{\mathrm{T}}\right)^{*}(\Delta\boldsymbol{A}\boldsymbol{\eta}(t) - \boldsymbol{B}\boldsymbol{u}(t) + (\boldsymbol{B} + \Delta\boldsymbol{B})(\mathrm{sat}(\boldsymbol{u}) + \boldsymbol{w}_0(t)))$$

where $(\boldsymbol{BB}^{\mathrm{T}})^{*}$ is the pseudo-inverse of matrix $\boldsymbol{BB}^{\mathrm{T}}$, $\boldsymbol{w}(t)$ is the comprehensive disturbance resulted from elliptical eccentricity, unknown mass, input saturation, and external disturbances.

Considering the coexistence of additive and multiplicative gain perturbations in the controller observer, the controller and observer perturbations $\Delta\boldsymbol{K}$ and $\Delta\boldsymbol{L}$ are assumed to be norm-bounded, satisfying

$$\Delta\boldsymbol{K} = \sigma_1\boldsymbol{M}_1\boldsymbol{F}_1(t)\boldsymbol{N}_1 + (1-\sigma_1)\boldsymbol{M}_2\boldsymbol{F}_2(t)\boldsymbol{N}_2\boldsymbol{K}, \quad \sigma_1 \in (0,1) \tag{10.3}$$

$$\Delta\boldsymbol{L} = \sigma_2\boldsymbol{M}_3\boldsymbol{F}_3(t)\boldsymbol{N}_3 + (1-\sigma_2)\boldsymbol{M}_4\boldsymbol{F}_4(t)\boldsymbol{N}_4\boldsymbol{L} \quad \sigma_2 \in (0,1) \tag{10.4}$$

where σ_1 and σ_2 are given constants, $\boldsymbol{M}_{1,2,3,4}$ and $\boldsymbol{N}_{1,2,3,4}$ are constant matrices reflecting the structure of uncertainty, and $\boldsymbol{F}_{1,2,3,4}(t)$ satisfy $||\boldsymbol{F}_{1,2,3,4}(t)|| \leq 1$. The first terms of Eqs. (10.3) and (10.4) both represent additive perturbations, while the second terms represent multiplicative perturbations. It is reasonable to assume that the perturbations are norm-bounded, as they stem from factors such as analog-to-digital and digital-to-analog conversion, truncation errors, round-off errors, and performance degradation (Keel & Bhattacharyya, 1997). Eqs. (10.3) and (10.4) represent one of the most widely used forms of gain perturbation in robust stability analysis.

To design a hybrid nonfragile disturbance observer, an auxiliary variable $\boldsymbol{z}(t)$ is innovated as

$$\boldsymbol{z}(t) = \hat{\boldsymbol{w}}(t) - (\boldsymbol{L} + \Delta\boldsymbol{L})\boldsymbol{\eta}(t) \tag{10.5}$$

In Eq. (10.5), the observer is treated as a subsystem, where $\boldsymbol{z}(t)$ is the state vector of the nonfragile observer, $\hat{\boldsymbol{w}}(t)$ is the estimation of the comprehensive disturbance, $\boldsymbol{L}$ represents the observer gain matrix. Then, a hybrid nonfragile disturbance observer is designed as

$$\begin{cases} \dot{\boldsymbol{z}}(t) = -(\boldsymbol{L} + \Delta\boldsymbol{L})(\boldsymbol{A}\boldsymbol{\eta}(t) + \boldsymbol{B}\boldsymbol{u}(t) + \boldsymbol{B}\hat{\boldsymbol{w}}(t)) \\ \hat{\boldsymbol{w}}(t) = \boldsymbol{z}(t) + (\boldsymbol{L} + \Delta\boldsymbol{L})\boldsymbol{\eta}(t) \end{cases} \tag{10.6}$$

Through the above hybrid nonfragile disturbance observer, a hybrid nonfragile control strategy is designed in the following form

$$\boldsymbol{u}(t) = -(\boldsymbol{K} + \Delta\boldsymbol{K})\boldsymbol{\eta}(t) - \hat{\boldsymbol{w}}(t) \tag{10.7}$$

where $\boldsymbol{K}$ is the controller gain matrix.

By importing Eqs. (10.6) and (10.7) into Eq. (10.2), one has

$$\begin{cases} \dot{\boldsymbol{\eta}}(t) = [\boldsymbol{A} - \boldsymbol{B}(\boldsymbol{K} + \Delta\boldsymbol{K})]\boldsymbol{\eta}(t) - \boldsymbol{B}\boldsymbol{e}(t) \\ \dot{\boldsymbol{e}}(t) = \dot{\boldsymbol{w}}(t) - (\boldsymbol{L} + \Delta\boldsymbol{L})\boldsymbol{B}\boldsymbol{e}(t) - \Delta\dot{\boldsymbol{L}}\boldsymbol{\eta}(t) \end{cases} \tag{10.8}$$

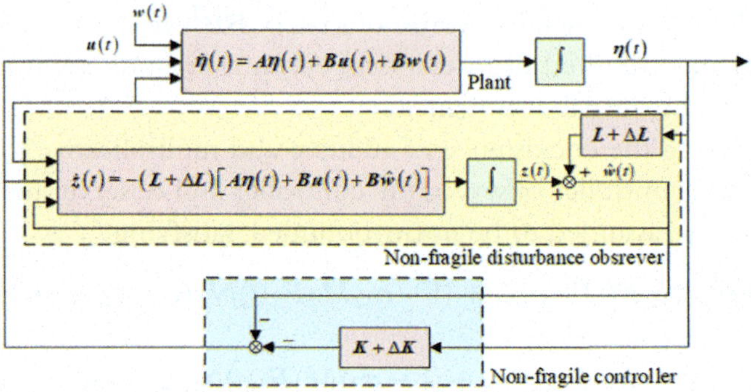

Figure 10.1 Structure of the disturbance observer-based dual nonfragile control system.

Thus, the state-space form of the electromagnetic docking system turns into

$$\begin{bmatrix} \dot{\boldsymbol{\eta}}(t) \\ \dot{\boldsymbol{e}}(t) \end{bmatrix} = \begin{bmatrix} \boldsymbol{A} - \boldsymbol{B}(\boldsymbol{K} + \Delta\boldsymbol{K}) & \boldsymbol{B} \\ -\Delta\dot{\boldsymbol{L}} & -(\boldsymbol{L} + \Delta\boldsymbol{L})\boldsymbol{B} \end{bmatrix} \begin{bmatrix} \boldsymbol{\eta}(t) \\ \boldsymbol{e}(t) \end{bmatrix} + \begin{bmatrix} 0 \\ \dot{\boldsymbol{w}}(t) \end{bmatrix} \tag{10.9}$$

where $\boldsymbol{e}(t) = \boldsymbol{w}(t) - \hat{\boldsymbol{w}}(t)$ is the error vector, $\dot{\boldsymbol{e}}(t)$ is the first derivative of $\boldsymbol{e}(t)$.

The structure of the close-loop system, whose controller and observer are suffered from gain perturbations, is displayed in Fig. 10.1.

The nonfragile controller continuously measures the system states and combines them with an estimated disturbance signal to compute the control input. The nonfragile disturbance observer operates by comparing the actual system state with the output of a nominal internal model. The meaning of the internal variable $\boldsymbol{z}$(t) represents the disturbance estimation error scaled by the $(\boldsymbol{L}+\Delta\boldsymbol{L})\boldsymbol{B}$ gain. This internal variable is then integrated to produce a real-time estimate $\hat{\boldsymbol{w}}(t)$ of the disturbance, which can be expressed as $\hat{\boldsymbol{w}}(t) = \boldsymbol{z}(t) = (\boldsymbol{L} + \Delta\boldsymbol{L})\boldsymbol{B}(\boldsymbol{w}(t) - \hat{\boldsymbol{w}}(t))$. The gain matrix $\boldsymbol{L}$ is the primary design parameter of the observer. It directly determines the convergence rate and bandwidth of the disturbance estimation by shaping how aggressively the observer corrects its error. By actively estimating and compensating for external disturbances and modeling uncertainties, including variations in feedback gains and system parameters, the structure can improve the overall robustness and nonfragility of the system.

The main perturbation forces experienced by a spacecraft in space include Earth's nonspherical gravitational force, atmospheric drag, electromagnetic forces from Earth's magnetic field, etc., but they are generally bounded. Since the physically real-world disturbances are always finite, and mathematically it is necessary to define a worst-case bound for stability analysis and controller design. In short, boundedness makes the robust control problem both physically meaningful and mathematically tractable, so the following assumption is raised.

Assumption 10.1 (Liu et al., 2018): The comprehensive disturbance $\boldsymbol{w}(t)$ and its first derivative $\dot{\boldsymbol{w}}(t)$ are bounded, i.e., $||\boldsymbol{w}(t)|| \le \Omega_1, ||\dot{\boldsymbol{w}}(t)|| \le \Omega_2$.

Based on the above closed-loop system and assumption, one can get the following theorem.

Theorem 10.1 The closed-loop system represented by Eq. (10.8) is uniformly ultimately bounded, if for given positive scalars $\xi_i (i = 1, 2, ..., 8)$, symmetric positive definite matrix $\widetilde{\boldsymbol{X}}_1$ and matrix $\boldsymbol{W}$ satisfying $||\boldsymbol{N}_4\boldsymbol{L}|| < ||\boldsymbol{W}||$, there exist symmetric positive definite matrices $\boldsymbol{X}_1, \boldsymbol{X}_2$ and matrix $\boldsymbol{H}$, such that the following LMIs hold:

$$\widetilde{\boldsymbol{X}}_1\boldsymbol{A}^{\mathrm{T}} + \boldsymbol{A}\widetilde{\boldsymbol{X}}_1 - \widetilde{\boldsymbol{X}}_1^{\mathrm{T}}\boldsymbol{K}^{\mathrm{T}}\boldsymbol{B}^{\mathrm{T}} - \boldsymbol{B}\boldsymbol{K}\widetilde{\boldsymbol{X}}_1 \le 0 \tag{10.10}$$

$$\begin{bmatrix} \boldsymbol{G}_{11} & \boldsymbol{G}_{12} \\ * & \boldsymbol{G}_{22} \end{bmatrix} \le 0 \tag{10.11}$$

where

$$\boldsymbol{G}_{11} = \begin{bmatrix} \boldsymbol{\Theta}_{11} & \boldsymbol{B}\boldsymbol{X}_2 \\ * & \boldsymbol{\Theta}_{22} \end{bmatrix}$$

$$\boldsymbol{G}_{12} = \begin{bmatrix} 0 & 0 & 0 & 0 & 0 & 0 & \boldsymbol{B}\boldsymbol{M}_1 & \boldsymbol{X}_1\boldsymbol{N}_1{}^{\mathrm{T}} & \boldsymbol{B}\boldsymbol{M}_2 & \boldsymbol{X}_1\boldsymbol{K}^{\mathrm{T}}\boldsymbol{N}_2{}^{\mathrm{T}} & \boldsymbol{X}_1\boldsymbol{N}_3{}^{\mathrm{T}} & \boldsymbol{X}_1\boldsymbol{W}^{\mathrm{T}} \\ \boldsymbol{M}_3 & \boldsymbol{M}_4 & \boldsymbol{M}_3 & \boldsymbol{X}_2\boldsymbol{B}^{\mathrm{T}}\boldsymbol{N}_3{}^{\mathrm{T}} & \boldsymbol{M}_4 & \boldsymbol{H}^{\mathrm{T}}\boldsymbol{N}_4{}^{\mathrm{T}} & 0 & 0 & 0 & 0 & 0 & 0 \end{bmatrix}$$

$$\boldsymbol{G}_{22} = \mathrm{diag}\left(\begin{bmatrix} -\sigma_2{}^{-1}\xi_5\boldsymbol{I}, -(1-\sigma_2)^{-1}\xi_6\boldsymbol{I}, -\sigma_2{}^{-1}\xi_7\boldsymbol{I}, -\sigma_2{}^{-1}\xi_7^{-1}\boldsymbol{I}, -(1-\sigma_2)^{-1}\xi_8\boldsymbol{I}, -(1-\sigma_2)^{-1}\xi_8^{-1}\boldsymbol{I}, \\ -\sigma_1{}^{-1}\xi_3\boldsymbol{I}, -\sigma_1{}^{-1}\xi_3^{-1}\boldsymbol{I}, -(1-\sigma_1)^{-1}\xi_4\boldsymbol{I}, -(1-\sigma_1)^{-1}\xi_4^{-1}\boldsymbol{I}, -\sigma_2{}^{-1}\xi_5^{-1}\boldsymbol{I}, -(1-\sigma_2)^{-1}\xi_6^{-1}\boldsymbol{I} \end{bmatrix}\right)$$

$$\boldsymbol{\Theta}_{11} = \boldsymbol{X}_1\boldsymbol{A}^{\mathrm{T}} - \boldsymbol{X}_1{}^{\mathrm{T}}\boldsymbol{K}^{\mathrm{T}}\boldsymbol{B}^{\mathrm{T}} + \boldsymbol{A}\boldsymbol{X}_1 - \boldsymbol{B}\boldsymbol{K}\boldsymbol{X}_1 + \xi_2\boldsymbol{X}_1$$

$$\boldsymbol{\Theta}_{22} = -\boldsymbol{H}_1{}^{\mathrm{T}} - \boldsymbol{H}_1 + \xi_1\boldsymbol{I} + \xi_2\boldsymbol{X}_2$$

then, one can obtain the controller gain matrix $\boldsymbol{K}$ and the observer gain matrix $\boldsymbol{L} = \boldsymbol{H}(\boldsymbol{B}\boldsymbol{X}_2)^*$, where $(\boldsymbol{B}\boldsymbol{X}_2)^*$is the pseudoinverse of $\boldsymbol{B}\boldsymbol{X}_2$.

Proof Choose a Lyapunov function candidate as

$$V(t) = \boldsymbol{\eta}(t)^{\mathrm{T}}\boldsymbol{P}_1\boldsymbol{\eta}(t) + \boldsymbol{e}(t)^{\mathrm{T}}\boldsymbol{P}_2\boldsymbol{e}(t) \tag{10.12}$$

where $\boldsymbol{P}_1$ and $\boldsymbol{P}_2$ are defined symmetric positive definite matrices, with $\boldsymbol{P}_1 = \boldsymbol{X}_1^{-1}, \boldsymbol{P}_2 = \boldsymbol{X}_2^{-1}$, and the first derivative of $V(t)$ is

$$\begin{aligned}
\dot{V}(t) &\leq \boldsymbol{\eta}^{\mathrm{T}}(t)\left[\boldsymbol{A}^{\mathrm{T}}\boldsymbol{P}_1 - (\boldsymbol{K}+\Delta\boldsymbol{K})^{\mathrm{T}}\boldsymbol{B}^{\mathrm{T}}\boldsymbol{P}_1 + \boldsymbol{P}_1\boldsymbol{A} - \boldsymbol{P}_1\boldsymbol{B}(\boldsymbol{K}+\Delta\boldsymbol{K})\right]\boldsymbol{\eta}(t) \\
&= \boldsymbol{\eta}^{\mathrm{T}}(t)\left[\boldsymbol{A}^{\mathrm{T}}\boldsymbol{P}_1 - (\boldsymbol{K}+\Delta\boldsymbol{K})^{\mathrm{T}}\boldsymbol{B}^{\mathrm{T}}\boldsymbol{P}_1 + \boldsymbol{P}_1\boldsymbol{A} - \boldsymbol{P}_1\boldsymbol{B}(\boldsymbol{K}+\Delta\boldsymbol{K})\right]\boldsymbol{\eta}(t) \\
&\quad + \boldsymbol{\eta}^{\mathrm{T}}(t)\left(\boldsymbol{P}_1\boldsymbol{B} - \Delta\dot{\boldsymbol{L}}^{\mathrm{T}}\boldsymbol{P}_2\right)\boldsymbol{e}(t) + \boldsymbol{e}^{\mathrm{T}}(t)\left(\boldsymbol{B}^{\mathrm{T}}\boldsymbol{P}_1 - \boldsymbol{P}_2\Delta\dot{\boldsymbol{L}}\right)\boldsymbol{\eta}(t) \\
&\quad - \boldsymbol{e}^{\mathrm{T}}(t)\left[\boldsymbol{B}^{\mathrm{T}}(\boldsymbol{L}+\Delta\boldsymbol{L})^{\mathrm{T}}\boldsymbol{P}_2 + \boldsymbol{P}_2(\boldsymbol{L}+\Delta\boldsymbol{L})\boldsymbol{B}\right]\boldsymbol{e}(t) \\
&\quad + \dot{\boldsymbol{w}}^{\mathrm{T}}(t)\boldsymbol{P}_2\boldsymbol{e}(t) + \boldsymbol{e}^{\mathrm{T}}(t)\boldsymbol{P}_2\dot{\boldsymbol{w}}(t)
\end{aligned} \tag{10.13}$$

According to Lemma 1.1, one has

$$\begin{aligned}
\dot{V}(t) &\leq \boldsymbol{\eta}^{\mathrm{T}}(t)\left[\boldsymbol{A}^{\mathrm{T}}\boldsymbol{P}_1 - (\boldsymbol{K}+\Delta\boldsymbol{K})^{\mathrm{T}}\boldsymbol{B}^{\mathrm{T}}\boldsymbol{P}_1 + \boldsymbol{P}_1\boldsymbol{A} - \boldsymbol{P}_1\boldsymbol{B}(\boldsymbol{K}+\Delta\boldsymbol{K})\right]\boldsymbol{\eta}(t) \\
&\quad + \boldsymbol{\eta}^{\mathrm{T}}(t)\left(\boldsymbol{P}_1\boldsymbol{B} - \Delta\dot{\boldsymbol{L}}^{\mathrm{T}}\boldsymbol{P}_2\right)\boldsymbol{e}(t) + \boldsymbol{e}^{\mathrm{T}}(t)\left(\boldsymbol{B}^{\mathrm{T}}\boldsymbol{P}_1 - \boldsymbol{P}_2\Delta\dot{\boldsymbol{L}}\right)\boldsymbol{\eta}(t) \\
&\quad - \boldsymbol{e}^{\mathrm{T}}(t)\left[\boldsymbol{B}^{\mathrm{T}}(\boldsymbol{L}+\Delta\boldsymbol{L})^{\mathrm{T}}\boldsymbol{P}_2 + \boldsymbol{P}_2(\boldsymbol{L}+\Delta\boldsymbol{L})\boldsymbol{B}\right]\boldsymbol{e}(t) \\
&\quad + \xi_1^{-1}\dot{\boldsymbol{w}}^{\mathrm{T}}(t)\dot{\boldsymbol{w}}(t) + \xi_1\boldsymbol{e}^{\mathrm{T}}(t)\boldsymbol{P}_2\boldsymbol{P}_2\boldsymbol{e}(t)
\end{aligned} \tag{10.14}$$

where ξ_1 is a given positive scalar.

With Assumption 10.1, Eq. (10.3)and (Eq. 10.4), and noticing that $\boldsymbol{L}$ is a constant matrix, one has

$$\begin{aligned}
\dot{V}(t) &\leq \boldsymbol{\eta}^{\mathrm{T}}(t)\begin{bmatrix}\boldsymbol{A}^{\mathrm{T}}\boldsymbol{P}_1 - (\boldsymbol{K}+\sigma_1\boldsymbol{M}_1\boldsymbol{F}_1(t)\boldsymbol{N}_1 + (1-\sigma_1)\boldsymbol{M}_2\boldsymbol{F}_2(t)\boldsymbol{N}_2\boldsymbol{K})^{\mathrm{T}}\boldsymbol{B}^{\mathrm{T}}\boldsymbol{P}_1 \\ +\boldsymbol{P}_1\boldsymbol{A} - \boldsymbol{P}_1\boldsymbol{B}(\boldsymbol{K}+\sigma_1\boldsymbol{M}_1\boldsymbol{F}_1(t)\boldsymbol{N}_1 + (1-\sigma_1)\boldsymbol{M}_2\boldsymbol{F}_2(t)\boldsymbol{N}_2\boldsymbol{K})\end{bmatrix}\boldsymbol{\eta}(t) \\
&\quad + \boldsymbol{\eta}^{\mathrm{T}}(t)\left[\boldsymbol{P}_1\boldsymbol{B} - \sigma_2\left(\boldsymbol{M}_3\dot{\boldsymbol{F}}_3(t)\boldsymbol{N}_3\right)^{\mathrm{T}}\boldsymbol{P}_2 - (1-\sigma_2)\left(\boldsymbol{M}_4\dot{\boldsymbol{F}}_4(t)\boldsymbol{N}_4\boldsymbol{L}\right)^{\mathrm{T}}\boldsymbol{P}_2\right]\boldsymbol{e}(t) \\
&\quad + \boldsymbol{e}^{\mathrm{T}}(t)\left[\boldsymbol{B}^{\mathrm{T}}\boldsymbol{P}_1 - \sigma_2\boldsymbol{P}_2\boldsymbol{M}_3\dot{\boldsymbol{F}}_3(t)\boldsymbol{N}_3 - (1-\sigma_2)\boldsymbol{P}_2\boldsymbol{M}_4\dot{\boldsymbol{F}}_4(t)\boldsymbol{N}_4\boldsymbol{L}\right]\boldsymbol{\eta}(t) \\
&\quad - \boldsymbol{e}^{\mathrm{T}}(t)\begin{bmatrix}\boldsymbol{B}^{\mathrm{T}}[\boldsymbol{L}+\sigma_2\boldsymbol{M}_3\boldsymbol{F}_3(t)\boldsymbol{N}_3 + (1-\sigma_2)\boldsymbol{M}_4\boldsymbol{F}_4(t)\boldsymbol{N}_4\boldsymbol{L}]^{\mathrm{T}}\boldsymbol{P}_2 \\ +\boldsymbol{P}_2[\boldsymbol{L}+\sigma_2\boldsymbol{M}_3\boldsymbol{F}_3(t)\boldsymbol{N}_3 + (1-\sigma_2)\boldsymbol{M}_4\boldsymbol{F}_4(t)\boldsymbol{N}_4\boldsymbol{L}]\boldsymbol{B}\end{bmatrix}\boldsymbol{e}(t) \\
&\leq \begin{bmatrix}\boldsymbol{\eta}(t) \\ \boldsymbol{e}(t)\end{bmatrix}^{\mathrm{T}}\begin{bmatrix}\boldsymbol{\Psi}_{11} & \boldsymbol{\Psi}_{12} \\ * & \boldsymbol{\Psi}_{22}\end{bmatrix}\begin{bmatrix}\boldsymbol{\eta}(t) \\ \boldsymbol{e}(t)\end{bmatrix} + \xi_1^{-1}\Omega_2^2
\end{aligned} \tag{10.15}$$

where

$$\begin{aligned}
\boldsymbol{\Psi}_{11} = {} & \boldsymbol{A}^{\mathrm{T}}\boldsymbol{P}_1 + \boldsymbol{P}_1\boldsymbol{A} - \boldsymbol{K}^{\mathrm{T}}\boldsymbol{B}^{\mathrm{T}}\boldsymbol{P}_1 - \boldsymbol{P}_1\boldsymbol{B}\boldsymbol{K} - \sigma_1\boldsymbol{N}_1^{\mathrm{T}}\boldsymbol{F}_1^{\mathrm{T}}(t)\boldsymbol{M}_1^{\mathrm{T}}\boldsymbol{B}^{\mathrm{T}}\boldsymbol{P}_1 \\
& - \sigma_1\boldsymbol{P}_1\boldsymbol{B}\boldsymbol{M}_1\boldsymbol{F}_1(t)\boldsymbol{N}_1 - (1-\sigma_2)\boldsymbol{K}^{\mathrm{T}}\boldsymbol{N}_2^{\mathrm{T}}\boldsymbol{F}_2^{\mathrm{T}}(t)\boldsymbol{M}_2^{\mathrm{T}}\boldsymbol{B}^{\mathrm{T}}\boldsymbol{P}_1 \\
& - (1-\sigma_1)\boldsymbol{P}_1\boldsymbol{B}\boldsymbol{M}_2\boldsymbol{F}_2(t)\boldsymbol{N}_2\boldsymbol{K}
\end{aligned}$$

$$\boldsymbol{\Psi}_{12} = \boldsymbol{P}_1\boldsymbol{B} - \sigma_2\boldsymbol{N}_3^{\mathrm{T}}\dot{\boldsymbol{F}}_3^{\mathrm{T}}(t)\boldsymbol{M}_3^{\mathrm{T}}\boldsymbol{P}_2 - (1-\sigma_2)\left(\boldsymbol{M}_4\dot{\boldsymbol{F}}_4(t)\boldsymbol{N}_4\boldsymbol{L}\right)^{\mathrm{T}}\boldsymbol{P}_2$$

$$\begin{aligned}\boldsymbol{\Psi}_{22} = & -\boldsymbol{B}^{\mathrm{T}}\boldsymbol{L}^{\mathrm{T}}\boldsymbol{P}_2 - \boldsymbol{P}_2\boldsymbol{L}\boldsymbol{B} - \sigma_2\boldsymbol{B}^{\mathrm{T}}\boldsymbol{N}_3^{\mathrm{T}}\boldsymbol{F}_3^{\mathrm{T}}(t)\boldsymbol{M}_3^{\mathrm{T}}\boldsymbol{P}_2 \\ & - \sigma_2\boldsymbol{P}_2\boldsymbol{M}_3\boldsymbol{F}_3(t)\boldsymbol{N}_3\boldsymbol{B} - (1-\sigma_2)\boldsymbol{B}^{\mathrm{T}}\boldsymbol{L}^{\mathrm{T}}\boldsymbol{N}_4^{\mathrm{T}}\boldsymbol{F}_4^{\mathrm{T}}(t)\boldsymbol{M}_4^{\mathrm{T}}\boldsymbol{P}_2 \\ & - (1-\sigma_2)\boldsymbol{P}_2\boldsymbol{M}_4\boldsymbol{F}_4(t)\boldsymbol{N}_4\boldsymbol{L}\boldsymbol{B} + \xi_1\boldsymbol{P}_2\boldsymbol{P}_2\end{aligned}$$

Premultiplying and postmultiplying both sides of inequality (Eq. 10.11) by diag $([\boldsymbol{P}_1\ \boldsymbol{P}_2\ \boldsymbol{I}_{12}])$, one has

$$\begin{bmatrix} \widetilde{\boldsymbol{G}}_{11} & \widetilde{\boldsymbol{G}}_{12} \\ * & \boldsymbol{G}_{22} \end{bmatrix} \le 0 \tag{10.16}$$

where

$$\widetilde{\boldsymbol{G}}_{11} = \begin{bmatrix} \boldsymbol{A}^{\mathrm{T}}\boldsymbol{P}_1 - \boldsymbol{K}^{\mathrm{T}}\boldsymbol{B}^{\mathrm{T}}\boldsymbol{P}_1 + \boldsymbol{P}_1\boldsymbol{A} - \boldsymbol{P}_1\boldsymbol{B}\boldsymbol{K} + \xi_2\boldsymbol{P}_1 & \boldsymbol{P}_1\boldsymbol{B} \\ * & -\boldsymbol{P}_2\boldsymbol{H}^{\mathrm{T}}\boldsymbol{P}_2 - \boldsymbol{P}_2\boldsymbol{H}\boldsymbol{P}_2 + \xi_1\boldsymbol{P}_2\boldsymbol{P}_2 + \xi_2\boldsymbol{P}_2 \end{bmatrix}$$

$$\widetilde{\boldsymbol{G}}_{12} = \begin{bmatrix} 0 & 0 & 0 & 0 & 0 & 0 & \boldsymbol{P}_1\boldsymbol{B}\boldsymbol{M}_1 & \boldsymbol{N}_1{}^{\mathrm{T}} & \boldsymbol{P}_1\boldsymbol{B}\boldsymbol{M}_2 & \boldsymbol{K}^{\mathrm{T}}\boldsymbol{N}_2{}^{\mathrm{T}} & \boldsymbol{N}_3{}^{\mathrm{T}} & \boldsymbol{W}^{\mathrm{T}} \\ \boldsymbol{P}_2\boldsymbol{M}_3 & \boldsymbol{P}_2\boldsymbol{M}_4 & \boldsymbol{P}_2\boldsymbol{M}_3 & \boldsymbol{B}^{\mathrm{T}}\boldsymbol{N}_3{}^{\mathrm{T}} & \boldsymbol{P}_2\boldsymbol{M}_4 & \boldsymbol{P}_2\boldsymbol{H}^{\mathrm{T}}\boldsymbol{N}_4{}^{\mathrm{T}} & 0 & 0 & 0 & 0 & 0 & 0 \end{bmatrix}$$

By using Lemma 1.3 and letting $||\boldsymbol{N}_4\boldsymbol{L}|| < ||\boldsymbol{W}||$ and $\boldsymbol{H} = \boldsymbol{L}\boldsymbol{B}\boldsymbol{X}_2$, one has

$$\begin{bmatrix} \boldsymbol{\Theta}_1 & \boldsymbol{P}_1\boldsymbol{B} \\ \boldsymbol{B}^{\mathrm{T}}\boldsymbol{P}_1 & \boldsymbol{\Theta}_2 \end{bmatrix} + \xi_2\begin{bmatrix} \boldsymbol{P}_1 & 0 \\ 0 & \boldsymbol{P}_2 \end{bmatrix} \le 0 \tag{10.17}$$

where

$$\begin{aligned}\boldsymbol{\Theta}_1 = & \boldsymbol{A}^{\mathrm{T}}\boldsymbol{P}_1 - \boldsymbol{K}^{\mathrm{T}}\boldsymbol{B}^{\mathrm{T}}\boldsymbol{P}_1 + \boldsymbol{P}_1\boldsymbol{A} - \boldsymbol{P}_1\boldsymbol{B}\boldsymbol{K} \\ & + \xi_6(1-\sigma_2)\boldsymbol{L}^{\mathrm{T}}\boldsymbol{N}_4^{\mathrm{T}}\boldsymbol{N}_4\boldsymbol{L} + \xi_5\sigma_2\boldsymbol{N}_3{}^{\mathrm{T}}\boldsymbol{N}_3 + \xi_4(1-\sigma_1)\boldsymbol{K}^{\mathrm{T}}\boldsymbol{N}_2{}^{\mathrm{T}}\boldsymbol{N}_2\boldsymbol{K} \\ & + \xi_4^{-1}(1-\sigma_1)\boldsymbol{P}_1\boldsymbol{B}\boldsymbol{M}_2\boldsymbol{M}_2{}^{\mathrm{T}}\boldsymbol{B}^{\mathrm{T}}\boldsymbol{P}_1 + \xi_3\sigma_1\boldsymbol{N}_1{}^{\mathrm{T}}\boldsymbol{N}_1 \\ & + \xi_3^{-1}\sigma_1\boldsymbol{P}_1\boldsymbol{B}\boldsymbol{M}_1\boldsymbol{M}_1{}^{\mathrm{T}}\boldsymbol{B}^{\mathrm{T}}\boldsymbol{P}_1\end{aligned}$$

$$\begin{aligned}\boldsymbol{\Theta}_2 = & -\boldsymbol{B}^{\mathrm{T}}\boldsymbol{L}^{\mathrm{T}}\boldsymbol{P}_2 - \boldsymbol{P}_2\boldsymbol{L}\boldsymbol{B} + \xi_1\boldsymbol{P}_2\boldsymbol{P}_2 \\ & + \xi_8(1-\sigma_2)\boldsymbol{B}^{\mathrm{T}}\boldsymbol{L}^{\mathrm{T}}\boldsymbol{N}_4{}^{\mathrm{T}}\boldsymbol{N}_4\boldsymbol{L}\boldsymbol{B} + \xi_8^{-1}(1-\sigma_2)\boldsymbol{P}_2\boldsymbol{M}_4\boldsymbol{M}_4{}^{\mathrm{T}}\boldsymbol{P}_2 \\ & + \xi_7\sigma_2\boldsymbol{B}^{\mathrm{T}}\boldsymbol{N}_3{}^{\mathrm{T}}\boldsymbol{N}_3\boldsymbol{B} + \xi_7^{-1}\sigma_2\boldsymbol{P}_2\boldsymbol{M}_3\boldsymbol{M}_3{}^{\mathrm{T}}\boldsymbol{P}_2 + \xi_6^{-1}(1-\sigma_2)\boldsymbol{P}_2\boldsymbol{M}_4\boldsymbol{M}_4{}^{\mathrm{T}}\boldsymbol{P}_2 \\ & + \xi_5^{-1}\sigma_2\boldsymbol{P}_2\boldsymbol{M}_3\boldsymbol{M}_3{}^{\mathrm{T}}\boldsymbol{P}_2\end{aligned}$$

With Lemma 1.1, one has

$$\begin{aligned}\begin{bmatrix} \boldsymbol{\Psi}_{11} & \boldsymbol{\Psi}_{12} \\ \boldsymbol{\Psi}_{21} & \boldsymbol{\Psi}_{22} \end{bmatrix} + \xi_1^{-1}\Omega_2^2 & \le \begin{bmatrix} \boldsymbol{\Theta}_1 & \boldsymbol{P}_1\boldsymbol{B} \\ \boldsymbol{B}^{\mathrm{T}}\boldsymbol{P}_1 & \boldsymbol{\Theta}_2 \end{bmatrix} + \xi_1^{-1}\Omega_2^2 \\ & \le -\xi_2\begin{bmatrix} \boldsymbol{P}_1 & 0 \\ 0 & \boldsymbol{P}_2 \end{bmatrix} + \xi_1^{-1}\Omega_2^2\end{aligned} \tag{10.18}$$

Then, one has

$$\begin{aligned}\dot{V}(t) &\le \begin{bmatrix}\boldsymbol{\eta}(t)\\ e(t)\end{bmatrix}^{\mathrm{T}}\begin{bmatrix}\boldsymbol{\Psi}_{11} & \boldsymbol{\Psi}_{12}\\ \boldsymbol{\Psi}_{21} & \boldsymbol{\Psi}_{22}\end{bmatrix}\begin{bmatrix}\boldsymbol{\eta}(t)\\ e(t)\end{bmatrix} + \xi_1^{-1}\Omega_2^2\\ &\le -\xi_2\begin{bmatrix}\boldsymbol{\eta}(t)\\ e(t)\end{bmatrix}^{\mathrm{T}}\begin{bmatrix}\boldsymbol{P}_1 & 0\\ 0 & \boldsymbol{P}_2\end{bmatrix}\begin{bmatrix}\boldsymbol{\eta}(t)\\ e(t)\end{bmatrix} + \xi_1^{-1}\Omega_2^2\end{aligned} \tag{10.19}$$

According to Lemma 1.2, one can guarantee the uniform ultimate boundness of the closed-loop system (Shi et al., 2020b). The proof is completed.

10.3 Simulation test

In order to illustrate the performance of the proposed dual hybrid nonfragile strategy, this section perform some simulations where the two spacecraft start docking from the initial state $\boldsymbol{\eta}(0) = [2\text{m}, 3\text{m}, 4\text{m}, 0\text{m/s}, 0\text{m/s}, 0\text{m/s}]^{\mathrm{T}}$ at 0 s to the docked state $\boldsymbol{\eta}_d = [0\text{m}, 0\text{m}, 0\text{m}, 0\text{m/s}, 0\text{m/s}, 0\text{m/s}]^{\mathrm{T}}$, and then separates from $\boldsymbol{\eta}_d$ at 100 s to the hovering state $\boldsymbol{\eta}_h = [2\text{m}, 3\text{m}, 4\text{m}, 0\text{m/s}, 0\text{m/s}, 0\text{m/s}]^{\mathrm{T}}$. The magnetic moment vector of target spacecraft is $\boldsymbol{\mu}_T = [-2.29, -8.17, -2.33]^{\mathrm{T}} \times 10^3\text{Am}^2$ while the upper bound of electromagnetic force is $U_{x,y,z} = 15\text{N}$.

The external disturbance $\boldsymbol{w}_0(t)$, which is chosen to be periodic or cyclic, usually includes earth magnetic force, earth nonspherical perturbation, solar radiation pressure, gravitational forces of the moon, etc. Since in common occasions the periodic disturbances can be expanded into Fourier series, which are composed of sin and cos functions, the combined external disturbance is chosen as

$$\boldsymbol{w}_0(t) = 5 \times 10^{-2} \times \begin{bmatrix}\cos(2\pi t) & \cos(2\pi t + \pi/2) & \cos(2\pi t + \pi)\end{bmatrix}^{\mathrm{T}}\text{N}$$

The matrices reflecting the structure of uncertainty are chosen as

$$\boldsymbol{M}_i = \rho_1 \times \begin{bmatrix}1 & 1 & 1\end{bmatrix}^{\mathrm{T}} (i = 1, 2, 3, 4)$$

$$\boldsymbol{N}_i = \begin{cases}\rho_1 \times \begin{bmatrix}1 & 1 & 1 & 1 & 1 & 1\end{bmatrix} (i = 1, 3)\\ \rho_1 \times \begin{bmatrix}1 & 1 & 1\end{bmatrix} (i = 2, 4)\end{cases}$$

$$F_i(t) = \sin(0.1\pi t)(i = 1, 2, 3, 4)$$

Table 10.1 Partial simulation parameters.

Item	a/km	e	μ/km^{3}s^{-2}	$\theta(0)$	$z(0)$	μ_0/NA^{-2}
Value	10^3	0.15	3.986×10^5	0	(0 0 0)	$4\pi \times 10^{-7}$
Item	$\boldsymbol{m_C}$**/kg**	$\boldsymbol{m_T}$**/kg**	$\boldsymbol{m_0}$**/kg**	σ_1	σ_2	ρ_1
Value	40	100	20	0.2	0.2	0.5

where ρ_1 is a coefficient of the constant matrixes. The matrices $\boldsymbol{M}$ and $\boldsymbol{N}$ are introduced to represent the range and structure of the perturbations, respectively, while $\boldsymbol{F}(t)$ represents the time-varying nature of these perturbations. The dimensions and magnitudes of perturbation matrices $\boldsymbol{M}$, $\boldsymbol{N}$, and $\boldsymbol{F}(t)$ are derived from previous works, which is a common practice in robust control literature. It is assume that each dimension experiences a similar range of perturbation, which is a reasonable simplification given the inherent random nature of these disturbances.

The matrix $\widetilde{\boldsymbol{X}}_1$'s selection is a user choice, but it can be determined through a systematic and rigorous LMI-based approach, which is illustrated as follows.

Let $\boldsymbol{Y} = \boldsymbol{K}\widetilde{\boldsymbol{X}}_1$. Then inequality (Eq. 10.10) becomes

$$\widetilde{\boldsymbol{X}}_1\boldsymbol{A}^{\mathrm{T}} + \boldsymbol{A}\widetilde{\boldsymbol{X}}_1^{\mathrm{T}} - \boldsymbol{Y}^{\mathrm{T}}\boldsymbol{B}^{\mathrm{T}} - \boldsymbol{B}\boldsymbol{Y} \leq 0 \ s.t.\ \widetilde{\boldsymbol{X}}_1 > 0$$

A feasible solution for $\widetilde{\boldsymbol{X}}_1$ can be obtained by solving the above LMI while some manual adjustment may be required. Through further numerical adjustment, the final value of $\widetilde{\boldsymbol{X}}_1$ is chosen as.

$$\widetilde{\boldsymbol{X}}_1 = 0.1 \times \begin{bmatrix} 2.3600 & 0.0333 & 0.0443 & -0.2639 & -0.0029 & -0.0053 \\ 0.0333 & 2.3871 & 0.0659 & -0.0029 & -0.2663 & -0.0057 \\ 0.0443 & 0.0659 & 2.4254 & -0.0024 & -0.0057 & -0.2696 \\ -0.2639 & -0.0029 & -0.0024 & 0.0499 & 0.0004 & 0.0006 \\ -0.0029 & -0.2663 & -0.0057 & 0.0004 & 0.0503 & 0.0009 \\ -0.0053 & -0.0057 & -0.2696 & 0.0006 & 0.0009 & 0.0508 \end{bmatrix}$$

The matrix $\boldsymbol{W}$ is chosen as

$$\boldsymbol{W} = \begin{bmatrix} 0 & 0 & 0 & 150 & 150 & 150 \end{bmatrix}$$

Some of the other simulation parameters are shown in Table 10.1.

The values of ξ are shown in Table 10.2.

The simulation environment is MATLAB R2021a (win64) on a Windows 10 computer with Intel Core i5 (10th Gen) and 16GB RAM. The LMIs are solved by YALMIP toolbox with the solver LMILAB. Other solver

Table 10.2 Other simulation parameters.

Item	ξ_1	ξ_2	ξ_3	ξ_4	ξ_5	ξ_6	ξ_7	ξ_8
Value	0.02	0.002	10^{-5}	10^{-7}	10^{-5}	10^{-6}	10^{-5}	10^{-6}

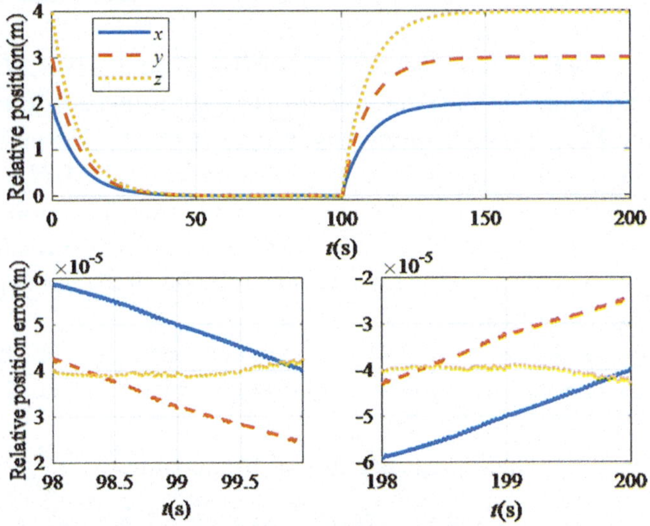

Figure 10.2 Time response of relative position.

for semidefinite programming in YALMIP, such as SDPT3, SEDUMI, and MOSEK, is also capable for this problem while the user may get different feasible solution of $\boldsymbol{K}$ and $\boldsymbol{L}$. The gains of the controller and observer calculated by Theorem 10.1 are

$$\boldsymbol{K} = 10^3 \times \begin{bmatrix} 0.5693 & -0.0060 & -0.0131 & 5.0860 & -0.0394 & -0.0700 \\ -0.0061 & 0.5626 & -0.0153 & -0.0394 & 5.0372 & -0.1075 \\ -0.0070 & -0.0154 & 0.5542 & -0.0702 & -0.1079 & 4.9798 \end{bmatrix}$$

$$\boldsymbol{L} = \begin{bmatrix} 0 & 0 & 0 & 25.5865 & 7.2696 & 7.2328 \\ 0 & 0 & 0 & 4.0680 & 22.3373 & 4.0338 \\ 0 & 0 & 0 & 0.8134 & 0.8160 & 19.1110 \end{bmatrix}$$

It is checked that $\boldsymbol{W}$ and L satisfy $||\boldsymbol{N}_4\boldsymbol{L}|| < ||\boldsymbol{W}||$.

Figs. 10.2 and 10.3 illustrate the relative position and velocity, which converge to a small neighborhood around the expected state $\boldsymbol{\eta}_d$ or $\boldsymbol{\eta}_h$ within 50s. The peak at 100s in Figs. 10.2 and 10.3 is due to the switching of the docking and separation mission. The steady-state error of the relative

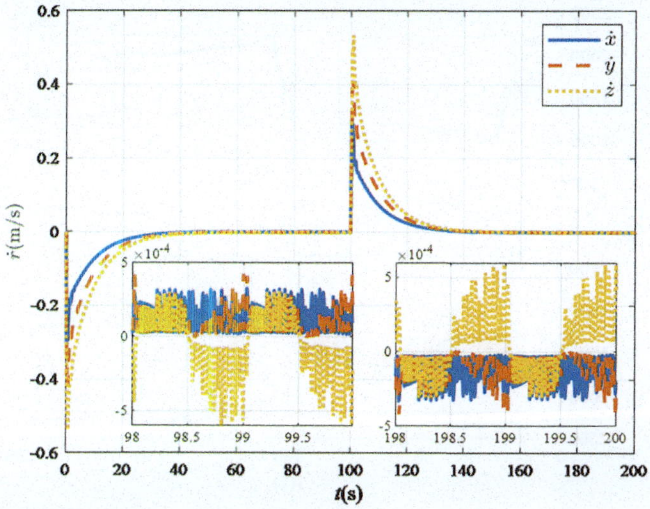

Figure 10.3 Time response of relative velocity.

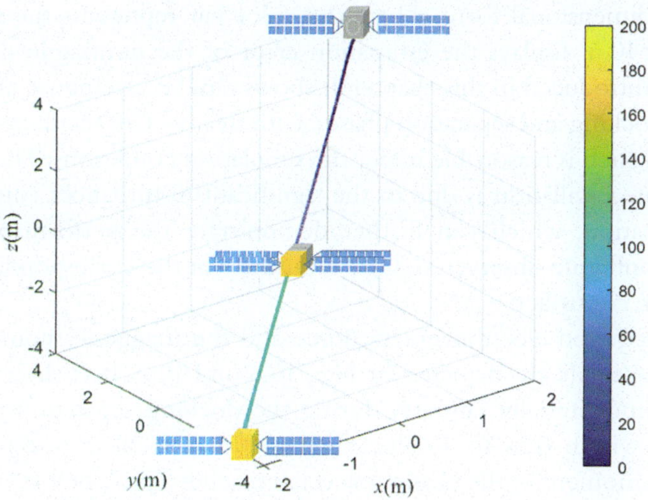

Figure 10.4 Spacecraft motion relative to "CM".

position is less than 6×10^{-5}m in both the docking phase and the separation phase per axis. Meanwhile, the steady-state error of the relative velocity is less than 6×10^{-4}m/s during both docking and separation phases per axis. In Fig. 10.4, the motion of the target spacecraft (the yellow one) and the chaser spacecraft (the grey one) in CM orbital coordinate system is demonstrated

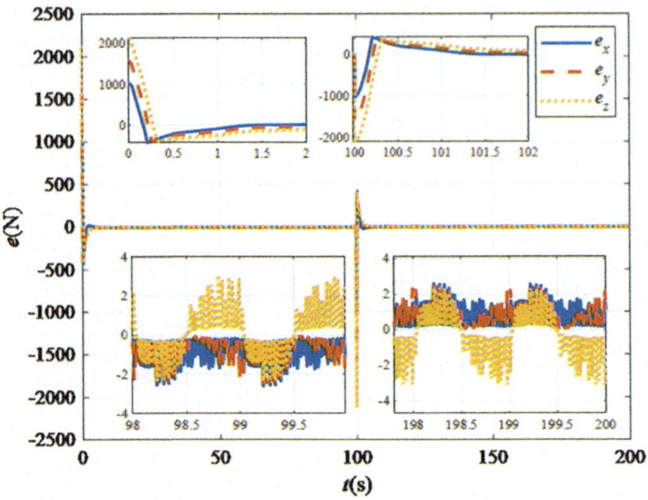

Figure 10.5 Time response of disturbance observer.

in three-dimensional form, where the color bar represents the simulation time. Fig. 10.5 displays the estimation error of the nonfragile disturbance observer introduced in this chapter. It shows rapid convergence at the onset of both docking and separation phases, with steady-state accuracy under 4N per axis, which is reasonable under the simulation conditions. The observed steady-state oscillation is due to the significant disturbances considered in the simulation, which can further demonstrate the performance of the hybrid nonfragile observer and the robustness of the corresponding hybrid nonfragile controller.

The required electromagnetic force and the magnetic moment of the chaser spacecraft are depicted in Figs. 10.6 and 10.7. It is shown that the control input initially saturates during the docking/separation phases but stabilizes within 0.5s to a value close to 0 across all axes. Similarly, the magnetic moment of the chaser spacecraft reaches a peak before 0.5s, on the order of $6\times10^6\,Am^2$, and then rapidly converges to a value very close to 0. The required magnetic moment is practical for a common electromagnetic mechanism on a 40kg chaser spacecraft.

To further illustrate the superiority of the dual hybrid nonfragile strategy, the following part gives comparisons of the following four disturbance observer-based cases, described in Table 10.3. Case 1 encompasses both additive and multiplicative perturbations in the controller and the observer. In contrast, Case 2 includes perturbations solely in the controller, while

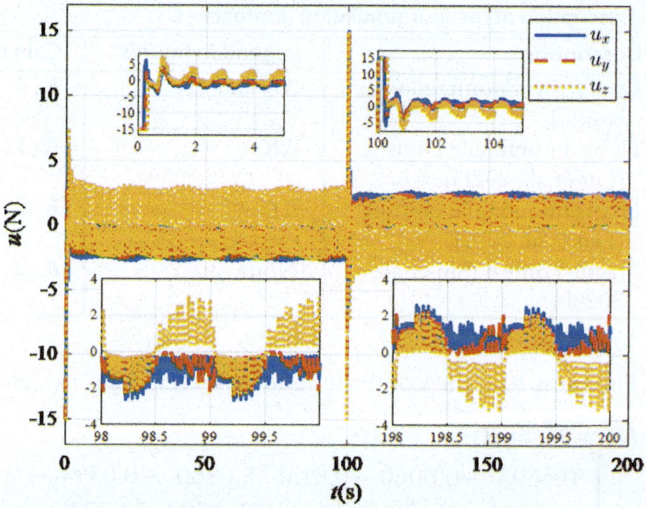

Figure 10.6 Time response of the required electromagnetic force.

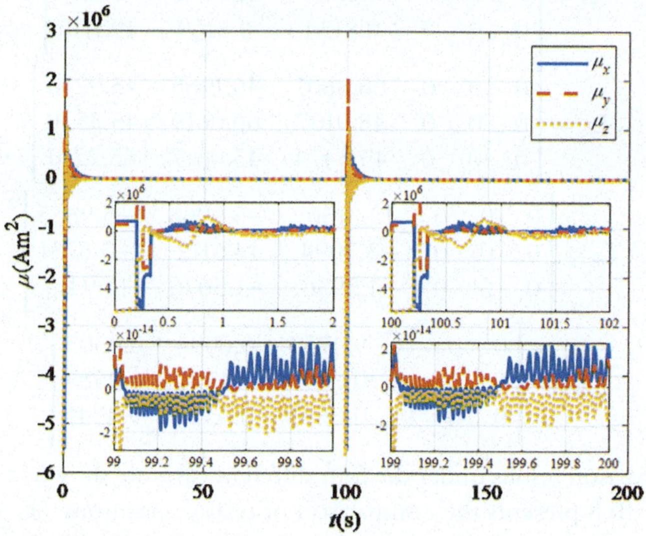

Figure 10.7 Magnetic moment of chaser spacecraft.

Case 3 includes them solely in the observer. Case 4 serves as the nominal case, devoid of any perturbations.

Table 10.3 Description of the four simulation approaches.

Case	Description	Legend in figures	Gain matrices
1	Dual hybrid nonfragile control	$\Delta\boldsymbol{K}$ and $\Delta\boldsymbol{L}$	$\boldsymbol{K}, \boldsymbol{L}$
2	Hybrid nonfragile control (Liu et al., 2021a)	$\Delta\boldsymbol{K}$	$\boldsymbol{K}_k, \boldsymbol{L}_k$
3	Hybrid nonfragile control (Lyu et al. 2022)	$\Delta\boldsymbol{L}$	$\boldsymbol{K}_l, \boldsymbol{L}_l$
4	Fragile control (Shi et al., 2020b)	None	$\boldsymbol{K}_f, \boldsymbol{L}_f$

With Theorem 10.1, the controller and observer gains are calculated as

$$\boldsymbol{K} = \boldsymbol{K}_k = \boldsymbol{K}_l = \boldsymbol{K}_f = 10^3 \times \begin{bmatrix} 0.5693 & -0.0060 & -0.0131 & 5.0860 & -0.0394 & -0.0700 \\ -0.0061 & 0.5626 & -0.0153 & -0.0394 & 5.0372 & -0.1075 \\ -0.0070 & -0.0154 & 0.5542 & -0.0702 & -0.1079 & 4.9798 \end{bmatrix}$$

$$\boldsymbol{L} = \begin{bmatrix} 0 & 0 & 0 & 25.5865 & 7.2696 & 7.2328 \\ 0 & 0 & 0 & 4.0680 & 22.3373 & 4.0338 \\ 0 & 0 & 0 & 0.8134 & 0.8160 & 19.1110 \end{bmatrix}$$

$$\boldsymbol{L}_k = \begin{bmatrix} 0 & 0 & 0 & 66.8801 & 46.1965 & 45.9246 \\ 0 & 0 & 0 & 45.9107 & 66.0619 & 45.3530 \\ 0 & 0 & 0 & 45.3474 & 45.0617 & 65.2210 \end{bmatrix}$$

$$\boldsymbol{L}_l = \begin{bmatrix} 0 & 0 & 0 & 12.3150 & -5.9968 & -5.9865 \\ 0 & 0 & 0 & -3.7094 & 14.6112 & -3.6731 \\ 0 & 0 & 0 & -1.3936 & -1.3676 & 16.9414 \end{bmatrix}$$

$$\boldsymbol{L}_f = \begin{bmatrix} 0 & 0 & 0 & 9.4753 & 0.0094 & 0.0303 \\ 0 & 0 & 0 & 0.0899 & 9.4747 & 0.0584 \\ 0 & 0 & 0 & 0.0683 & 0.0394 & 9.4741 \end{bmatrix}$$

Comparison results under the four different cases are shown in Figs. 10.8–10.11. Fig. 10.8 presents the comparison of relative positions, i.e., $r = ||\boldsymbol{r}||$, with Case 2 demonstrating the best performance, followed by Case 1. In Fig. 10.9 which compares relative velocity, i.e., $\dot{r} = ||\dot{\boldsymbol{r}}||$, Case 1 exhibits the best performance, while Case 2 shows significant oscillations, which is caused by the high observer gain. Figs. 10.10 and 10.11 illustrate the comparison of disturbance estimation errors, with Fig. 10.11 using a logarithmic scale with base 10 on the vertical axis. It is demonstrated that Case 2 still oscillates

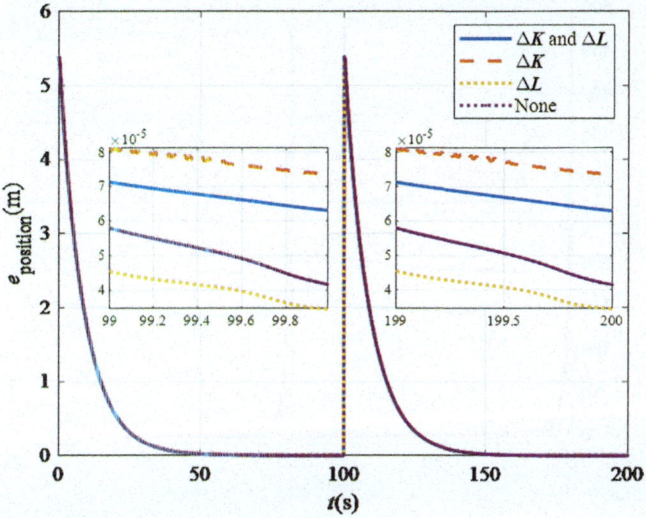

Figure 10.8 Comparison of relative position.

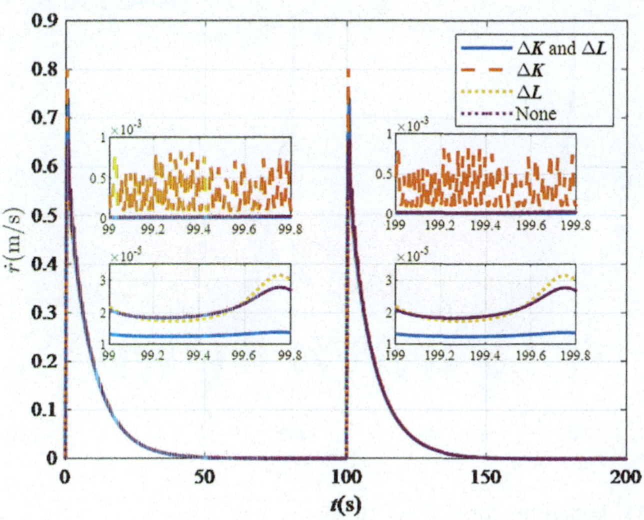

Figure 10.9 Comparison of relative velocity.

significantly, while Cases 1, 3, and 4 show similar good performance. It is shown that Case 1 integrates the advantages of the other cases with a balanced observe gain.

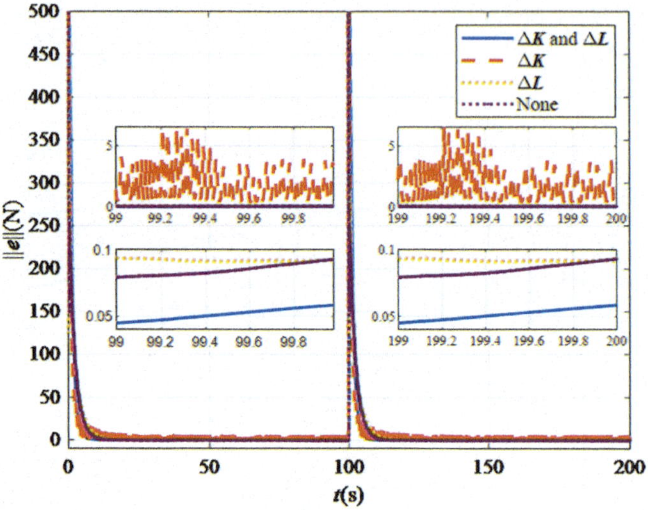

Figure 10.10 Comparison of estimation error.

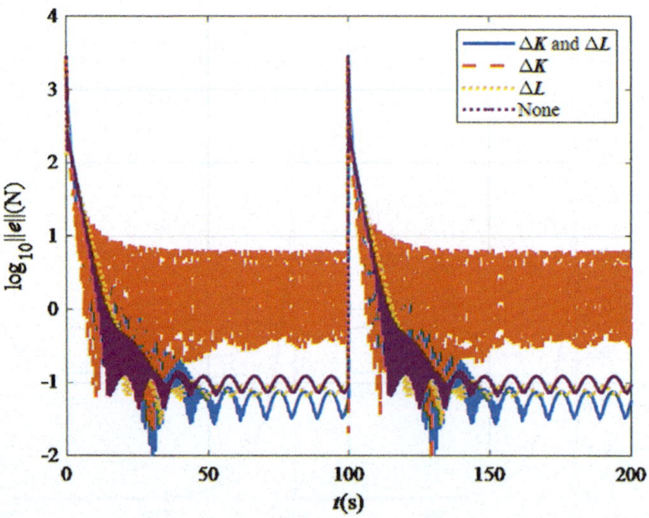

Figure 10.11 Logarithm form of Fig. 10.10.

Considering energy consumption during simulations, an optimal cost function-based energy index function is defined as $E = \frac{1}{2}\int_0^T ||\boldsymbol{u}||dt$, in which E is the energy index and T is the simulation time. Fig. 10.12 displaces the comparison of the required electromagnetic force. It can be observed that the

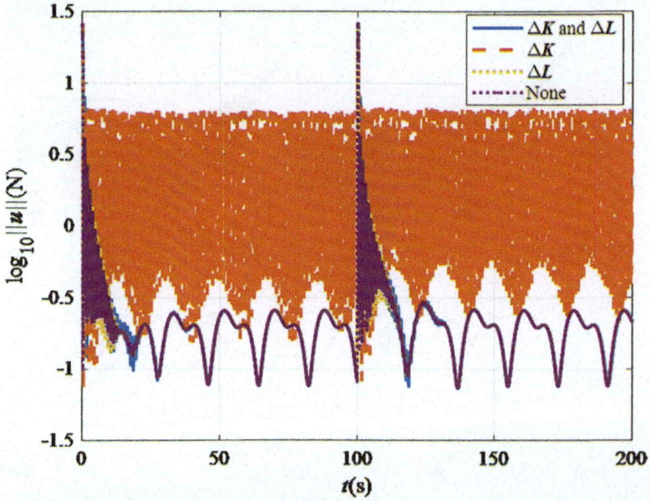

Figure 10.12 Logarithm form of the required electromagnetic force comparison.

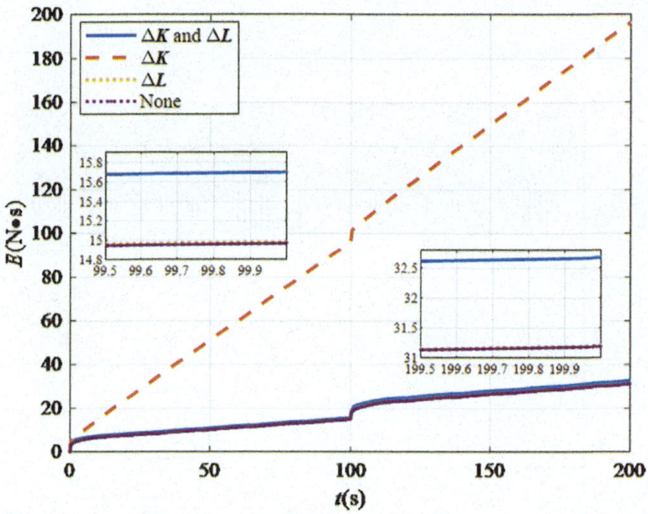

Figure 10.13 Comparison of energy index.

electromagnetic force required for Case 2 is significantly greater than that of the other cases, while the required electromagnetic forces for the other cases show minimal differences. Fig. 10.13 compares the energy index, indicating that Case 2 consumes more than 10 times the amount of energy compared to the other cases. Case 1 has the second-highest energy consumption, while the

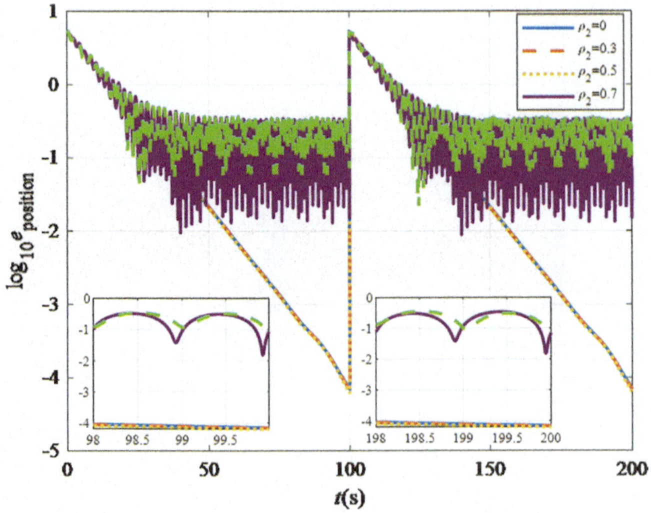

Figure 10.14 Relative position error under different perturbation.

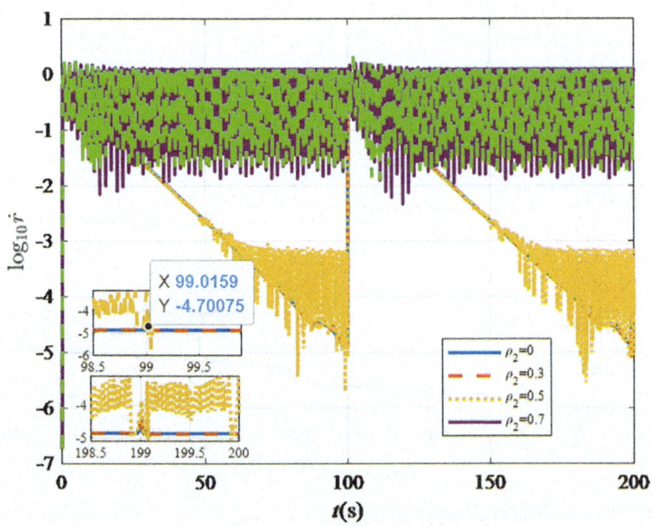

Figure 10.15 Relative velocity error under different perturbation.

difference in energy consumption among Cases 1, 3, and 4 is not significant. A comparative analysis of control accuracy and the energy consumption index reveals that the proposed strategy yields a more optimized control gain.

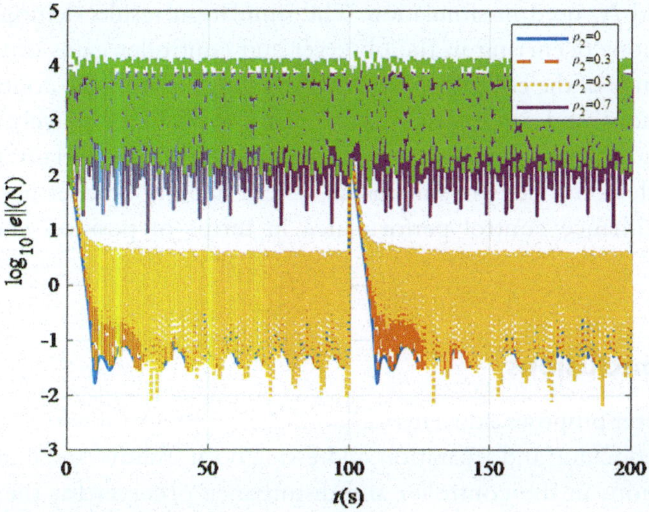

Figure 10.16 Logarithm form of estimation error.

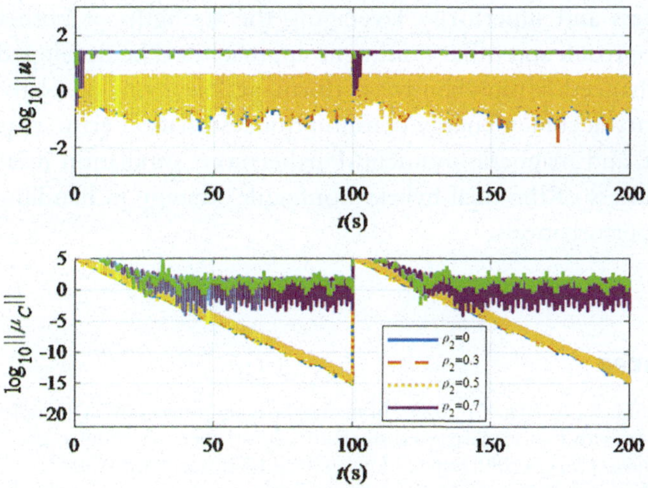

Figure 10.17 Logarithm form of required electromagnetic force and magnetic moment.

This ensures high control precision while maintaining a moderate increase in the energy consumption coefficient.

Figs. 10.14–10.17 give the comparisons of the developed dual hybrid nonfragile control strategy under different perturbations, and ρ_1 is the coefficient of $\boldsymbol{M}_i$ and $\boldsymbol{N}_i$ used in the gains' calculation while ρ_2 is the coefficient of $\boldsymbol{M}_i$ and $\boldsymbol{N}_i$ used in simulation. The simulation results demonstrate that perturbations occurring in the observer and controller gains can affect the performance of the control system. By designing a nonfragile controller, gain perturbations under the design specifications can be effectively addressed. Furthermore, even for larger perturbations within a certain range, the non-fragile controller can still ensure system stability. Smaller perturbations notably enhance control performance in terms of position and velocity accuracy.

10.4 Conclusions

This chapter proposes a dual hybrid nonfragile strategy for spacecraft electromagnetic docking and separation, addressing both additive and multiplicative perturbations in the controller and disturbance observer for the first time. Utilizing Lyapunov stability theory, a theorem is formulated to ensure the uniform ultimate boundedness of the closed-loop system, which can be effectively solved by LMI approach. Simulation results demonstrate the effectiveness and superiority, leveraging the strengths of both the fragile control approach and other nonfragile approaches. The developed approach shows enhanced comprehensive performance in terms of relative position and velocity accuracy, energy consumption, estimation error, required control input, and magnetic moment. Furthermore, simulation results validate the robustness of the dual hybrid nonfragile strategy in handling a certain range of perturbations.

References

Dong, Q., Shi, S., & Ma, Y. (2020). Non-fragile synchronization of complex dynamical networks with hybrid delays and stochastic disturbance via sampled-data control. *ISA Transactions, 105*, 174–189. https://doi.org/10.1016/j.isatra.2020.05.047.

Fang, H., Cheng, Q., & Sun, C. (2018). A method of the non-fragile controller based on LMI. *Proceedings of 2018 IEEE 7th data driven control and learning systems conference, DDCLS 2018* (pp. 102–106). https://doi.org/10.1109/DDCLS.2018.8516102.

Fauré, M., Henry, D., Cieslak, J., Lachevre, P., & Ankersen, F. (2023). Optimization of spacecraft thrusters configuration under fault diagnosability and recoverability constraints. *IEEE Transactions on Aerospace and Electronic Systems, 59*(5), 5275–5286. https://doi.org/10.1109/TAES.2023.3256976.

Hafezi, A., & Khandani, K. (2022). Non-fragile exponential H∞ observer-based event-triggered control of uncertain systems with disturbance input: Theoretical studies and application investigations. *ISA Transactions, 129*, 13–23. https://doi.org/10.1016/j.isatra.2021.12.015.

Han, T. J., & Kim, H. S. (2023). Disturbance observer-based nonfragile fuzzy tracking control of a spacecraft. *Advances in Space Research, 71*(9), 3600–3612. https://doi.org/10.1016/j.asr.2022.11.043.

Kaviarasan, B., Kwon, O. M., Park, M. J., & Sakthivel, R. (2021). Stochastic faulty estimator-based non-fragile tracking controller for multi-agent systems with communication delay. *Applied Mathematics and Computation, 392*. https://doi.org/10.1016/j.amc.2020.125704.

Keel, L. H., & Bhattacharyya, S. P. (1997). Robust, fragile, or optimal? *IEEE Transactions on Automatic Control, 42*(8), 1098–1105. https://doi.org/10.1109/9.618239.

Li, W. J., Cheng, D. Y., Liu, X. G., Wang, Y. B., Shi, W. H., Tang, Z. X., & ..., Gao, Z.-L. (2019). On-orbit service (OOS) of spacecraft: A review of engineering developments. *Progress in Aerospace Sciences, 108*, 32–120. https://doi.org/10.1016/j.paerosci.2019.01.004.

Liu, C., Ma, Y., Yue, X., & Shi, K. (2024). Improved sliding mode tracking control for spacecraft electromagnetic separation supporting on-orbit assembly. *Advances in Space Research, 73*(9), 4711–4720. https://doi.org/10.1016/j.asr.2024.01.034.

Liu, C., Sun, Z., Shi, K., & Wang, F. (2017). Robust dynamic output feedback control for attitude stabilization of spacecraft with nonlinear perturbations. *Aerospace Science and Technology, 64*, 102–121. https://doi.org/10.1016/j.ast.2017.01.016.

Liu, C., Vukovich, G., Sun, Z., & Shi, K. (2018). Observer-based fault-tolerant attitude control for spacecraft with input delay. *Journal of Guidance, Control, and Dynamics, 41*(9), 2039–2051. https://doi.org/10.2514/1.g003555.

Liu, C., Yang, Z., Yue, X., & Wang, X. (2021a). Hybrid non-fragile attitude stabilization for post-capture spacecraft. *Proceedings of the 33rd Chinese control and decision conference, CCDC 2021* (pp. 6090–6095). https://doi.org/10.1109/CCDC52312.2021.9601750.

Liu, C., Yue, X., & Yang, Z. (2021b). Are nonfragile controllers always better than fragile controllers in attitude control performance of post-capture flexible spacecraft? *Aerospace Science and Technology, 118*, 107053. https://doi.org/10.1016/j.ast.2021.107053.

Lyu, B., Liu, C., & Yue, X. (2022). Hybrid nonfragile intermediate observer-based T-S fuzzy attitude control for flexible spacecraft with input saturation. *Aerospace Science and Technology, 128*, 107753. https://doi.org/10.1016/j.ast.2022.107753.

Rajavel, S., Samidurai, R., Cao, J., Alsaedi, A., & Ahmad, B. (2017). Finite-time non-fragile passivity control for neural networks with time-varying delay. *Applied Mathematics and Computation, 297*, 145–158. https://doi.org/10.1016/j.amc.2016.10.038.

Safana, M. A., Sakthivel, R., Devi, N. B., & Anthoni, S. M. (2024). Disturbance rejection-based non-fragile hybrid-triggered control design for repeated scalar nonlinear systems. *International Journal of Systems Science, 55*(5), 1005–1020. https://doi.org/10.1080/00207721.2023.2301491.

Schweighart, S. A. (2005). *Electromagnetic formation flight dipole solution planning*. Massachusetts Institute of Technology.

Shi, K., Liu, C., Biggs, J. D., Sun, Z., & Yue, X. (2020a). Observer-based control for spacecraft electromagnetic docking. *Aerospace Science and Technology, 99*, 105759. https://doi.org/10.1016/j.ast.2020.105759.

Shi, K., Liu, C., & Sun, Z. (2019). Constrained fuel-free control for spacecraft electromagnetic docking in elliptical orbits. *Acta Astronautica, 162*, 14–24. https://doi.org/10.1016/j.actaastro.2019.05.016.

Shi, K., Liu, C., Sun, Z., & Yue, X. (2020b). Disturbance observer-based attitude stabilization for rigid spacecraft with input MRCs. *Advances in Space Research, 66*(3), 689–701. https://doi.org/10.1016/j.asr.2020.04.016.

Shi, K., Liu, C., Sun, Z., & Yue, X. (2022). Coupled orbit-attitude dynamics and trajectory tracking control for spacecraft electromagnetic docking. *Applied Mathematical Modelling, 101*, 553–572. https://doi.org/10.1016/j.apm.2021.08.030.

Wang, H., Deng, J., Zhang, L., Bao, Q., & Mao, Y. (2024). Enhanced disturbance observer-based hybrid cascade active disturbance rejection control design for high-precise tracking system in application to aerospace satellite. *Aerospace Science and Technology, 146*, 108939. https://doi.org/10.1016/j.ast.2024.108939.

Wu, C., Yang, J., Li, S., Li, Q., & Guo, L. (2018). Disturbance observer based model predictive control for accurate atmospheric entry of spacecraft. *Advances in Space Research, 61*(9), 2457–2471. https://doi.org/10.1016/j.asr.2018.02.010.

Xia, K., & Zou, Y. (2021). Adaptive saturated fault-tolerant control for spacecraft rendezvous with redundancy thrusters. *IEEE Transactions on Control Systems Technology, 29*(2), 502–513. https://doi.org/10.1109/tcst.2019.2950399.

Zhang, H., Zheng, Y., & Wang, Y. (2024). Event-triggered fault-tolerant attitude tracking control for spacecraft with fixed-time controller and disturbance observer under input constraints. *Advances in Space Research, 73*(6), 3148–3165. https://doi.org/10.1016/j.asr.2023.12.048.

Zhang, Y. (2021). Piece-wise control of constant electromagnetic moments for spacecraft self- and soft-docking exploiting dynamics conservation. *Asian Journal of Control, 23*(5), 2459–2473. https://doi.org/10.1002/asjc.2359.

Zhang, Y. W., & Chen, P. L. (2023). Self-alignment capability analysis and self-attraction dipole solution of spacecraft electromagnetic docking. *AIAA Journal, 61*(8), 3247–3256. https://doi.org/10.2514/1.j062361.

Zhang, Y. W., Yang, L. P., Zhu, Y. W., Ao, H. J., & Qi, D. W. (2016). Self-docking analysis and velocity-aimed control for spacecraft electromagnetic docking. *Advances in Space Research, 57*(11), 2314–2325. https://doi.org/10.1016/j.asr.2016.03.022.

Zhang, Y., Yang, L., Zhu, Y., & Huang, H. (2013). Dynamics and nonlinear control of space electromagnetic docking. *Journal of Systems Engineering and Electronics, 24*(3), 454–462. https://doi.org/10.1109/jsee.2013.00054.

Zhu, X., Chen, J., & Zhu, Z. H. (2021). Adaptive sliding mode disturbance observer-based control for rendezvous with non-cooperative spacecraft. *Acta Astronautica, 183*, 59–74. https://doi.org/10.1016/j.actaastro.2021.03.005.

Index

Page numbers followed by "*f*" and "*t*" indicate, figures and tables respectively.

Printed and bound by CPI Group (UK) Ltd, Croydon, CR0 4YY
14/09/2026
02216410-0001